Clinical management of bladder cancer

Edited by

Reginald R. Hall MS, FRCS

A member of the Hodder Headline Group
LONDON • SYDNEY • AUCKLAND
Co-published in the United States of America by
Oxford University Press Inc., New York

First published in Great Britain in 1999 by
Arnold, a member of the Hodder Headline Group,
338 Euston Road, London NW1 3BH

http://www.arnoldpublishers.com

Co-published in the United States of America by
Oxford University Press Inc.,
198 Madison Avenue, New York, NY10016
Oxford is a registered trademark of Oxford University Press

British Library Cataloguing in Publication Data
A catalogue record for this book is available from the British Library

Library of Congress Cataloging-in-Publication Data
A catalog record for this book is available from the Library of Congress

ISBN 0 340 74092 2

1 2 3 4 5 6 7 8 9 10

Typeset in 11 on 13pt Times by Genesis Typesetting, Rochester, Kent
Printed and bound in Great Britain by St Edmundsbury Press, Suffolk
and MPG Books Ltd, Bodmin

Dedicated to Dr E. R. Wide of Baringa Hospital, The Congo Balolo Mission, who watched over my first breath in 1939 and my first attempts at surgery in 1963.

Contents

Contributors

Peter R. Auriemma MD
Assistant Professor, Department of Urology, WVU Health Sciences Center,
PO Box 9251, Morgantown, WV 26506–9251, USA
Fax: 00 1 304 293 2807

Rob G. Bristow MB, Ph.D., FRCPC
Associate Professor, Department of Radiation Oncology, University of Toronto
and Princess Margaret Hospital, 610 University Avenue, Toronto,
Ontario M5G 2M9, Canada
Tel: 00 1 416 946 4421; Fax: 00 1 416 946 6556

Sarah Burdett
Meta-analysis Group Research Assistant, MRC Cancer Trials Office,
5 Shaftesbury Road, Cambridge CB2 2BW, United Kingdom
Tel: 01223 311110; Fax: 01223 311844; e-mail via Max Parmar

Margaret M. Charlton BA, M.Sc., RGN
Chemotherapy Research Sister, Department of Urology, Freeman Hospital,
High Heaton, Newcastle upon Tyne NE7 7DN, United Kingdom
Tel: 0191 284 3111 ext. 26115; Fax: 0191 223 1317; e-mail via R. R. Hall

Suzanne Cholerton B.Sc., Ph.D.
Lecturer in Pharmacology, Department of Pharmacological Sciences,
The Medical School, University of Newcastle, Newcastle upon Tyne NE2 4AA,
United Kingdom
Tel: 0191 222 8511; Fax: 0191 222 7230; e-mail: suzanne.cholerton@ncl.ac.uk

Mary K. Gospodarowicz MD, FRCPC
Department of Radiation Oncology, University of Toronto and Princess Margaret
Hospital, 610 University Avenue, Toronto, Ontario M5G 2M9, Canada
Tel: 00 1 416 946 4421; Fax: 00 1 416 946 6556; e-mail:
marykg@pmh.toronto.on.ca

Reginald R. Hall MS, FRCS
Macmillan Lead Clinician, Northern Cancer Network, Visiting Professor at the University of Newcastle, Northern Cancer Network, Derwent Court, Freeman Hospital, Newcastle upon Tyne NE7 7DN, United Kingdom
Tel: 0191 223 1313; Fax: 0191 223 1317; e-mail:r.r.hall@ncl.ac.uk

Harry W. Herr MD
Attending Surgeon, Department of Surgery, Memorial Sloan-Kettering Cancer Center, 1275 York Avenue, New York 10021, USA
Tel: 00 1 212 639 8193; Fax: 00 1 212 717 3175

Alan Horwich Ph.D., FRCP, FRCR
Chair, Academic Unit of Radiotherapy and Oncology, The Royal Marsden Hospital, Downs Road, Sutton, Surrey SM2 5PT, United Kingdom
Tel: 0181 642 6011 ext. 3274; Fax: 0181 643 8809

Gerhard Jakse MD
Professor and Chairman, Clinic of Urology, University Clinic of Rheinisch-Westfälische Technische Hochschule, Medical Faculty, Aachen, Germany
Tel: 00 49 241 8089377; Fax: 00 49 241 8888441

Saad Khoury MD
Professor of Urology, Clinique Urologique, Hôpital de la Pitié, 83 bd. De l'Hôpital, 75634 Paris Cedex 13, France
Tel: 00 33 1 42 177120; Fax: 00 33 1 42 177122; e-mail: khoury@pratique.fr

Karlheinz Kurth Ph.D., MD
Professor and Chairman, Department of Urology, Academic Medical Center/University of Amsterdam, Meibergdreef 9, 1105 AZ Amsterdam, The Netherlands
Tel: 00 31 30 5663775/5666030; Fax: 00 31 20 6911389; e-mail: Urologie@AMC.UVA.NL

Donald L. Lamm MD
Chair, Department of Urology, WVU Health Sciences Center, PO Box 9251, Morgantown, WV 26505–9251, USA
Tel: 00 1 304 293 2706; Fax: 00 1 304 293 2807

Christopher J. Logothetis MD
Chair, Department of GU Medical Oncology, Internist/Professor of Medicine, M. D. Anderson Cancer Center, The University of Texas, 1515 Holcombe Boulevard, Houston, Texas 77030, USA
Tel: 00 1 713 7922 830; Fax: 00 1 713 792 7573

Luc M. F. Moonen MD
The Netherlands Cancer Institute, Antoni van Leeuwenhoek Huis,
Plesmanlaan 121, 1066 CX Amsterdam, The Netherlands
Tel: 00 31 20 512 2117; Fax: 00 31 20 669 1101

Patricia Ongena
Clinical Data Manager, AZ Middelheim, Lindendreef 1, 202 Antwerpen, Belgium
Tel: 00 32 3 2803226; Fax: 00 32 3 2184696

Mahesh K. B. Parmar D.Phil., M.Sc.
Acting Chief Medical Statistician, MRC Cancer Trials Office, 5 Shaftesbury
Road, Cambridge CB2 2BW, United Kingdom
Tel: 01223 311110; Fax: 01223 311844; e-mail: mp@mrc-cto.cam.ac.uk

Derek Raghavan MBBS, Ph.D., FRACP, FACP
Professor of Medicine and Urology, Chief, Division of Medical Oncology,
Associate Director, USC-Norris Comprehensive Cancer Center, University of
California, 1441 Eastlake Avenue, Room 3450, Los Angeles, CA 90033, USA
Tel: 00 1 323 865 3900; Fax: 00 1 323 865 0061; e-mail: draghavan@hsc.usc.edu

J. Trevor Roberts MBBS (Lon), FRCP (UK), FRCR
Consultant Clinican Oncologist, Northern Centre for Cancer Treatment,
Newcastle General Hospital, Westgate Road, Newcastle upon Tyne NE4 6BE,
United Kingdom
Tel: 0191 219 4200 ext. 4245; Fax: 0191 272 4236

Mary C. Robinson FRCPath, MBBS
Consultant Histopathologist, Freeman Hospital, High Heaton,
Newcastle upon Tyne NE7 7DN, United Kingdom
Tel: 0191 284 3111 ext. 26537; Fax: 0191 223 1317; e-mail via R. R. Hall

Donald G. Skinner MD
Professor and Chairman, Department of Urology, University of Southern
California, 1441 Eastlake Avenue, #7414, Los Angeles, CA 90033, USA
Tel: 00 1 213 764 3707; Fax: 00 1 213 764 0120

Philip H. Smith MB, FRCS
Medical Director, St James's University Hospital, Beckett Street, Leeds LS9 7TF,
United Kingdom
Tel: 0113 206 5431; Fax: 0113 206 4954

Urs E. Studer MD
Professor and Chairman, University of Berne, Department of Urology, Inselspital/Anna Seiler-Haus, 3010 Bern, Switzerland
Tel: 00 41 31 632 3641; Fax: 00 41 31 632 2180

Richard J. Sylvester D.Sc.
Assistant Director Biostatistics, European Organisation for Research and Treatment of Cancer Data Center, 83 Avenue E Mounier, Box 11, 1200 Brussels, Belgium
Tel: 00 32 2 774 1613; Fax: 00 32 2 772 3545

William H. Turner MD, FRCS (Urol)
56 Thorp Arch Park, Thorp Arch, West Yorkshire LS23 7AN, United Kingdom
Tel: 0113 243 2799; Fax: 0113 292 6490

Adrian P. M. van der Meijden Ph.D., MD
Chairman, Genito-Urinary Tract Cancer Co-operative Group, Groot Ziekengasthuis's Hertogenbosch, Postbuus 90153, NL – 5200 ME Den Bosch, The Netherlands
Tel: 00 31 736 862434; Fax: 00 31 616 2373

Padraig Warde MB, MRCPI, FRCPC
Associate Professor, Department of Radiation Oncology, University of Toronto and Princess Margaret Hospital, 610 University Avenue, Toronto, Ontario M5G 2M9, Canada
Tel: 00 1 416 946 4421; Fax: 00 1 416 946 6556

Preface

'Come and see this fascinating case of bladder cancer' was David Wallace's greeting on my first visit to the Royal Marsden Hospital in 1971. His fascination was matched by his enthusiasm to discover the cause and improve the treatment of bladder cancer. Both were infectious. Nearly 30 years later we know much more about the disease and have refined its treatment considerably but the urothelium remains an enigma, we do not yet understand the cause of bladder cancer nor can we cure it. Surgery removes the cancer, radiotherapy and chemotherapy interfere with its growth. All three control the disease with variable success. None reverses the underlying neoplastic process and even the need for randomized trials is a sign of our therapeutic inadequacy.

In this book colleagues of different opinions and backgrounds have described and discussed what we have to offer our patients: clinical examination, endoscopy, light microscopy, imaging, organ preserving and radical surgery, radiotherapy and chemotherapy. In recent years a series of international conferences has sought to reach consensus about the management of bladder cancer, and in large measure these conferences have succeeded. Inevitably, however, differences of opinion and practice persist and an awareness of these differences can be educational. To make the most of this wider experience, comment has been invited from international colleagues, and where they have considered this to be desirable and appropriate their commentary has been added to the main chapter.

Prevention may be better than cure but, at present, is equally difficult. None the less, the role of cigarette smoking in bladder cancer has been presented in detail to draw attention to its clinical importance and to encourage further research. How many patients will be persuaded to stop smoking is uncertain; how many can be prevented from starting is beyond clinical management. For the foreseeable future, it seems that improvements in the treatment of bladder cancer will occur only as small steps. Given that benefit always bears a cost, it is important to be sure that improvements are real rather than apparent. Hence the chapter on statistics. Until somebody stumbles on that crucial aetiological factor that enables us to prevent or cure bladder cancer, all our patients with this disease will suffer inconvenience, discomfort, indignity and anxiety. Many will die of their cancer despite our expertise. Talking with patients, sharing their fears, restoring confidence in their bodies and themselves, exploring the positive side of life without a bladder, these tend not to be strong points of urology or non-surgical oncology. It seems appropriate to give the last words to our patients.

Acknowledgements

I wish to thank the colleagues who have contributed to this book, each in their own inimitable way. We meet infrequently, so the faxes, letters, courier packages and e-mails have been awaited eagerly, proved thought-provoking and have enriched my own understanding of bladder cancer. Writing books of this nature is always an extra to busy professional lives, at the expense of precious time with family or other activities. To all our partners, and to Val in particular, thank you.

Floppy discs and e-mail have simplified communication, but manuscripts still require a lot of work. Judie Graham's impeccable secretarial skills are greatly appreciated. The book has been a long time in the writing; I am grateful to Joanna Koster of Arnold for her considerable patience.

The management of patients with bladder cancer is a learning experience shared with many professional colleagues. As I take leave of clinical urology I wish to record my gratitude to those who have made it fun and, I hope, worthwhile for those in our care. Especially to clinical oncologist, Trevor Roberts; nurses Margaret Charlton, Sally Burton and Frances Errington; urologist, Philip Powell; histopathologist, Mary Robinson; data manager, Janice Reading; statisticians, Max Parmar and Richard Sylvester, and the many friends in the MRC Bladder Cancer Working Parties and the EORTC Genito-Urinary Group.

1

The diagnosis of bladder cancer

Reginald R. Hall

INTRODUCTION

The diagnosis of most bladder cancers is very easy. Cystoscopy shows something abnormal, the resectoscope is used to remove it, or at least take generous biopsies, and the pathologist provides the answer. Some cases are a little more difficult, and that is where the suspicion of the skilled urologist – call it clinical acumen if you will – comes into play. Something in the bladder or prostatic urethra looks not quite right and the biopsy, or bladder washing cytology makes the diagnosis. The difficult part of diagnosing bladder cancer is bringing the patient to the point of cystoscopy.

HAEMATURIA

If somebody has visible haematuria they will usually seek medical help immediately and any delay in diagnosis is likely to be the fault of the medical profession or health-care system. Thirty-five per cent of people more than 50 years old who complain of painless haematuria have bladder cancer[73]. A few family doctors still do not appreciate this and continue to treat haematuria as a urinary infection. If the patient is a woman and she has frequency and dysuria as well, bacterial urinary infection is most probably the correct diagnosis. None the less proof is required; a mid-stream urine sample should always be collected for culture before starting antibiotics. To ensure that the infection is not secondary to an underlying tumour, if the urine is not sterile following appropriate antibiotics or if the infection recurs, cystoscopy is advised. In men, urinary infection with or without bleeding should be investigated and any woman or man should be cystoscoped without delay if haematuria is not accompanied by other symptoms. Even under 50 years of age, 10% of patients with painless haematuria have bladder cancer[73].

Some urologists make the sweeping statement that any patient with haematuria, painless or not, should be cystoscoped. This is not realistic. For each bladder cancer seen by a family doctor tens or scores of women will be seen with haematuria, albeit usually accompanied by other symptoms. If the simple principles set out above are followed, the diagnosis of no bladder tumour should be delayed.

Dipstick haematuria

'Dipstick' haematuria, a positive reaction for blood on urine-reagent-strip testing of asymptomatic people, is an entirely different problem, created by the growth in popularity of routine health checks. Should dipstick haematuria in an otherwise healthy person be investigated? At present the answer has to be a reluctant 'yes'.

Mohr *et al.*[48] found a prevalence rate of 13% for microscopic haematuria and only 2.3% with significant urological disease in a large healthy population and concluded that investigation of microscopic haematuria may not be necessary or that investigation should follow only repeated or persistent microscopic bleeding. However, most previous[11,20,21] and subsequent reviews[9,14,67,74] have concluded that microscopic haematuria should be investigated because the yield in terms of significant urological disease was high.

Microscopy of the urine has been replaced by 'dipstick' testing in most routine health examinations. Britton *et al.*[8] found that 20% of 2356 men over 60 years of age had dipstick haematuria on a single or repeated testing for 10 weeks. Not all agreed to be investigated. Seventeen bladder cancers were diagnosed, a pick-up rate of 0.7% overall. None were invading muscle. The number of bladder cancers diagnosed was equivalent to the expected incidence over 5 years in this group of men. Messing *et al.*[44,45] reported a 1.3% incidence of bladder cancer in a similar population-based study of regular dipstick testing in men over 50 years of age. Continued testing for 1 year, rather than 10 weeks probably explains the higher diagnosis rate.

From the urologist's viewpoint Sultana *et al.*[73] reported their experience of asymptomatic dipstick haematuria investigated in a special haematuria clinic. Urine microscopy, culture and cytology were performed together with flexible cystoscopy and intravenous urography (IVU) or an ultrasound scan in all patients. No cancer was found in 131 patients under 50 years, but bladder cancer was confirmed in 4.4% (11/250) over 50 years of age. In addition three transitional cell carcinomas (TCC) of the upper tract (0.8%) and four renal cancers were diagnosed. Patients investigated for visible haematuria during the same period at the same clinic had a 10% bladder cancer incidence under the age of 50 (6/66) and 31% over 50 (53/173). The sensitivity of urine cytology was only 36% overall. Most recently Pickard *et al.*[55] confirmed the likelihood of finding a significant number of bladder cancers. Of 474 patients referred to a haematuria clinic with only dipstick haematuria, 30 had bladder cancer (6.3%). Bullock[9] and Schroder[67] have stated that investigation of any dipstick haematuria is mandatory, but it is not known how much importance is attached by family doctors to a single positive urine test. Many probably do not refer these patients for urological investigations unless there are other symptoms. Ritchie *et al.*[59] reported that 39% of men found to have dipstick haematuria at health check examinations were not referred for investigation, and Fraser *et al.*[18] had found previously that microscopic haematuria detected in hospital outpatient clinics had been investigated in only 41% of cases. The difference in the incidence of bladder cancer in population-based studies and reports from special haematuria clinics suggests that a similar degree of selection still occurs, but by which criteria is not known.

It is common practice to examine the urine microscopically or to repeat the dipstick test as part of the investigation of dipstick haematuria and, in some clinics, to proceed with investigation only if haematuria is confirmed. Lynch *et al.*[42] warned against this as they found malignancy in 4% of patients with initial dipstick haematuria and 10% of patients with initial visible haematuria which was not confirmed by repeat testing: 'Repeating the urine analysis when haematuria has been detected, to determine "a high risk group" for investigation is misleading, as a negative result leads to false reassurance and does not rule out underlying urological pathology.'[42] Schroder[67] recommended several examinations of the urinary sediment and that urine culture was mandatory if white cells were present. This seems unnecessary. Whether haematuria is confirmed or not, cystoscopy is still required and totally asymptomatic bacterial infection needing treatment is most unlikely.

The age of the individual with dipstick haematuria has influenced the decision concerning investigation in some urologists' practice. Bullock[9] recommended investigation irrespective of the age of the patient, but Sultana *et al.*[73] suggested that dipstick haematuria below 50 years of age probably does not require cystoscopy. Jones *et al.*[30] were of a clear opinion that it does not. A prospective study of 100 men with dipstick haematuria aged 16–40 years, found no malignancy, and showed that a careful history and clinical examination were usually sufficient to indicate the underlying diagnosis. If a confident clinical diagnosis could be reached without cystoscopy this final examination could be avoided. Similarly, a study of haematuria in young people with HIV in the US Air Force (mean age 32 for men, 30 for women)[13] revealed that visible or microscopic haematuria occurred in 25%, but cancer was never a cause. The opposite advice was given by Sparwasser *et al.*[71] who found two urothelial cancers in 157 men aged 18–53. However, asymptomatic dipstick haematuria had been investigated only if confirmed by repeat testing, which would have selected cases considerably. These authors also detected glomerulonephropathy in 16.5% (by renal biopsy), which emphasizes the need to be aware of a possible 'nephrological' cause if urological investigations are negative, particularly in younger patients.

Patients on long-term anti-coagulant therapy may be more prone than others to developing haematuria and immediate investigation may not always be considered necessary. However, Savage and Fried[62] found two invasive bladder cancers in 24 patients with visible haematuria, and significant urological disease in 30% of 32 patients with any haematuria while on anti-coagulant therapy. Antolak and Mellinger[3] reported a similar high incidence of underlying pathology revealed by haematuria while on anti-coagulants.

In countries where *Schistosoma haematobium* is endemic, haematuria is the presenting symptom of the primary infection with this parasite. The bladder cancers that arise subsequently tend to be diagnosed late because there is little to distinguish the developing cancer from the symptoms caused by the recurring infections associated with schistosomiasis.

In summary, haematuria is a finding or symptom that should always be noted with concern. **Any** dipstick haematuria in a person over the age of 40 should be investigated, as should painless visible haematuria at any age. If thorough investigation is negative, some colleagues have recommended repeat investigations for up to 3 years, following the proposal of Golin and Howard in 1980[20]. However, when these authors reviewed the outcome after 10 years and found that no cancers had developed, they abandoned follow-up investigation and advised re-examination only if symptoms occurred[27].

Other symptoms

Knowing that 80% of bladder cancers present with haematuria is not particularly helpful, except to emphasize that 20% do not. **Carcinoma in situ** (CIS) usually causes frequency, nocturia, urgency, sometimes suprapubic or penile pain, dysuria and haematuria; symptoms that used to associated with tuberculosis but are now more likely to suggest bladder instability. In a man, these symptoms should therefore arouse suspicion and in the absence of bladder outflow obstruction, cystoscopy, random biopsies and bladder washing cytology should be performed.

Bladder outflow obstruction caused by bladder cancer is rare, but both diseases are common in the same age group. Men presenting with prostatism should be asked specifically about haematuria and a urine sample tested for blood with dipstick. This is particularly important if non-surgical treatment is planned or a long wait anticipated before prostatectomy. For the few men with very large prostates undergoing open prostatectomy, the old warning to perform preliminary cystoscopy is still important.

Recurrent frequency and dysuria, with or without proven infection, is very common in young women. Fortunately most do not have haematuria, but those who do should be cystoscoped. Urinary infections that are difficult to eradicate or recur in older women should also arouse suspicion and indicate cystoscopy.

Most muscle invasive bladder cancers arise *de novo* with no previous history of superficial bladder tumour[33]. When questioned, many will admit to an episode of haematuria that was ignored several years before, or repeated prescriptions for antibiotics for urinary infections that had not been investigated. The average family doctor sees a new patient with bladder cancer every 3–4 years, just one of several hundred consultations about bladder symptoms. Vigilance, common sense and caution are essential.

CYSTOSCOPY

Anticipating a cystoscopy for the first time causes apprehension for most patients. Even if the discomfort of a rigid cystoscopy is minimized by a skilled urologist using topical anaesthetic or by a general anaesthetic, it is an experience to be avoided if possible. The introduction of the flexible cystoscope 14 years ago[56] transformed this situation[16]. Passing the flexible instrument through the external sphincter and prostatic urethra is uncomfortable for some men, but universal experience has confirmed the flexible cystoscope as the single diagnostic investigation of choice for haematuria. Any area of suspicious mucosa can be biopsied and small lesions vapourized or coagulated with diathermy without anaesthesia. If indicated by preliminary ultrasound or IVU, retrograde imaging can be performed as well. Thus for any patient with haematuria the definitive diagnostic process comprises a detailed medical history, careful clinical examination (including rectal or vaginal pelvic examination), measurement of blood pressure, urine testing for protein and flexible cystoscopy. Any further debate concerns the choice between IVU or plain X-ray and an ultrasound scan of the upper urinary tract, and the contribution, if any, of cytology and other urine tests.

IMAGING OF THE UPPER URINARY TRACT

In the study of haematuria by Sultana *et al.*[73], 302 patients underwent both IVU and plain X-ray with ultrasound scan. The latter combination failed to detect two of three cases with isolated upper-tract TCC and the IVU did not detect one of four patients with renal cancer. As expected, ultrasound was much better at distinguishing renal cysts.

Mokulis *et al.*[49] suggested that ultrasound is unnecessary if an IVU has been performed and is normal. While this may be true for the diagnosis of upper-tract TCC, the superiority of ultrasound in diagnosing small renal parenchymal tumours and renal cysts, together with its low cost and zero morbidity, make it difficult to justify its exclusion. Conversely, many colleagues consider that these factors mandate the use of ultrasound instead of IVU. They are supported by the findings of Starling *et al.*[72] who used both in the investigation of 238 patients with haematuria. The accuracy, sensitivity and specificity were equivalent, but the positive predictive value was greater for ultrasound (72.7 versus 50.0). Given the potential for adverse contrast reactions, significant cost and potential for missing renal tumours, ultrasound was preferred by these authors. Inevitably neither investigation is perfect. If a choice has to be made, plain X-ray and ultrasound would appear to be the most efficient overall. As suggested by Ohja *et al.*[52], IVU need be performed only in those patients in whom a cause for haematuria has not been found by cystoscopy, plain X-ray and ultrasound scan.

URINE CYTOLOGY AND OTHER DIAGNOSTIC TESTS

Cells are shed continually from the surface of the urothelium. Urothelial cancers, especially CIS, shed more cells than normal because of their decreased cell adhesion, and it has long been recognized that 'exfoliated cells provide a rich source for cytologic analysis'[37]. This in turn underlies the search for a non-invasive technique for diagnosing bladder cancer that avoids cystoscopy. Papanicolaou staining of the centrifuged urine sediment has been a routine investigation for many years. Most patients suspected of bladder cancer will be asked to pass a urine specimen for cytology and many urologists rely on cytology for their follow-up of bladder cancer patients.

Cytology is especially valuable in carcinoma in situ[58]. It took a long time for urologists to realize that 'denuding cystitis' was in fact CIS and that the diagnosis was floating in the urine rather than in the bladder wall biopsy that contained little or nothing of diagnostic value. For CIS, bladder washing cytology has become a standard investigation for initial diagnosis and follow-up.

Voided urine cytology

For papillary Ta and T1 tumours, the usefulness of cytology is limited. Exfoliated cells from well-differentiated TCC look normal on light microscopy and cells from moderately differentiated tumours are often reported as 'atypical', not even 'suspicious' and certainly not diagnostic. For this reason the **sensitivity** of cytology is poor although its **specificity** is good, as summarized in Table 1.1. Just how poor

Table 1.1 Voided urine cytology: sensitivity and specificity

Author	No. of patients	Sensitivity (%)	Specificity (%)
Bhargava *et al.*[6]	75	47	97
Johnston *et al.*[29]	100	39	86
Leyh *et al.*[40]	414	25	100
Miyanaga *et al.*[47]	40	40	100
Pickard *et al.*[55]	474	27	99
Sultana *et al.*[73]	460	36	99

Table 1.2 Voided urine cytology: sensitivity according to tumour grade and T category

No. of patients	T category	Sensitivity (%)	Grade	Sensitivity (%)
30	Ta	7	1	0
11	T1	64	2	18
6	T2–4	50	3	57
10	TIS	40	–	–

From Leyh *et al.*[41]

the detection of low-grade and low-stage tumours can be is shown in Table 1.2. It is rare for voided urine cytology to diagnose a papillary or invasive cancer of the bladder, urethra, ureter or renal pelvis that is not apparent on cystoscopy, bimanual palpation, ultrasound or IVU. On the basis of recent experience, it is difficult to justify any expenditure on routine voided urine cytology for either the diagnosis or follow-up of Ta or T1 bladder cancer[79]. As with IVU in the diagnosis of haematuria, cytology may be reserved only for those patients in whom genuine doubt about the diagnosis remains after the use of the other more sensitive investigations.

Bladder washing cytology

Barbotage of the bladder provides a more cellular specimen for cytology[76] and has higher sensitivity than voided urine[4]. Its disadvantage is that vigorous washing of the bladder is uncomfortable and is therefore usually performed at the same time as rigid cystoscopy. Barbotage of the bladder through a flexible cystoscope is not possible because the irrigation channel is too narrow. Although bladder washings are used extensively for research purposes, they do not contribute to the diagnosis or follow-up of most bladder cancers; their clinical usefulness is limited to patients suspected of having CIS. Gentle 'washing' of the ureter or renal pelvis is feasible using a 10 ml syringe attached to a rigid ureteroscope. This may be useful if good access to the renal pelvis is not possible. However, it should always be remembered that a false positive cytology result may arise from the trauma of instrumentation, indwelling catheters, stones, urinary infection, radiotherapy, intravesical therapy, ileal conduits or bowel substitution.

Regular cytology is still used for the screening of some high-risk populations exposed to industrial carcinogens, such as men who worked in the past in rubber manufacturing or who continue to handle MbOCA in paint manufacture. This concern for workers' health should be maintained and possibly extended to other industries, but in view of the inefficiency of cytology as a screening test, there is a strong case for its replacement by one of the other tests described below.

There can be not doubt that a reliable, non-invasive test that could replace cystoscopy would be welcomed by patients and the providers of health care, although the loss of income could be a problem for some urologists. In the search for such a test a variety of cytological and behavioural characteristics of bladder cancer have been exploited.

Other tests

Flow cytometry (FCM) measures the DNA content of cell nuclei and thereby gives an indication of the number of cells synthesizing DNA and their proliferative activity. The normal urothelium displays a range of DNA activity; Melamed[43] proposed that the presence of more than 15% aneuploid cells (compared to diploid) should define malignancy in the urothelium. In general, diploid tumours tend to be G1, Ta or T1 and tetraploid or aneuploid tumours are more likely to be poorly differentiated or invading muscle[5,22,75]. Badalament *et al.*[5] demonstrated in 228 patients that flow cytometry varied according to T category; it was greater in TIS and T1 than in Ta. Comparison with cytology in a subset of 103 patients showed that the overall sensitivity of a single FCM examination was 78%, and that the sensitivity was significantly greater than voided urine cytology for all three categories of superficial bladder cancer. Cytology was positive when FCM was negative in 3/103 patients.

De Vere White and Baker[15] adopted a very practical approach and demonstrated that flow cytometry could provide equally satisfactory results with voided urine as with bladder washings. However, 14% of control subjects had aneuploidy so that DNA flow cytometry could not be recommended as a diagnostic test for bladder cancer. Flow cytometry can also be used simultaneously to measure **cytokeratin**[35] or **blood group antigen expression**, and its specificity may be improved by doing so[17]. DNA content can also be measured in individual cells by using automated **quantitative fluorescent image analysis**[12] or fluorescence in situ hybridization (**FISH**)[65,68].

The **nuclear matrix protein (NMP) 22** immunoassay measures a nuclear matrix protein released from urothelial cells into the urine. Nuclear matrix proteins determine nuclear morphology and regulate DNA replication and gene expression[19]. NMP 22 has been developed as a quantitative test[10,47] to predict the presence of TCC or the likelihood of recurrence[70]. Miyanaga *et al.*[47] found raised NMP 22 levels (>10.0 U/ml) in 81% (38/47) of patients with bladder cancer, compared with 36% (74/207) of patients following endoscopic resection of their tumour, or benign controls. Soloway *et al.*[70] reported that 86% of patients with post-transurethral resection (post-TUR) levels below 10 U/ml had no recurrence at the first 3-month cystoscopy, compared with levels of NMP 22 greater than 10 U/ml in post-TUR urine that suggested a high risk of recurrence. These authors have suggested that this may 'prompt more aggressive management if NMP 22 levels are elevated following surgery'. As a diagnostic test, NMP 22 was reported to have a sensitivity of 61% compared with 33% for voided urine cytology[1]. Further published data are awaited, but with positive tests in 36% of controls and only 61% sensitivity for tumour diagnosis, this test in its present form does not appear promising.

The bladder tumour-associated analyte (**BTA**) test has been evaluated more extensively. This is a latex agglutination test that detects the presence of basement membrane complexes in urine. The overall sensitivity and specificity, and the sensitivity according to tumour grade and category are summarized in Tables 1.3 and 1.4. Although more sensitive than cytology, the BTA test has been superseded by the **BTA STAT** test by the same manufacturers. This test measures complement factor H-related proteins in the urine[34]. Leyh *et al.*[41] reported overall sensitivity of the new test for diagnosing bladder cancer to be 72% compared with 28% for cytology. Sarosdy *et al.*[61] compared BTA STAT with BTA in 181 archived urine samples and found superior sensitivity for BTA STAT (66% versus 59%) but equal specificity.

Table 1.3 BTA diagnostic urine test: sensitivity and specificity compared with urine cytology

Author	No. of patients	BTA		Cytology	
		Sensitivity (%)	Specificity (%)	Sensitivity (%)	Specificity (%)
Bhargava *et al.*[6]	75	33	90	47	97
Johnston *et al.*[29]	100	30	85	39	92
Leyh *et al.*[40]	414	70	90	25	100
UK and Eire Study[77]	272	58	86	–	–

Table 1.4 BTA diagnostic urine test: sensitivity according to tumour grade and T category

T category	Sensitivity (%)	Grade	Sensitivity (%)	
Ta	38	G1	17†	41*
T1	87	G2	64	58
T2–4	89	G3	92	70
TIS	75			

* From Ishak et al.[28]
† From Leyh *et al.*[41]

Ishak *et al.*[28] evaluated another alternative, the **BTA TRAK** test, which is a quantitative enzyme immunoassay using monoclonal antibodies to bind bladder tumour antigen in urine. Comparison with BTA in 180 archived samples showed improved sensitivity for BTA TRAK: sensitivity for grades 1, 2 and 3 tumours were 55%, 67% and 85%, compared with 41%, 58% and 70% for BTA.

Measurement of fibrinogen degradation products in urine is the basis of the **Aura-Tek FDP** dipstick rapid immunoassay. Two studies have shown the test to be significantly better than cytology for the detection of bladder cancer. Overall sensitivity of 69–84% and specificity of 78–79% compared with 33–39% sensitivity for cytology has been reported[29,66].

Telomerase is a ribonucleoprotein enzyme that is present in almost all cancer cells but not in benign tissue[39,50]. Telomerase can be measured in the urine or bladder washings[32,38,39] by a fluorescein-labelled PCR (polymerase chain reaction) based technique, or the highly sensitive telomeric repeat amplification protocol (**TRAP assay**)[38,50]. These preliminary studies detected telomerase activity in 90% of 100 voided urine or bladder wash samples from patients with proven bladder cancer, including 17/18 well-differentiated (Grade 1) tumours. Urine from all of the few CIS patients was also positive. Of the six new urine diagnostic tests, telomerase would seem to be the most promising because preliminary results suggest that it is able to detect well-differentiated tumours, something that the other tests and cytology do poorly.

From the practical point of view, a decision has to be made concerning the use of these innovative diagnostic tests in routine practice. The published data suggest that the sensitivity of each test is higher than cytology except for carcinoma in situ. (The specificity of cytology for CIS is very high and probably cannot be improved upon.) Dipstick testing is quicker and less labour intensive than cytology. Given the improved detection rates and current costs of BTA, FDP and NMP 22, these tests are almost certainly more cost effective than cytology. Those who include cytology in their investigations for haematuria would be well advised to consider replacing cytology by one of these tests. However, the diagnostic ability of none compares with cystoscopy and it has yet to be shown that any provide worthwhile additional information. In keeping with cytology they may be excluded from the investigation of haematuria and the diagnosis and follow-up of bladder cancer unless there is residual doubt about the underlying condition after flexible cystoscopy has been performed, or if CIS is suspected or involved.

MUCOSAL BIOPSIES

Early publications that drew attention to invisible carcinoma in situ and dysplasia in the normal-looking bladder mucosa of patients with TCC reported a high incidence of this finding[2,36,63,64]. These reports were based on step sections of cystectomy specimens usually removed for multiple or invasive bladder cancers. None the less they led to the practice of 'bladder mapping' in which multiple random mucosal biopsies have been taken as part of the initial assessment of Ta and T1 bladder tumours, both as an aid to prognosis and as a guide to treatment[25,51,69]. There was some concern that unnecessary biopsies might cause tumour recurrence by implantation[26], but evidence of this has not been forthcoming[69].

The validity of random biopsies was questioned by Richards *et al.*[57] who reported considerable inter- and intra-pathologist variability in their interpretation. The most extreme was the diagnosis of severe dysplasia or CIS in 27% by one pathologist compared with 7% by another, in a review of 92 biopsies. At the final consensus discussion the five pathologists agreed a diagnosis of severe dysplasia/CIS in four of the 92 specimens.

The patients studied by Richards *et al.*[57] all had newly diagnosed, mainly Ta tumours. EORTC protocol 30863[53] recruited similarly good prognosis patients: solitary and mainly Ta. Biopsies of normal mucosa far from the tumour revealed CIS in 2.0%, Ta tumours in 1.5% and T1 in 0.5% of 393 patients. In a subsequent trial of 602[78] multiple or recurrent Ta, T1 bladder tumours, CIS was found in 3.5%, Ta tumours in 8.6% and T1 in 2.8% of normal-looking random biopsies. This information, summarized in Table 1.5, shows that the incidence of CIS or papillary tumour in random biopsies is very low in good-risk patients, and certainly not the serious problem suggested by previous authors. Unsuspected abnormalities are more common in association with multiple and recurrent tumours, although the overall finding in 14.9% is not great. Bollina *et al.*[7] clarified the situation further by analysing the incidence of mucosal abnormalities according to the grade of the primary tumour (Table 1.6). CIS was infrequent with G1 and G2 tumours but very common with G3 cancers, confirming many previous reports of the frequent association of CIS with pT1 G3 bladder cancers.

Richards *et al.*[57] noted that the finding of CIS in a few patients had no impact on subsequent tumour progression, although numbers were very small. The largest study of random biopsies reported to date was by Witjes *et al.*[81]. Univariate analysis

Table 1.5 Random mucosal biopsies: incidence of abnormalities in biopsies from normal-looking bladder mucosa distant from tumour

Author	CIS (%)	Ta (%)	T1 (%)
Richards *et al.*[57] newly diagnosed Ta, T1	4.3	0	0
EORTC 30863[53] solitary Ta, T1	2.0	1.5	0.5
EORTC 30911[78] multiple, recurrent Ta, T1	3.5	8.6	2.8

Table 1.6 Random mucosal biopsies: the incidence of carcinoma in situ and abnormalities in random mucosal biopsies at first follow-up cystoscopy, according to grade of primary tumour

Grade	No. of patients	CIS at diagnosis (%)	Abnormal random biopsy at first follow-up cystoscopy (%)
G1	85	5.8	7.1
G2	95	5.3	12.6
G3	67	47.8	34.0

From Bollina *et al.*[7]

of 1026 patients with Ta and T1 bladder cancer showed a slightly higher risk of recurrence for patients with mucosal abnormalities (47.5% versus 44.5% for abnormal biopsies). However, multivariate analysis demonstrated that the outcome of random biopsies was of no additional prognostic value. In view of this, it is relevant that Parmar *et al.*[54] were unable to include the result of mucosal biopsies in their analysis of prognostic factors because of the variability of their histological interpretation, reported previously by Richards *et al.*

Carcinoma in situ of the bladder is frequently multifocal and the advisability of gaining an approximate estimate of its extent by taking multiple biopsies before starting treatment is generally recognized. However, pT1 G3 tumours are also a cause for clinical concern, mainly because of the high risk of local progression. What is not often acknowledged is that 50% of these patients have carcinoma in situ elsewhere in the bladder and the prostatic urethra[31]. It may be the frequent association with CIS, as much as the invasive potential of the primary tumour, that explains the poor prognosis of so many of these tumours. Some urologists recommend immediate cystectomy; others favour conservative treatment with transurethral resection (TUR) and bacille Calmette–Guérin (BCG), or possibly radiotherapy. For the former, knowledge of the state of the prostatic urethra is important before cystectomy[24]. For the latter, knowledge of the presence and extent of CIS will almost certainly determine the choice of BCG and radiotherapy.

The frequency of mucosal abnormalities in association with **muscle invasive bladder cancer** has been poorly documented. Wolfe *et al.*[82] reported that the presence of concomitant CIS predicted failure of radiotherapy. Hardeman and Soloway[24] drew attention to the need for random **biopsy of the prostate** to identify TCC in the prostatic urethra or prostatic ducts before embarking on cystectomy. Their review of 86 cystectomies found that 37% (11/30) of patients with prostatic TCC developed urethral relapse, compared with 4% (2/56) when the prostate was free of urothelial cancer. One of the latter patients had multiple tumours and CIS in the bladder, the other a single tumour. Several other authors have reported the frequency with which prostatic TCC can occur in conjunction with multiple Ta, T1 tumours in the bladder, particularly poorly differentiated, carcinoma in situ and muscle invasive bladder cancer[23]. A decade ago this was of little importance as surgical fashion favoured cysto-prostato-urethrectomy in most patients, but since the advent of orthotopic bladder substitution urologists have sought every opportunity to leave the urethra *in situ*. If this is the preferred option, it is important to know beforehand if there is TCC in the prostatic urethra. This information is obtained by taking several TUR biopsies from the bladder neck to the distal end of the verumontanum at 5 and 7 o'clock, making sure that underlying prostate stroma is included[60,83].

The relevance of all this information for routine practice would seem to be as follows.

- Random biopsies are not necessary in patients with newly diagnosed or recurrent Ta, T1 tumours unless carcinoma in situ is suspected.
- Patients found on histology to have poorly differentiated (G3) tumours merit repeat cystoscopy assessment including multiple random biopsies.
- Patients with multiple, recurrent tumours will usually receive intensive intravesical therapy and the finding of an abnormality on random biopsy is unlikely to influence the choice of treatment.
- Radiotherapy may not be the appropriate treatment for muscle invasive bladder cancer if CIS is present. A study to determine the incidence of CIS and its response to radiotherapy is needed.
- TUR biopsies of the prostatic urethra should be taken in all patients planned for orthotopic bladder substitution after cystectomy.

SCREENING FOR BLADDER CANCER

The question that arises from the population studies of dipstick haematuria by Britton *et al.*[8] and Messing *et al.*[44] is whether dipstick screening for bladder cancer should be recommended. Woolhandler *et al.*[84] reviewed the literature in 1989 and concluded that fewer than 2% of young adults with a positive dipstick test for blood had a serious and treatable urinary tract disease, which was too few to justify screening and the risks of subsequent work-up. For older populations they considered the evidence to be contradictory.

The only evidence in favour of such a proposal is that provided by a follow-up report by Messing *et al.*[46]. On the same premise as any screening programme that early detection may contribute to better outcome, 1575 healthy men, 50 years of age or older, tested their urine regularly for up to 1 year with clinical reagent strips. Twenty-one bladder cancers were diagnosed, a pick-up rate of 1.3%[44,45]. This finding, and the results of the similar United Kingdom study[8], stirred little public health interest despite detection rates similar to those of breast cancer from mammography.

The significance of these findings lies in the nature and prognosis of the tumours diagnosed. Messing *et al.* also analysed the symptomatic bladder cancers that were diagnosed over the subsequent 18 months in men who had declined to join the screening study[46]. A similar proportion (1.2%) occurred. Compared with these symptomatic cases and all bladder cancers registered during the period of the study, those diagnosed by dipstick screening were less likely to be poorly differentiated or invading muscle, and no patient died of bladder cancer, compared with 16.4% bladder cancer deaths within 2 years in the unscreened cancers (Table 1.7). Furthermore, only two of the 19 screen-detected cancers required cystectomy and among the screened population no additional bladder cancers were diagnosed during the subsequent 18 months. The good prognosis of screen-detected bladder tumours was confirmed by Whelan *et al.*[80] who reported that none of the 17 tumours detected in the British screening study[8] were invading muscle at the time of diagnosis, and only one progressed to muscle invasion during 3 years of follow-up. Of the 2339 men who were either dipstick negative or normal on investigation, only one is known to have developed bladder cancer in 3 years; symptomatic haematuria developed in this patient 6 months after screening negative in the study.

From this evidence, routine urine dipstick testing in men over 50 years of age appears to diagnose bladder cancers at an early stage that are amenable to bladder conserving treatment, and reduces deaths from bladder cancer. For each cancer diagnosed 18 (28) men were worried unnecessarily and 12 (19) underwent unnecessary IVU, cystoscopy and bladder lavage, in the Wisconsin and British

Table 1.7 Dipstick haematuria screening for bladder cancer. Number of bladder cancers diagnosed, their T category and grade, and death from bladder cancer in 1575 men older than 50 years undergoing repeated dipstick urine testing for 1 year compared with the same data for men of the same age who declined screening plus all bladder cancers registered in the Wisconsin area during the same period

	Screened	Unscreened
Total number of new bladder cancers	21	511
Ta, T1: G1, G2	11 (52.4%)	290 (56.8%)
G3 Ta, T1; CIS	9 (42.9%)	99 (19.4%)
T2–4; N+; M+	1 (4.8%)	122 (23.9%)
Deaths from bladder cancer	0	84 (16.4%)

From Messing *et al.*[46]

studies, respectively. However, in addition to bladder cancers a significant number of prostate and kidney cancers were detected as well as benign conditions that merited treatment. Compared with existing breast and cervical cancer screening programmes, and the pressure for prostate-specific antigen (PSA) screening for prostate cancer, the Wisconsin findings provide a compelling rationale for conducting a prospective randomized trial to determine the efficacy of bladder cancer screening.

REFERENCES

1. Akaza H, Miyanaga N, Tsukamoto T *et al.* (1997) Evaluation of nuclear matrix protein 22 (NMP 22) as a diagnostic marker for bladder cancer: a multicentre trial in Japan. *J Urol* **157**: 337, Abstract 1315.
2. Althausen AF, Prout GR and Dalley JJ (1976) Non-invasive papillary carcinoma of the bladder associated with carcinoma in situ. *J Urol* **116**: 575–580.
3. Antolak SJ and Mellinger GT (1969) Urologic evaluation of haematuria occurring during anticoagulant therapy. *J Urol* **101**: 111–113 .
4. Badalament RA, Hermansen DK, Kimmel M *et al.* (1987) The sensitivity of bladder wash flow cytometry, bladder wash cytology and voided cytology in the detection of bladder carcinoma. *J Urol* **137**: 215A, Abstract 447.
5. Badalament RA, Kimmel M, Gay H *et al.* (1987) The sensitivity of flow cytometry (FCM) compared with conventional cytology in the detection of superficial bladder carcinoma. *J Urol* **137**: 215A, Abstract 448.
6. Bhargava V, Singh S, Shankar D *et al.* (1997) Prospective evaluation of cytology and the bladder tumour antigen (BTA) test in the evaluation of patients with haematuria or history of bladder cancer. *J Urol* **157**: 338, Abstract 1319.
7. Bollina PR, Akhtar NR, Labiad AM and Grigor KM (1996) Random mucosal biopsies in superficial bladder carcinoma: when and in which patients. *Eur Urol* **30** (Suppl. 2): 232, Abstract 86.
8. Britton JP, Dowell AC, Whelan P and Harris CM (1992) A community study of bladder cancer screening by the detection of occult urinary bleeding. *J Urol* **148**: 788–790.
9. Bullock KN (1986) Asymptomatic microscopical haematuria. *BMJ* **292**: 645.
10. Carpinito GA, Stadler WM, Briggman JV *et al.* (1996) Urinary nuclear matrix protein as a marker for transitional cell carcinoma of the urinary tract. *J Urol* **156**: 1280–1285.
11. Carson CC, Segura JW and Greene LF (1979) Clinical importance of micro-haematuria. *J Am Med Ass* **241**: 149–150.
12. Carter HB, Amberson JB, Bander MH *et al.* (1987) Newer diagnostic techniques for bladder cancer. *Urol Clin N Am* **14**: 763–769.

13. Cespedes RD, Peretsman SJ and Blatt SP (1995) The significance of haematuria in patients infected with the human immunodeficiency virus. *J Urol* **154**: 1455–1456.
14. Corwin HL and Silverstein MD (1988) The diagnosis of neoplasia in patients with asymptomatic microscopic haematuria: a decision analysis. *J Urol* **139**: 1002–1006.
15. De Vere White RW and Baker WC (1987) Use of flow cytometric analysis of urine in the diagnosis of bladder cancer. *J Urol* **137**: 216A, Abstract 451.
16. Flannagan GM, Gelister JSK, Noble JG and Milroy EJG (1988) Rigid versus flexible cystoscopy. A controlled trial of patient tolerance. *Br J Urol* **62**: 537–540.
17. Fradet Y, Tardif M, Bourget L *et al.* (1990) Clinical cancer progression in urinary bladder tumours evaluated by multiparameter flow cytometry with monoclonal antibodies. *Cancer Res* **50**: 432–437.
18. Fraser CG, Smith BC and Peake MJ (1977) Effectiveness of an outpatient urine screening programme. *Clin Chem* **23**: 2216–2218.
19. Getzenberg RH, Konety BR, Nguyen TT *et al.* (1997) Characterisation of bladder cancer associated nuclear matrix proteins. *J Urol* **157**: 49, Abstract 186.
20. Golin AL and Howard RS (1980) Asymptomatic microscopic haematuria. *J Urol* **124**: 389–391.
21. Greene LF, O'Shaughnessey EJ and Hendricks ED (1956) Study of 500 patients with asymptomatic micro-haematuria. *J Am Med Ass* **161**: 610–613.
22. Griffiths TRL, Mellon JK, Pyle GA *et al.* (1995) p53 and ploidy assessed by flow cytometry in bladder washings. *Br J Urol* **76**: 575–579.
23. Hall RR and Robinson MC (1998) Transitional cell carcinoma in the prostate. European Board of Urology Update Series. *Eur Urol Update Series* **7**: 1–7.
24. Hardeman SW and Soloway MS (1990) Urethral recurrence following radical cystectomy. J Urol **144**: 666–669.
25. Heney NM, Knocks BN, Daly JJ *et al.* (1982) Ta and T1 bladder cancer: location, recurrence and progression. *Br J Urol* **54**: 152–157.
26. Hicks RM (1992) The development of bladder cancer, in *Scientific Foundations of Urology*, 2nd edn (eds GD Chisholm and DI Williams), Heinemann, London, pp. 711–722.
27. Howard RS and Golin AL (1991) Long-term followup of asymptomatic microhaematuria. *J Urol* **145**: 335–336.
28. Ishak LM, Enfield DL, Sarosdy MF *et al.* (1997) Detection of recurrent bladder cancer using a new quantitative assay for bladder tumour antigen. *J Urol* **157**: 337, Abstract 1317.
29. Johnston B, Morales A, Emerson L and Lundy M (1997) A comparative evaluation of point-of-care urine tests for the detection of transitional cell carcinoma (TCC) of the bladder. *J Urol* **157**: 342, Abstract 1336.
30. Jones DJ, Langstaff RJ, Holt SD and Morgans BT (1988) The value of cystourethroscopy in the investigation of microscopic haematuria in adult

males under 40 years: a prospective study in 100 patients. *Br J Urol* **62**: 541–545.

31. Kakizoe T, Matumoto K, Nishio Y *et al.* (1985) Significance of carcinoma in situ and dysplasia in association with bladder cancer. *J Urol* **133**: 395–398.
32. Kavaler E, Schu WP, Chang Y *et al.* (1997) Detection of human bladder cancer cells in voided urine samples by assaying the presence of telomerase activity. *J Urol* **157**: 338, Abstract 1321.
33. Kaye KW and Lange PH (1982) Mode of presentation of invasive bladder cancer: reassessment of the problem. *J Urol* **128**: 31–33.
34. Kinders RJ, Root R, Jones T and Hass GM (1997) Complement factor H-related proteins are expressed in bladder cancers. *Proc Am Ass Can Res* **38**: 29, Abstract 189.
35. Klein A, Zemer R, Buchumensky V *et al.* (1997) Detection of bladder carcinoma: a urine test, based on cytokeratin expression. *J Urol* **157**: 339, Abstract 1326.
36. Koss LG (1979) Mapping of the urinary bladder: its impact on the concepts of bladder cancer. *Human Pathol* **10**: 533–548.
37. Kurth KH (1997) Diagnosis and treatment of superficial transitional cell carcinoma of the bladder: facts and perspectives. *Eur Urol* (Suppl. 1): 10–19.
38. Lance RS, Aldous WK, Blaser J and Brantley Thrasher J (1997) Telomerase activity in solid transitional cell carcinoma (TCC) and bladder washings. *J Urol* **157**: 338, Abstract 1320.
39. Lee D, Yang S, Hong SJ *et al.* (1997) Telomerase activity in bladder wash cytology: is it reliable? *J Urol* **157**: 338, Abstract 1322.
40. Leyh H, Hall RR, Mazeman E and Blumenstein BA (1997) Comparison of the BARD BTA test with voided urine and bladder wash cytology in the diagnosis and management of cancer of the bladder. *Urology* **50**: 49–53.
41. Leyh H, Marberger M, Pagano F *et al.* (1997) Results of a European multicentre trial comparing the BTA STAT test to urine cytology in patients suspected of having bladder cancer. *J Urol* **157**: 337, Abstract 1316.
42. Lynch TH, Waymont B, Dunn JA *et al.* (1994) Repeat testing for haematuria and underlying urological pathology. *Br J Urol* **74**: 730–732.
43. Melamed MR (1984) Flow cytometry of the urinary bladder. *Urol Clin N Am* **11**: 599–608.
44. Messing EM, Young TB, Hunt VB *et al.* (1987) The significance of symptomatic micro-haematuria in men 50 or more years old: findings of a home screening study using urinary dipsticks. *J Urol* **137**: 919–922.
45. Messing EM, Young TB, Hunt VB *et al.* (1992) Home screening for haematuria: results of a multi-clinic study. *J Urol* **148**: 289–292.
46. Messing EM, Young TB, Hunt VB *et al.* (1995) Comparison of bladder cancer outcome in men undergoing haematuria home screening versus those with standard clinical presentations. *Urology* **45**: 387–397.

47. Miyanaga N, Akaza H, Ishikawa S *et al.* (1997) Clinical evaluation of nuclear matrix protein 22 (NMP 22) in urine as a novel marker for urothelial cancer. *Eur Urol* **31**: 163–168.
48. Mohr DN, Offard KP, Owen RA and Melton LJ (1986) Asymptomatic microhaematuria and urologic disease. A population-based study. *J Am Med Ass* **256**: 224–229.
49. Mokulis JA, Arndt WF, Downey JR *et al.* (1995) Should renal ultrasound be performed in the patient with microscopic hematuria and a normal excretory urogram? *J Urol* **154**: 1300–1301.
50. Muller M, Heine B, Heicappell R *et al.* (1997) Telomerase activity in bladder cancer: bladder washings and in urine. *J Urol* **157**: 50, Abstract 187.
51. Murphy WM, Nagy GK, Rad MK *et al.* (1979) Normal urothelium in patients with bladder cancer. A preliminary report from the National Bladder Cancer Collaborative Group A. *Cancer* **44**: 1050–1058.
52. Ohja H, Fletcher CJ and Kadow CK (1997) An ultrasonography-based rapid-access haematuria clinic. *Br J Urol* **79** (Suppl. 4): 31, Abstract 118.
53. Oosterlink W, Kurth KH, Schroder F *et al.* (1993) A prospective European Organisation for Research and Treatment of Cancer, Genitourinary Group randomised trial comparing transurethral resection followed by a single intravesical instillation of epirubicin or water in single stage Ta, T1 papillary carcinoma of the bladder. *J Urol* **149**: 749–752.
54. Parmar MKB, Friedman LS, Hargreave TB and Tolley DA (1989) Prognostic factors for recurrence and follow up policies in the treatment of superficial bladder cancer: report from the British Medical Research Council Subgroup on superficial bladder cancer (Urological Cancer Working Party). *J Urol* **142**: 284–288.
55. Pickard R, Charlton M, Thorpe J and Neal DE (1997) Examination of the urinary sediment in dipstick haematuria. *Br J Urol* **79** (Suppl. 4): 33, Abstract 127.
56. Powell PH, Manohar V, Ramsden PD and Hall RR (1994) A flexible cystoscope. *Br J Urol* **56**: 622–624.
57. Richards B, Parmar MKB, Anderson CK *et al.* (1991) Interpretation of biopsies of 'normal' urothelium in patients with superficial bladder cancer. *Br J Urol* **67**: 369–375.
58. Rife CC, Farrow GM and Utz DC (1979) Urine cytology of transitional cell neoplasms. *Urol Clin N Am* **6**: 599–612.
59. Ritchie CD, Bevan EA and Collier S (1986) Importance of occult haematuria found at screening. *BMJ* **292**: 681–683.
60. Sakamoto N, Tsuneyoshi M, Naito S and Kumazawa J (1993) An adequate sampling of the prostate to identify prostatic involvement by urethral carcinoma in bladder cancer patients. *Urology* **149**: 318–321.
61. Sarosdy MF, Hudson MA, Ellis WJ *et al.* (1997) Detection of recurrent bladder cancer using a new one-step test for bladder tumour antigen. *J Urol* **157**: 337, Abstract 1318.

62. Savage JGV and Fried FA (1995) Anticoagulant associated haematuria: a prospective study. *J Urol* **153**: 1594–1596.
63. Schade ROK and Swinney J (1968) Pre-cancerous changes in bladder epithelium. *Lancet* **2**: 943–948.
64. Schade ROK and Swinney J (1973) The association of urothelial atypism with neoplasia. *J Urol* **109**: 619–622.
65. Schapers RF, Ploem-Zaaijer JJ, Pauwels RP *et al.* (1993) Image cytometric DNA analysis in transitional cell carcinoma of the bladder. *Cancer* **72**: 182–189.
66. Schmetter S, Habicht KK, Lamm DL *et al.* (1996) Results of a multi-centre trial evaluation of Aura-tek FDP: an aid in the management of bladder cancer patients. *J Urol* **155**: 492A, Abstract 725.
67. Schroder FH (1994) Microscopic haemturia requires investigation. *BMJ* **309**: 70–72.
68. Shankey V, Stankiewicz J, Waters B *et al.* (1997) Fluorescence in-situ hybridisation analysis of bladder washings as a means to detect genetic changes in the urothelium in transitional cell bladder cancers. *J Urol* **157**: 339, Abstract 1325.
69. Smith G, Elton RA, Beynon LL *et al.* (1993) Prognostic significance of biopsy results of normal-looking mucosa in cases of superficial bladder cancer. *Br J Urol* **55**: 665–669.
70. Soloway MS, Briggman JV, Carpinito GA *et al.* (1996) Use of new tumour marker, urinary NMP 22, in the detection of occult or rapidly recurring transitional cell carcinoma of the urinary tract following surgical treatment. *J Urol* **156**: 363–367.
71. Sparwasser C, Cimniak HU, Treiber U and Pust RA (1994) Significance of the evaluation of asymptomatic microscopic haematuria in young men. *Br J Urol* **74**: 723–729.
72. Starling JF, Camerer A, Tiwari A *et al.* (1997) Ultrasonography vs intravenous pyelography in the initial evaluation of patients with haematuria. *J Urol* **157**: 124, Abstract 485.
73. Sultana SR, Goodman CM, Byrne DJ and Baxby K (1996) Microscopic haematuria: urological investigation using a standard protocol. *Br J Urol* **78**: 691–698.
74. Thompson IM (1987) The evaluation of microscopic haematuria: a population based study. *J Urol* **138**: 1189–1190.
75. Tribukait B, Gustafson H and Espositi PL (1982) The significance of ploidy and proliferation in the clinical and biological evaluation of bladder tumours: a study of 100 untreated cases. *Br J Urol* **54:** 130–135.
76. Trott PA and Edwards L (1973) Comparison of bladder washings and urine cytology in the diagnosis of bladder cancer. *J Urol* **110**: 664–666.
77. United Kingdom and Eire Bladder Tumour Antigen Study Group (1997) The use of the bladder-tumour associated analyte test to determine the type of

cystoscopy in the follow up of patients with bladder cancer. *Br J Urol* **79**: 362–366.

78. van der Meijden APM, Oosterlink W, Sylvester R and members of the EORTC GU Group Superficial Bladder Committee (1998) The significance of bladder biopsies in Ta, T1 bladder tumours. Report from the EORTC GU Group, in press.
79. Vogeli TA, Thiel R, Grimm MO and Ackermann R (1997) Does urine cytology contribute to decision making for r-TUR of transitional cell carcinoma of the bladder. *J Urol* **157**: 382, Abstract 1497.
80. Whelan P, Britton JP and Dowell AC (1993) Three-year follow-up of bladder tumours found on screening. *Br J Urol* **72**: 893–896.
81. Witjes JA, Kiemeney LALM, Verbeek ALM *et al.* (1992) Random biopsies and the risk of recurrent superficial bladder cancer: a prospective study in 1026 patients. *World J Urol* **10**: 231–234.
82. Wolf H, Olsen PR and Hojgaard K (1985) Urothelial dysplasia concomitant with bladder tumours: a determinant for future new occurrences in patients treated by full-course radiotherapy. *Lancet* **1**: 1005–1008.
83. Wood DP, Montie JE, Pontes JE and Levin HS (1989) Identification of transitional cell carcinoma of the prostate in bladder cancer patients: a prospective study. *J Urol* **142**: 83–85.
84. Woolhandler S, Pels RJ, Bor DH *et al.* (1989) Dipstick urinalysis screening of asymptomatic adults for urinary tract disorders. (1) Hematuria and proteinuria. *J Am Med Ass* **262**: 1215–1220

Commentary

Harry W. Herr

Professor Hall provides a particularly lucid and complete treatise covering all aspects of the diagnosis of bladder cancer. The observations stem from solid interpretation of relevant studies and translating their lessons into practical recommendations. Perhaps most compelling is his emphasis on 'clinical acumen' which is the basis of accurate diagnosis and follow-up of bladder tumours. All of these sophisticated tests and methods currently available or evolving do not replace common sense and experience!

The diagnosis of bladder cancer is made by cystoscopy and biopsy. There is no reliable substitute. This procedure is indicated for haematuria, dysuria, or for that matter any lower urinary tract symptoms in adult patients, especially among smokers and women. Smokers (including prior smokers) have a higher incidence of bladder cancer than individuals who never smoked. Women present with higher-stage bladder cancer and tend to fare worse, stage for stage, than men[3].

The onus of cystoscopy as an uncomfortable, morbid procedure has largely been eliminated by the flexible cystoscope. Flexible cystoscopy can be accomplished with minimal discomfort in both men and women. Topical anaesthetic jelly often facilitates the ease of the procedure both mentally and physically, by relaxing the external sphincter in men and reducing discomfort during visual passage of the instrument. This probably owes more to adequate lubrication than to its local anaesthetic effect. Over the past year our Urology Department performed 2696 outpatient flexible cystoscopies of which 21% (566 patients) included fulguration of small papillary tumours[4]. Only one patient was unable to tolerate the procedure.

URINE CYTOLOGY

The point about cytology is that a positive result detects reliably dangerous high-grade Ta, T1 and TIS tumours, whereas a negative cytology is often associated with relatively innocuous papillomas or low-grade Ta tumours. Prompt diagnosis of a Ta G1 tumour is not compelling but cystoscopy provides sure diagnosis. Other tests such as FCM, NMP 22, BTA or Aura-Tek FDP, are unlikely to show sufficiently superior sensitivity and specificity to become practical substitutes for voided urine cytology, especially when used with judicious cystoscopy[5].

TELOMERASE

It is suggested that telomerase is a promising tumour marker for bladder cancer detection. We evaluated telomerase prospectively with the developers of the TRAP assay[2]. In 66 consecutive patients with a history of bladder tumour, the TRAP assay identified correctly 50% of the patients with pathologically confirmed bladder tumours, while the sensitivity of voided urine cytology was 71%. The sensitivity of voided urine cytology was 50% for Ta, 92% for T1 and 62% for T2+ and 100% TIS tumours, compared to 46%, 50%, 18% and 20% respectively for the TRAP assay. The TRAP assay is reproducible and is not dependent on the expertise of the cytopathologist. Despite these disappointing preliminary results, telomerase should be investigated further in bladder cancer patients.

MUCOSAL BIOPSIES

Cystoscopy aims to discover tumours and mucosal abnormalities in the bladder and urethra. Transurethral resection (TUR) aims to remove all of these abnormalities completely for histological evaluation. Such philosophy suggests that random or selected-site biopsies of normal mucosa are of little or no use in the evaluation of bladder tumours. A post-TUR urine cytology is more accurate in indicating the presence of diffuse neoplasia. Besides, I have yet to see a biopsy of normal-appearing mucosa show frank carcinoma in situ, either adjacent to or remote from a bladder tumour.

TCC IN THE PROSTATIC URETHRA

The importance of prostatic urethral biopsy to detect occult disease in this extravesical site is stressed correctly. Extension of bladder cancer to involve the prostatic epithelium has important implications for prognosis and management, especially in favour of cystectomy rather than continued conservative therapy. However, are we to infer that a man with CIS involving the prostatic urethra and ducts is not a candidate for orthotopic urinary diversion after cystectomy? I think not. Prostatic disease is usually limited to the surgical specimen, as confirmed by negative distal urethral margins. The quality of life afforded and the low incidence of urethral recurrence even with prostatic tumour makes internal diversion applicable in most men.

BLADDER CANCER SCREENING

Hall makes a case for prospective bladder cancer screening. I agree, especially if conducted in heavy smokers and patients exposed to known bladder cancer carcinogens. The value is to detect not only bladder tumours at an early stage but also other primary neoplasms, such as prostate or kidney[1]. A secondary benefit of screening may be to enrol patients in smoking cessation programmes.

COMMENTARY REFERENCES

1. Begg CB, Zhang Z-F, Sun M, Herr HW and Schantz SP (1995) Methodology for evaluating the incidence of second primary cancers with application to smoking-related cancer. *Am J Epidemiol* **142**: 635–655.
2. Dalbagni G, Han W, Herr HW *et al.* (1998) Evaluation of the telomeric repeat amplification protocol (TRAP) assay for telomerase as a diagnostic modality in bladder cancer. *Clin Can Res,* in press.
3. Fleshner NE, Herr HW and Stewart AK (1996) The National Cancer Data Base report on bladder cancer. *Cancer* **78**: 1505–1513.
4. Herr HW (1990) Outpatient flexible cystoendoscopy with fulguration in the management of recurrent bladder tumours. *J Urol* **144**: 1365–1366.
5. Schwalb DM, Herr HW and Fair WR (1993) The management of clinically unconfirmed positive urinary cytology. *J Urol* **150**: 1751–1756.

2

Histopathology of urothelial cancer: consensus or controversy?

Mary C. Robinson and Reginald R. Hall

The paramount importance of histological examination in the diagnosis of bladder cancer has been recognized for many years and has been confirmed by each successive edition of the TNM classification. However, this important principle has not always been followed and many surgeons have yielded to the temptation of destroying papillary lesions of the bladder without taking a biopsy. Despite claims to the contrary, no urologist can guess the histological grade of a bladder tumour from its cystoscopic appearance, nor be absolutely certain of the depth of tumour invasion. The histopathologist is vital to the management of **all** patients with bladder cancer.

Much of this work may be regarded as routine, but a number of features are contentious or ill understood and are worthy of further discussion. The term 'papilloma' is still used by some colleagues for well-differentiated non-invasive papillary tumours. Is this appropriate, or should almost all papillary tumours be considered to be malignant and therefore called carcinoma? Clinicians expect tumours to be divided according to a grading system and often rely on this information to guide treatment decisions. It is accepted that no system is perfect but how reliable is the separation of one grade from another in bladder cancer, and how accurate is the prognostic information this provides? Carcinoma in situ is an uncommon but serious clinical problem. For many pathologists CIS means only high-grade flat, non-invasive carcinoma, but others are of the opinion that less severe degrees of dysplasia should be included in the diagnosis. Are urologists aware of this; do lower grades of dysplasia require treatment? A number of publications have drawn attention to the fact that pathologists do not always agree about the grade or depth of invasion of papillary bladder cancers. Is this only of relevance to randomized clinical trials or does the subjectivity of light microscopy have an impact on clinical management?

PAPILLOMA OR CARCINOMA?

The term 'papilloma' of the bladder, together with euphemisms such as 'warts in the bladder' were common parlance a decade and more ago. The latter is heard rarely today but the use of 'papilloma' is varied and its correctness debatable. For some authors papilloma has been synonymous with the common, well-differentiated, non-infiltrating papillary Ta G1 transitional cell tumour[47,64,128], while others restrict their use of the term to very rare papillary lesions that are considered to be benign[60]. In a comprehensive review of the literature, Eble and Young[29] document the development of the papilloma–papillary carcinoma controversy which has 'troubled pathologists since the earliest days of our specialty'.

According to the World Health Organization (WHO) classification[77] a transitional cell papilloma is:

> a papillary tumour with a delicate fibrovascular stroma covered by regular transitional cell epithelium indistinguishable from that of the normal bladder and not more than six layers in thickness. The individual cells are slender, parallel to each other and at right angles to the basement membrane. The nuclei are uniform with normal distribution of chromatin. Mitotic figures are absent or rare, and if present, they are in a basal location.

This was qualified with the statement that 'the designation transitional cell papilloma is based on purely histological considerations and is not intended to imply innocent biological behaviour, particularly as such epithelial tumours may recur and behave in a malignant manner'. In practice transitional cell papilloma as restrictively defined is a very uncommon tumour, accounting for less than 2% of large series of primary bladder tumours[40,72].

Histological definitions of 'papilloma' have varied and attitudes have developed from denying the existence of benign papillary lesions, accepting their reality but conceding an inability to recognize them reliably, to an acceptance of rigidly defined papillary lesions as papillomas[97]. Koss[60] allowed normal or nearly normal urothelium composed of not more than seven layers and without nuclear abnormalities. Pugh[101] additionally suggested that papilloma should be restricted to a tumour with cylindrical shape and parallel sides with 4/5 layers indistinguishable from those of normal bladder. Olsen[88] recorded that papillary tumours with the slightest degree of epithelial anaplasia, epithelial thickening above seven layers, increased numbers of mitoses or nuclear atypia should be excluded from the diagnosis and called carcinoma. Ayala and Ro[13] did not find the distinction between a papilloma and a Ta G1 papillary tumour to be of major clinical significance, because even Grade 1 papillary carcinomas, particularly when solitary, are managed conservatively. They rarely applied the term papilloma to lesions in patients above the fifth decade, but found the term convenient in young patients with a solitary papillary tumour with the bland features noted. The intention was to discourage patient anxiety and overly assiduous follow-up.

It is undoubtedly useful to have a category called 'papilloma', to describe the rare papillary lesions covered by normal urothelium that do occasionally arise. The more contentious problem (than whether a benign papilloma exists) is whether WHO Grade 1 carcinomas should be referred to as papillomas. This choice of terminology is important from the point of view of clinical management because the use of the term papilloma suggests a benign tumour that will neither recur nor metastasize, and will therefore need neither adjuvant treatment nor follow-up. The debate has been reopened by the proposal that Ta G1 tumours should be called papilloma in the grading system described in the most recent Armed Forces Institute of Pathology, Fascicle 11[81]. This represents a change from the terminology of the previous series[60] in which the term 'papilloma' was applied only to papillary tumours without significant cytological abnormalities and which were considered to have a much better prognosis than papillary carcinoma Grade 1. Murphy[79] has previously cited his reasons for regarding Ta G1 tumours as papillomas, namely, almost 50% of patients with a Ta G1 tumour have no recurrence after primary resection, neither invasion nor metastases are well documented, and new tumours that occur in 50% of patients are almost always low-grade, non-invasive papillary lesions, similar to the primary tumour.

Jordan *et al.*[58], with Murphy, studied a series of patients with long-term follow-up of transitional cell neoplasms. Ten- and 20-year survivals for 91 patients with Grade 1 tumours were 98% and 93% respectively. Eighty-four patients had disease which did not progress and had normal life expectancies regardless of the number of recurrences. However, 4.4% of patients died of bladder cancer 9–15 years after diagnosis, and a further 3.3% progressed to papillary TCC Grade 3 but did not die of bladder cancer. Despite this overall progression of 7.7%, the authors concluded that Grade 1 TCC was not aggressive and should be redesignated transitional cell papilloma.

In a more recent study, Prout *et al.*[100] reported 178 patients with Grade 1 Ta bladder tumours that were followed for 1–10 years. There were 419 recurrent tumours in 109 patients (61%). There was progression to a more advanced grade in 16% of patients, and invasion developed in 4.5%, involving subepithelial connective tissue in five patients and detrusor muscle in three patients. Only one patient died of bladder cancer. Prout *et al.*[100] concluded that in general, Grade 1 tumours rarely behave as malignancies, but that their presence suggests that in some patients there is a gradually expressed perturbation of the vesical urothelium with subsequent development of higher grade and sometimes invasive cancer. The authors suggested that Grade 1 tumours might best be referred to as Stage Ta Grade 1 transitional cell **tumours**, as in the grading system of Berqvist, rather than carcinomas, but did not support the use of the term papilloma. Heney *et al.*[53] reported that seven of 116 Grade 1 tumours invaded lamina propria **at diagnosis** and that muscle invasion or metastasis occurred in 2%, the majority within 2 years.

The National Bladder Cancer Collaborative Group A[84] concluded that recurrences are actually new occurrences of tumour unless the primary lesion has been resected or destroyed incompletely. Therefore it is as much the behaviour of the rest of the bladder mucosa as that of the primary lesion that confirms that the underlying process is malignant rather than benign. While some Ta G1 tumours will fulfil the expectations of a benign lesion, some will not, and to apply the term 'papilloma' to well-differentiated tumours as a whole is to instil a sense of false security for some of them. As, at present, light microscopy can distinguish only rarely the tumour that over a lifetime of follow-up proves to be genuinely benign and non-recurring from those that will recur or demonstrate frankly malignant characteristics, TCC is the safest name to use to describe this disease, retaining the term 'papilloma' as defined by the WHO classification for the very small number of tumours conforming to this strict definition, and for which confident predictions of cure can be made. Having established the appropriate measure of caution thereby, the pathologist and clinician can utilize other factors described in Chapter 4 to refine both prognosis and choice of treatment for Ta G1 disease, until such time as routinely applicable and reproducible histological and non-histological methods of defining risk for an individual tumour are developed.

TUMOUR GRADE

For a grading system to be successful certain criteria need to be met[36]:

- A range of grades (cellular morphology) must be evident in different patients.
- There should be a uniform grade in all samples from the same tumour.
- The separation into different grades must have prognostic relevance.
- There should be a significant number of tumours in each of the separate grades.

Several grading systems have been described, the best known of which are those of Berqvist *et al.*[15] WHO[77] and Murphy *et al.*[81]. These are summarized in Table 2.1.

Table 2.1 Comparison of grading sytems for bladder tumours

WHO[77]	Murphy[81]	Bergvist[15]
Papilloma		Grade 0 urothelial tumour
	Papilloma	
Grade 1 TCC		Grade 1 urothelial tumour
	Low-grade carcinoma	
Grade 2 TCC		Grade 2 urothelial tumour
	High-grade carcinoma	
Grade 3 TCC		Grade 3 urothelial tumour
Undifferentiated carcinoma		Grade 4 urothelial tumour

Exact extrapolation between individual grading sytems is difficult to achieve and equivalents are approximate.

Exact extrapolation between individual grading systems is difficult to achieve and to facilitate the sharing of experience in publications and the analysis of clinical trials, the use of a single universally accepted grading system would be an advantage. The features common to the different grading systems are brought together in the WHO classification and this is the one that we recommend[77]. It has three grades of TCC. **Grade 1** applies to tumours with the least degree of cellular anaplasia compatible with a diagnosis of malignancy. **Grade 3** applies to tumours with the most severe degree of cellular anaplasia. **Grade 2** lies in between. The criteria given for anaplasia are increased cellularity, nuclear crowding, altered cell polarity, irregular cell size, nuclear pleomorphism, altered chromatin pattern, displaced or abnormal mitotic figures and giant cells. The drawback is that no specific histological criteria are described to distinguish individual grades, although the latter were well illustrated in the original WHO publication[77].

To some extent grading is artificial, in that there is a spectrum of change within urothelial tumours rather than discrete steps in differentiation[19], and the criteria used for individual grades are open to variation of interpretation between pathologists[89]. Some use intermediate grades, e.g. 'Grade 1/2' for a tumour with differing areas of differentiation, while others apply only the highest (worst) grade[21,77]. The features of individual grades in the WHO system have been defined by Ayala and Ro[13] who provide a useful practical reference, as follows.

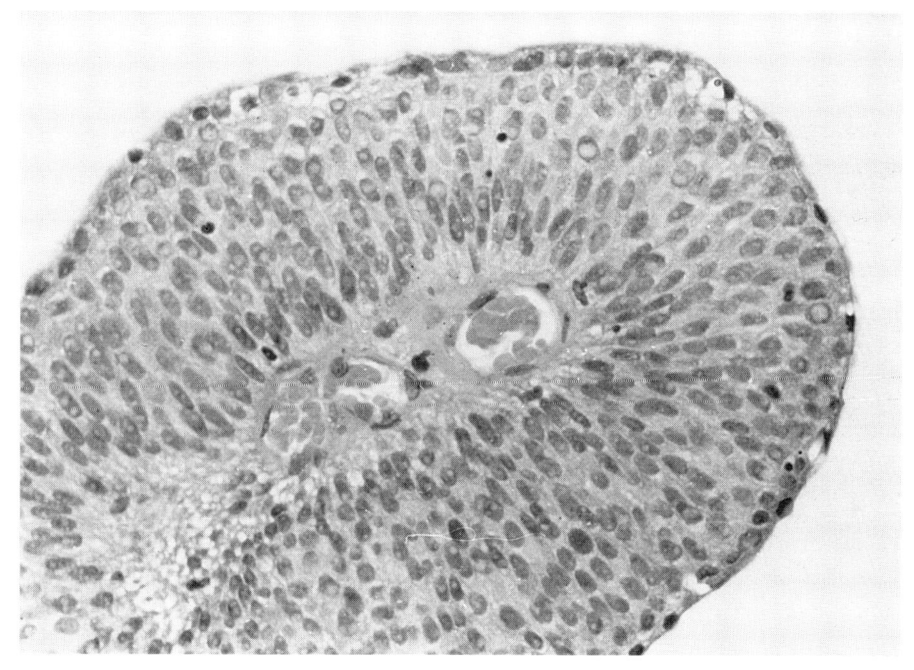

Figure 2.1 Grade 1 papillary transitional cell carcinoma.

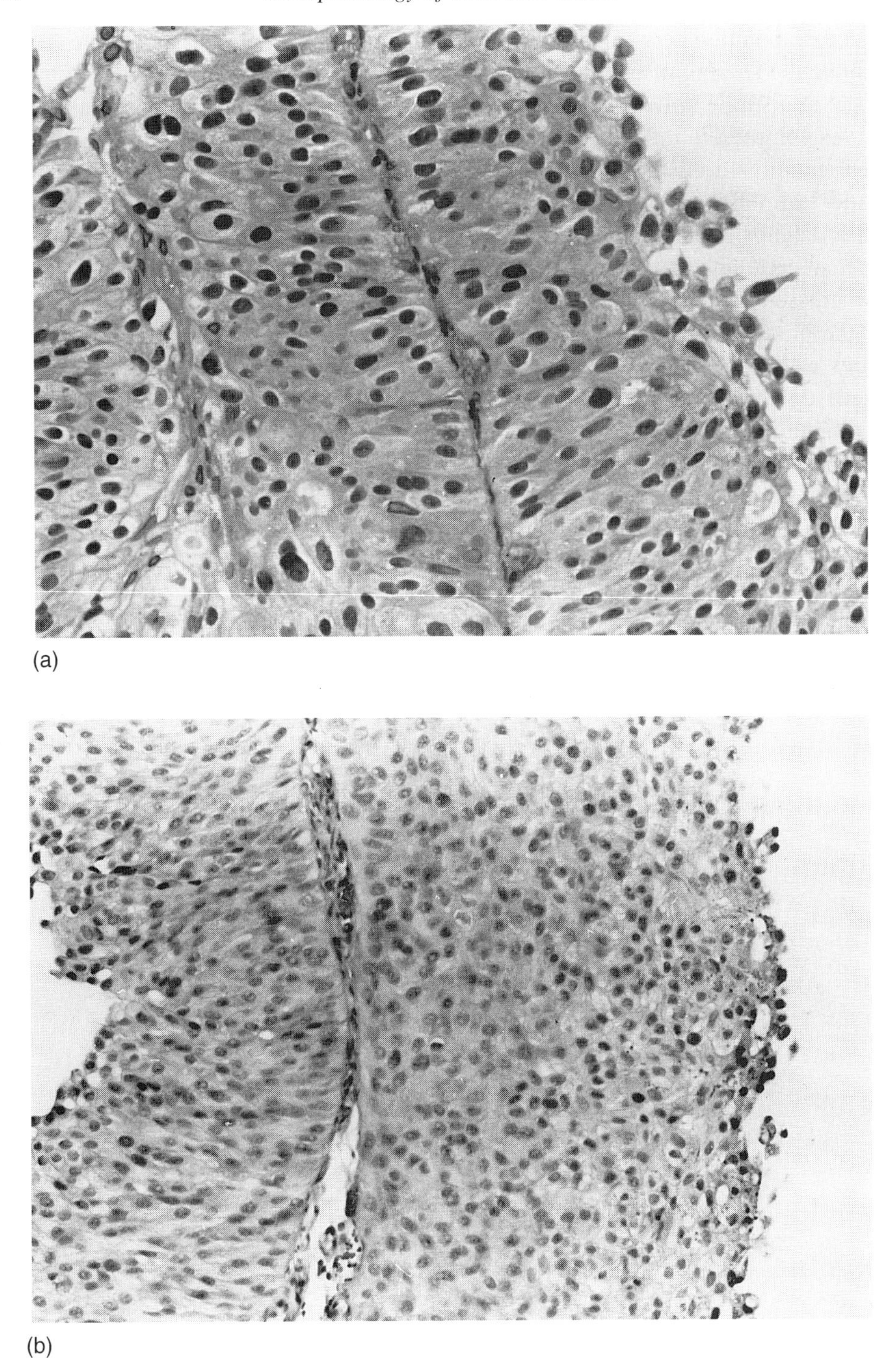

(a)

(b)

Figure 2.2 Grade 2 papillary transitional cell carcinoma (see text).

Grade 1 papillary TCC (Figure 2.1) shows an increased number of cell layers, superficial cells are usually present, there is reduced or absent cytoplasmic clearing, increased nuclear size, slight nuclear pleomorphism and slightly abnormal nuclear polarization, slight hyperchromatism, absent or rare mitoses and nuclear grooves are present.

Grade 2 papillary TCC (Figure 2.2a, b) shows a variable number of cell layers, absent superficial cells, often absent cytoplasmic clearing, increased nuclear size, moderate nuclear pleomorphism, abnormal nuclear polarization, moderate hyperchromatism, mitotic figures and nuclear grooves are present.

Grade 3 papillary TCC (Figure 2.3) shows a variable number of cell layers, absent superficial cells, absent cytoplasmic clearing, greatly increased nuclear size, marked nuclear pleomorphism, absent nuclear polarization, marked hyperchromatism, prominent mitoses and absent nuclear grooves.

As already noted in the criteria for a successful grading system, the separation into different grades must have prognostic relevance. Herein lies a problem with WHO Grade 2 tumours which are a heterogeneous group clinically and histologically[21,22], as illustrated in Figure 2.2a and b. As a result, various attempts have been made to subdivide them into more useful prognostic categories[22,96,113]. Pauwels *et*

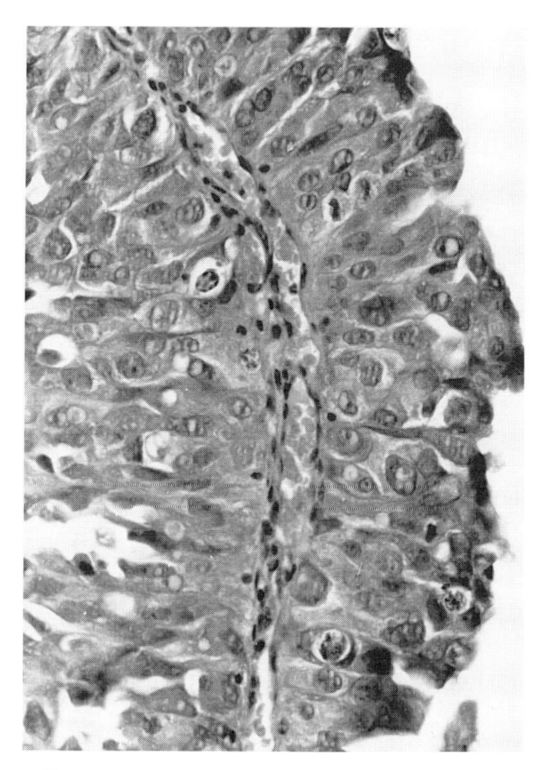

Figure 2.3 Grade 3 papillary transitional cell carcinoma.

al.[96] reclassified WHO Grade 2 tumours into Grade 2a (Figure 2.2a), tumours showing slight cellular variation, but normal cellular polarity, and Grade 2b tumours which show clear cytological deviation and a tendency to lose normal polarity (Figure 2.2b). Progression was seen in 4% of Grade 2a patients compared with 33% of Grade 2b. Pauwels recommended combining WHO Grade 1 TCC with 2a tumours as low grade, retaining 2b as the intermediate grade, and WHO Grade 3 tumours remained as high grade. Carbin *et al.*[21] subdivided WHO Grade 2 tumours with reference to the degree of nuclear atypia and the number of mitoses. Grade 2a showed slight to moderate variation in nuclear size with less than twofold variation in each diameter at the same axis among adjacent cells, smooth nuclear membranes, even, fine chromatin pattern and few mitoses. Grade 2b showed moderate to strong variation in nuclear size with at least a twofold variation in each diameter at the same axis, prominent irregular nuclear membranes, coarse chromatin and frequent mitoses. The 5-year survival was 92% for patients with Grade 2a tumours and 43% for Grade 2b[22].

Schapers *et al.*[113] developed a simplified three-tier grading system to include all urothelial tumours; namely, 'papilloma', 'low-grade' and 'high-grade' carcinomas:

- **Papilloma**: urothelium fewer than seven layers in thickness, normal nuclear polarity, no pleomorphism.
- **Low-grade carcinoma**: urothelium seven layers or thicker, nuclear polarity in 95% or more of the tumour and no or only slight pleomorphism, in 95% or more of the tumour.
- **High-grade carcinoma**: loss of nuclear polarity and moderate or prominent pleomorphism.

In a collaborative evaluation of this grading system, in comparison with the WHO system, four histopathologists examined 88 consecutive bladder tumours[113]. The group Kappa value for the four pathologists using the Schapers grading system was estimated at 0.78, indicating excellent inter-observer agreement, compared with only fair agreement using the WHO system (group Kappa 0.48). A significant difference was found in survival between the low and high grades ($P < 0.001$). However, in a multivariate analysis incorporating this grading system with other prognostic factors the above simplified grading system provided no significant additional independent discrimination[113].

In general terms, each of the established grading systems has been shown by numerous studies to correlate with prognosis. Grade is inseparably linked with stage (depth of invasion) but within individual T categories poorly differentiated tumours (Grade 3) relapse and progress more frequently than Grade 1 or Grade 2. Despite this, grade is not a very good predictor of either superficial recurrence or muscle invasion[93]. For example, G3 T1 tumours have a notorious prognosis compared with other T1 tumours, with progression to muscle invasion in up to 50% within 2 years[1,16,56]. However, 50% do not progress and light microscopy is incapable of making this crucial distinction. Furthermore, Grade 2 tumours include an even wider range of morphology within which a small but unidentifiable proportion will relapse as invasive cancers at the same site and prove fatal. As illustrated above, alternative grading systems have not resolved the problem. A variety of alternative assays have sought to remedy the deficiency but none has yet proved sufficiently accurate, reproducible and practicable to be able to add to histology in routine practice.

In addition to the division of bladder tumours into prognostic groups that have distinct morphological characteristics, it is important that there are adequate numbers of patients in each grading group. If nearly all tumours are of the same grade, little is achieved by grading. Like all other cancers, the appreciation of subtle differences in bladder cancer morphology is inevitably prone to some subjective interpretation, but it is a skill that is acquired by the training and experience made possible only by specialization. The inexperienced or generalist histopathologist may tend to classify most tumours as Grade 2, thereby denying the patient the benefits of the grading system that has important implications for treatment.

INVERTED TRANSITIONAL CELL PAPILLOMA

This uncommon tumour comprising 1% of urothelial tumours[79] was originally described by Paschkis in 1927[94] but was named inverted papilloma by Potts and Hirst[99] in 1963 because of its histological similarity to inverted papilloma of the nose and paranasal sinuses. The majority of patients are men aged 50–75 years, but there is a wide age range and inverted papilloma has been reported in a 14-year-old boy[39]. Inverted papilloma is most often situated at the trigone or bladder neck and appears as a sessile or pedunculated solitary polyp with a smooth or lobulated surface. The tumour is usually less than 3 cm in size, but some have been reported measuring up to 8 cm[79].

As the name implies it looks, microscopically, like a bladder papilloma that has developed upside down, growing into the lamina propria instead of the bladder lumen (Figure 2.4a). The microscopic appearance of inverted papilloma is characteristic (Figure 2.4b)[13]. The smooth surface is composed of cytologically unremarkable urothelium from which anastomosing cords, nests and columns of transitional cells push down as an expansile mass to invaginate the underlying lamina propria, but do not invade detrusor muscle. The basement membrane of these columns is continuous with that of the surface urothelium, and the peripheral cells lie perpendicular to the basement membrane, with a palisaded appearance. The inner cells are uniform and slightly spindle-shaped with minimal cytological atypia and lie in a horizontal position, parallel to the basement membrane. Mitoses are absent or rare. Non-keratinizing squamous metaplasia is common[48] and cyst formation is frequent. Kunze described trabecular and glandular variants[61]. The trabecular type shows the typical appearances described above. The glandular

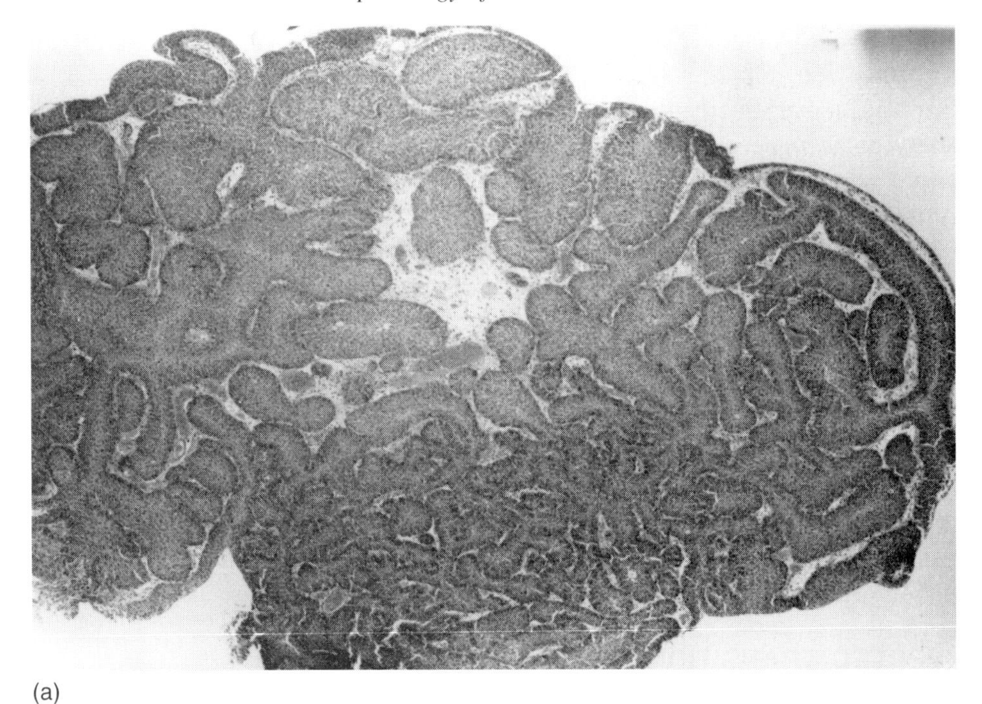

(a)

(b)

Figure 2.4 Inverted papilloma. (a) Low-power view to show overall architecture; (b) high-power view, showing normal attenuated surface epithelium and intact basement membrane.

variant shows either pseudoglandular urothelial lined spaces within the cell masses or true glandular differentiation, with a lining of columnar, mucus-secreting cells. This histological similarity to cystitis glandularis prompted speculation that inverted papilloma might develop by neoplastic transformation of cystitis glandularis[61]. Indeed, it is sometimes difficult to differentiate between these two entities.

Inverted papilloma is commonly considered to be benign; however, rare cases of recurrence of inverted papilloma have been recorded in the literature[26,104]. Coexistence of inverted papilloma with TCC has been described at separate locations[8] or growing together as a single lesion[63,104,120]. In one case[127], the TCC was considered to have developed by neoplastic transformation of the surface epithelium of the inverted papilloma. In view of this association with TCC, it has been suggested by some workers that inverted papilloma may not be a totally benign lesion, and therefore should be followed up regularly[120].

When considering these variations in possible behaviour, it should be realized that some transitional cell carcinomas may develop an **endo**phytic (i.e. inverted) growth pattern instead of the usual **exo**phytic appearance. Amin *et al.*[6] have detailed a series of 18 cases of TCC with an endophytic growth pattern, and suggested that the previous cases of combined inverted papilloma and TCC may have represented examples of inverted, endophytic TCC. Distinction of inverted papilloma from an endophytic growth of TCC may be difficult. Cytological atypia, including nuclear pleomorphism, nuclear membrane irregularity, clumped chromatin, prominent nucleoli and appreciable mitotic activity, are not seen in inverted papilloma, although a minor degree of focal cytological atypia is acceptable[6]. In TCC, the cell columns are thicker and more variable in width, with transition into solid areas. The peripheral palisading and orderly maturation seen in inverted papilloma is not conspicuous in TCC. Amin *et al.* recommended that an inverted tumour with cytological features equating to Grade 2 TCC is inconsistent with the diagnosis of inverted papilloma and should be considered as TCC with an inverted growth pattern[6].

In the nose, where these tumours are more common, inverted papillomas may show varying degrees of dysplasia and do, on occasions, develop into carcinomas. In the bladder it does seem illogical to follow up all Ta G1 exophytic tumours but not do the same for similar endophytic tumours. Thus, if the strict cytological criteria for a benign exophytic papilloma, described previously in this chapter, are not met in an inverted tumour, annual flexible cystoscopy would be a wise precaution.

PROBLEMS OF HISTOPATHOLOGICAL STAGING

In terms of histological detail, there are several features which may confuse the assessment of the presence and depth of tumour invasion, even when the specimen is well orientated and the detrusor muscle is included.

Assessment of lamina propria invasion

Recognition of invasion of the lamina propria may be difficult and interpretation tends to be subjective with considerable variation between pathologists[1,6]. Cross-cutting of the base of the fronds of non-invasive papillary tumours may result in islands of tumour cells apparently stranded within the lamina propria, simulating invasion. Careful examination will show an intact basement membrane and further levels may show continuity with the overlying papillary tumour. Non-invasive papillary tumours may push into the underlying lamina propria as complex, papillary infoldings which may be difficult to distinguish from superficial invasion[13,19]. The presence of true invasion requires breaching of the basement membrane, which is seen as small, irregularly shaped nests or strands of invasive tumour cells[135] which are sometimes of a higher grade than cells of the overlying tumour (Figure 2.5)[48], and sometimes show a surrounding desmoplastic stromal response. Amin *et al.*[6] have described urothelial TCC with an endophytic growth pattern, with broad-front extension pushing into the lamina propria, which may approach the muscularis propria (detrusor). They considered that a diagnosis of invasion was not warranted as long as the basement membrane was intact at light microscopic level.

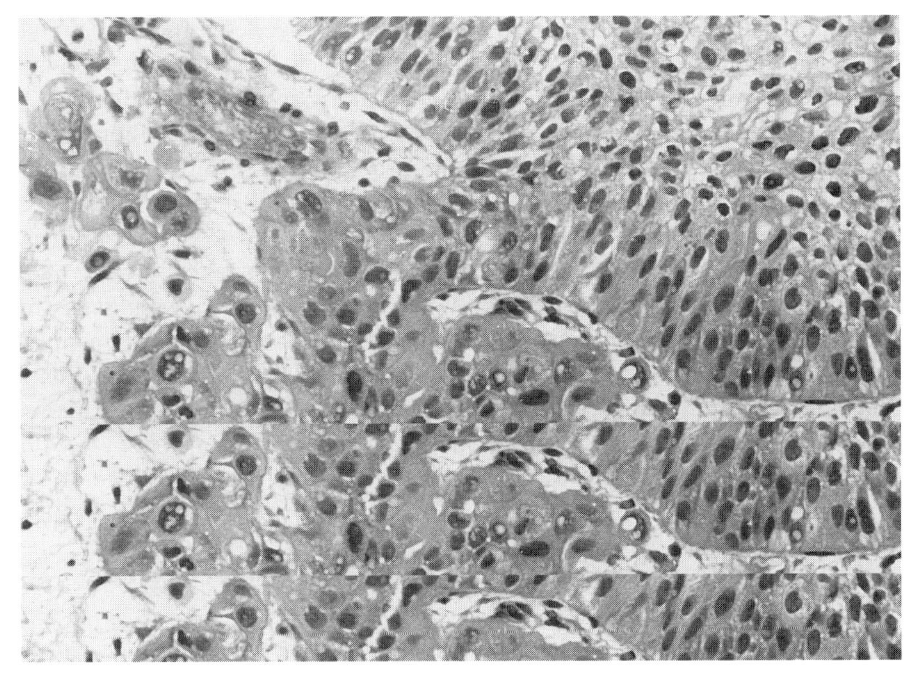

Figure 2.5 Early invasion of the lamina propria of a frond of a G2 papillary TCC. Note that some invasive cells are of higher grade.

Immunohistochemical staining of basement membrane components laminin and type IV collagen has been suggested to be helpful in the detection of breaches of the basement membrane and identification of early micro-invasion[112]. On the other hand, Deen and Ball[25] reported that focal loss of reactivity for collagen type IV was also found in some cases of inflammation, dysplasia and non-invasive papillary TCC, making this an unreliable feature.

Occasionally, TCCs may have invasive foci with deceptively bland cytological features, with small nests of transitional cells that are very difficult to distinguish from von Brunn's nests[137] and may lead to an erroneous benign diagnosis. The presence of angulated nests of epithelial cells with spiked projections arising from the periphery of these nests, sometimes giving rise to small clusters of infiltrating cells, is helpful in reaching the correct diagnosis[137]. Other infiltrative tumours that may be confused with benign conditions, such as cystitis glandularis, cystitis cystica and nephrogenic adenoma in biopsy material are the microcystic[136] and nested[81] variants of transitional cell carcinoma.

The muscularis mucosae

In the past there has been debate as to the presence of a muscularis mucosae within the bladder. Webb[123] considered that the development of smooth muscle fibres in the lamina propria probably occurs as a response to inflammation, previous surgery or irradiation. Ro *et al.*[106] were able to identify thin bundles of smooth muscle fibres, sometimes as a discontinuous layer and usually associated with a layer of medium-sized dilated vessels running parallel to the mucosa in all but six of 100 cystectomy specimens studied. The muscularis mucosae may be less easy to identify in transurethral resection biopsies partly because of its discontinuity, but also due to misorientation of the biopsy fragments, and the artefactual changes associated with transurethral resection (TUR). Angulo *et al.*[11] were able to identify the level of the muscularis mucosae in 65% of cases, as distinct muscle bundles in 39%, or by using the level of the layer of large vessels in the upper half of the lamina propria as a morphological marker of the muscularis mucosae in a further 26% of cases. It is extremely important that the pathologist assessing the depth of invasion into the bladder wall does not confuse these thin, wispy, smooth muscle bundles of muscularis mucosae (Figure 2.6) with the larger, more rounded muscle bundles of the detrusor muscle (Figure 2.7) or a serious error of overstaging will result. In some cases with a very prominent desmoplastic stromal response around tumour islands, there may be hypertrophy and lack of parallel orientation of the fibres of the muscularis mucosae, so that it is impossible to distinguish this from the muscularis propria in a small biopsy. The pathologist should draw attention to this problem in the biopsy report[6].

Various workers have suggested that subdivision of the depth of invasion within the connective tissue between the mucosa and the detrusor muscle may yield

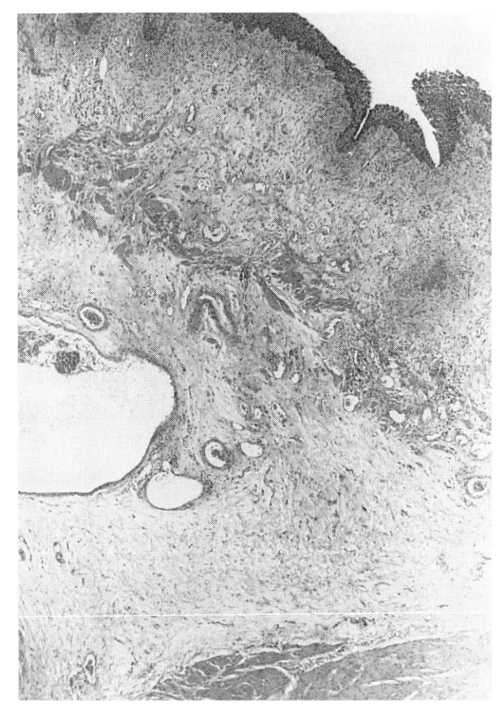

Figure 2.6 Muscularis mucosae. A bladder biopsy illustrating **muscularis mucosae demarcating** lamina propria (above) from submucosa (below). Note the detrusor muscle at the lower margin.

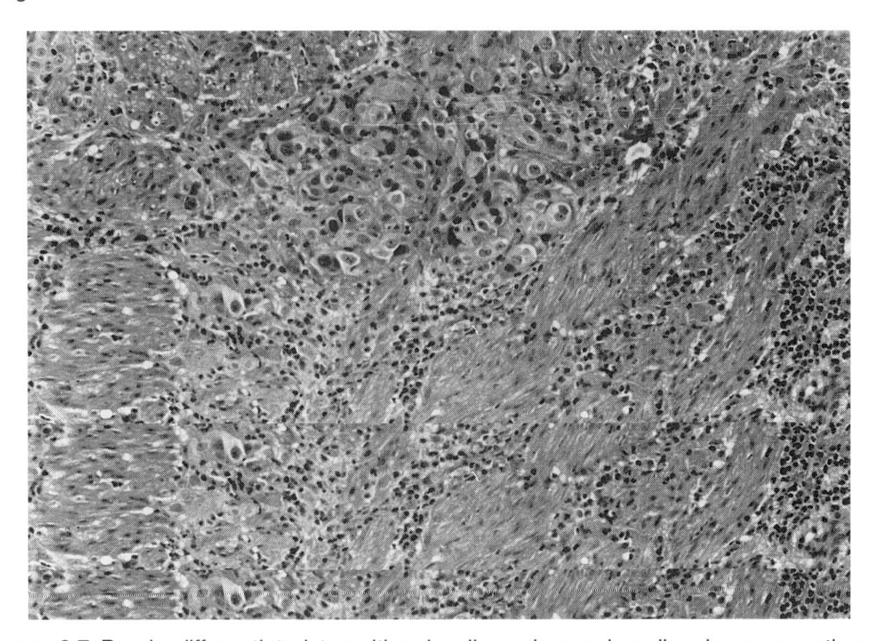

Figure 2.7 Poorly differentiated transitional cell carcinoma invading large smooth muscle bundles of the detrusor muscle.

further prognostic information[11,133]. Younes *et al.*[133] reported that high-grade tumours invading the lamina propria, i.e. the connective tissue between mucosa and muscularis mucosae, and tumours reaching the level of the muscularis mucosae had a 75% 5-year survival, whereas tumours invading through the muscularis mucosae into submucosa but not into the detrusor, had an 11% 5-year survival. The latter was comparable with the survival of patients with tumours invading the detrusor. Angulo *et al.*[11] also found significantly different survivals between patients with tumours above (pT1A) and below (pT1B) the muscularis mucosae, although this distinction could only be made in 58% of specimens. Eighty-six per cent of patients with tumours infiltrating the lamina propria but not invading through the muscularis mucosae survived 5 years, compared with 52% of those whose tumours invaded the submucosa. For patients with pT1A G2 and pT1B G2 bladder cancer, a statistically significant difference in survival was detected, with 5-year survivals of 75% for pT1A G2 patients and 41% for pT1B G2 tumours. Multivariate analysis identified the levels of subepithelial connective tissue invasion as independent predictors of survival. Hasui *et al.*[52] in a review of 88 bladder cancers invading lamina propria found that the tumour progression rate was significantly higher (53.5%) in pT1b cancer, defined as tumours with involvement of the lamina propria under the stalk of the tumour, with invasion to or near the muscularis mucosae, in comparison with a progression rate of 6.7% in pT1a cancer, defined as tumours without cancer invasion to or near the muscularis mucosae[52].

The very mixed prognosis of T1 and Grade 2 tumours is a continuing source of clinical uncertainty, and the findings of the above authors[11,52,133] appear worthy of further evaluation in larger patient numbers.

Invasion of detrusor muscle (Figure 2.7)

Pathologists must state in the biopsy report whether detrusor muscle is present or not, and whether it is invaded by tumour. In a transurethral resection, pathologists are not able to distinguish between superficial detrusor muscle and deep muscle unless the biopsy is fortuitously orientated to include the mucosal surface, with tumour extending directly through the subepithelial connective tissue into detrusor muscle. Invasion of adipose tissue in a small transurethral resection biopsy does not always represent perivesical extension. Although not widely recognized or seen frequently, fat may be present within detrusor muscle and even in lamina propria[80].

The diagnosis of superficial muscle invasion is a crucial decision which cannot be based on cystoscopy, bimanual palpation or any imaging technique. Confirmation of invasion of the detrusor muscle is the responsibility of the pathologist. In case of doubt, re-resection of the tumour site is recommended to obtain further biopsy material.

Angiolymphatic invasion

Angiolymphatic invasion has been reported to be an important predictor of tumour progression in both T1 TCC[7,38,66] and invasive bladder cancer[54,57]. Blood vessel invasion is unusual in low-grade tumours and almost always occurs in association with other poor prognostic indicators such as muscle invasion[80]. Lopez and Angulo[66] studied angiolymphatic (vascular) invasion in 170 T1 bladder carcinomas and identified vascular invasion, verified by immunohistochemistry, in 10% of patients, most frequently in high-grade tumours. There was a statistically significant difference in the 5-year disease-free survival for cases without (81%) compared with those with vascular invasion (44%; logrank $P = 0.004$). In multivariate analysis, vascular invasion was an independent prognostic factor of survival. Anderström *et al.*[7] also regarded lymphatic invasion as an important predictor for prognosis in patients with T1 tumours, regardless of tumour grade.

Figure 2.8 illustrates submucosal lymphatic invasion. It is important when identifying vascular spaces to be certain that this is not a retraction artefact, commonly seen around small groups of tumour cells in the lamina propria. Larsen *et al.*[62] used the endothelial marker *Ulex europaeus* agglutinin (UEA-1) to

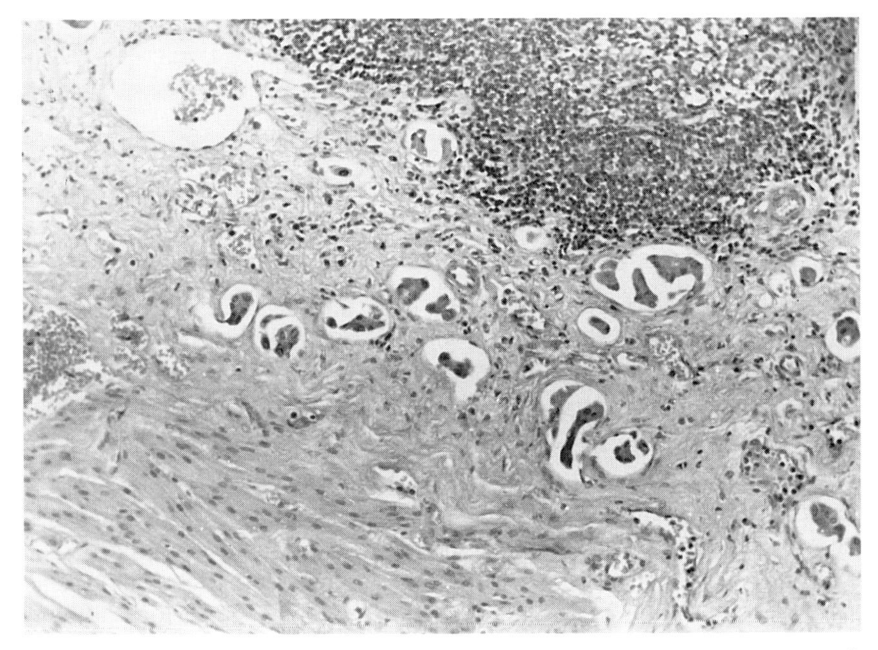

Figure 2.8 Lymphatic invasion. Submucosal lymphatic invasion at some distance from a macroscopic tumour (not shown). Note the endothelial cells lining lymphatic spaces which are permeated by tumour.

confirm vascular spaces. Ramani *et al.*[102] additionally used antibodies to von Willebrand's factor (VWF) and the monoclonal antibody QBEND/10 to distinguish small vessel invasion from retraction artefact in cases of T1 bladder cancer. Larsen identified angiolymphatic invasion in 7% and Ramani in 12.5% of T1 tumours, an incidence similar to that found by Lopez and Angulo[66], but in both series the numbers of cases were too small to allow prognostic evaluation.

Tumour may spread to adjacent areas of the bladder wall by tunnelling under intact mucosa, and lymphatic invasion should also be sought in apparently uninvolved fragments of bladder wall in a transurethral resection. It should be noted that the depths of lymphatic and/or vascular invasion are not taken into account in determining tumour category[77].

Prostatic involvement by transitional cell carcinoma

The exact incidence of prostatic involvement by TCC has ranged from 12 to 43% in cystoprostatectomy specimens[114,131]. In the TNM classification (Chapter 3), stage pT4a has usually been taken to mean TCC invading the prostate stroma from the bladder base or neck. There is no provision in this category for different degrees of prostatic involvement by either papillary TCC or CIS, which may be limited to the prostatic urethral mucosa and/or involving the periurethral prostatic ducts and acini, or infiltrating from these locations into the prostate stroma. Various classifications of prostatic involvement have been suggested[24,51,92,129]. Hardeman and Soloway[51] proposed a classification based on the integrity of the basement membrane surrounding the prostatic urethra and periurethral ducts and acini, with three subgroups:

1. Tumour confined to prostatic urethral urothelium.
2. Tumour invasion (extension) into ducts and acini, but limited by the basement membrane.
3. Stromal invasion.

The classification did not differentiate between primary TCC of the prostate and secondary prostatic TCC in a patient with previous or concurrent bladder cancer.

Esrig *et al.*[35] studied survival of patients with prostate involvement by TCC in radical cystoprostatectomy specimens. Five-year overall survival rates for prostate stromal involvement by direct extension of TCC from the bladder was 21%, which compared unfavourably with survival in patients with prostatic involvement arising within the prostatic urethra of 55%. Five-year overall survival rates for prostatic urethral or ductal tumours and for CIS in the urethra or ducts, when there was no evidence of stromal invasion by tumour were similar: 71%, compared with 36% when stromal invasion arising from tumour in the ducts or urethra was present. Esrig *et al.*[35] concluded that prostatic urethral or ductal transitional cell carcinoma without stromal invasion does not alter survival predicted by primary bladder stage alone and should not be classified as pT4a, but prostatic stromal invasion arising intraurethrally significantly decreases survival.

Other workers have confirmed that non-invasive TCC of the prostatic urethra, ducts or acini has a generally good prognosis which is similar to that of CIS or T1 tumour in the bladder[24,115]. TCC invading the prostatic stroma has a poor prognosis, whether it has arisen *de novo* in the prostate, secondary to or concurrently with superficial tumours of the bladder or by direct extension of a T3 cancer of the bladder neck[114,119,129]. However, Reese *et al.*[103], reviewing the clinical outcome in patients with prostatic involvement by TCC, found that prostatic invasion either into the stroma or involving ducts and acini only, had no adverse effect on outcome and that lymph node status and bladder stage, and not prostatic invasion, were the determining factors in outcome.

It is clear that future revision of the T4a category will be necessary to separate these different prognostic groups and that histological assessment is critical in making this distinction. TUR biopsy of the prostate is recommended for staging and should be sufficiently deep to include prostate ducts and suburethral connective tissue[70]. Sakamoto *et al.*[111] reported that biopsies from the 5 and 7 o'clock positions of the verumontanum portion of the urethra, substantially improved detection of ductal and acinar involvement by transitional cell carcinoma and carcinoma in situ. It may be necessary to examine several levels through the tissue block to identify this; often the urethral surface is severely traumatized or denuded of epithelium.

Papillary TCC in the urethra may involve and distend the orifices of the periurethral ducts, resulting in a compressed papillary pattern, with fine stromal cores evident. Urothelial CIS in prostatic ducts is usually accompanied by CIS in the prostatic urethra[67]. Large polygonal tumour cells, with pleomorphic enlarged and hyperchromatic nuclei and eosinophilic cytoplasm, insinuate themselves and spread between the intact prostatic luminal epithelial cells and the basal cells, eventually completely filling and distending the ducts and acini[67]. The cell masses filling the ducts may develop central necrosis, similar to the comedonecrosis seen in intraductal breast carcinoma[33]. Single tumour cells may spread in a pagetoid fashion into adjacent ductal epithelium. While tumour remains *in situ*, the nests of malignant transitional cells have the distribution and contours of prostatic ducts and acini, with smooth, rounded and circumscribed edges (Figure 2.9). There should be no surrounding desmoplastic stromal reaction. Early invasion may be recognized as small, spiky projections or cords of cells, extending into the surrounding stroma from the edges of the cell masses, eliciting a fibrous stromal response[33]. Larger nests of invasive tumour have overrun the ducts and show irregular borders with areas of necrosis and often surrounding chronic inflammation and desmoplastic fibrosis. Differentiation from high-grade prostatic adenocarcinoma may be aided by remnants of an acinar pattern in prostate cancer and by immunohistochemical staining for prostate-specific antigen and prostatic acid phosphatase. It should be noted, however,

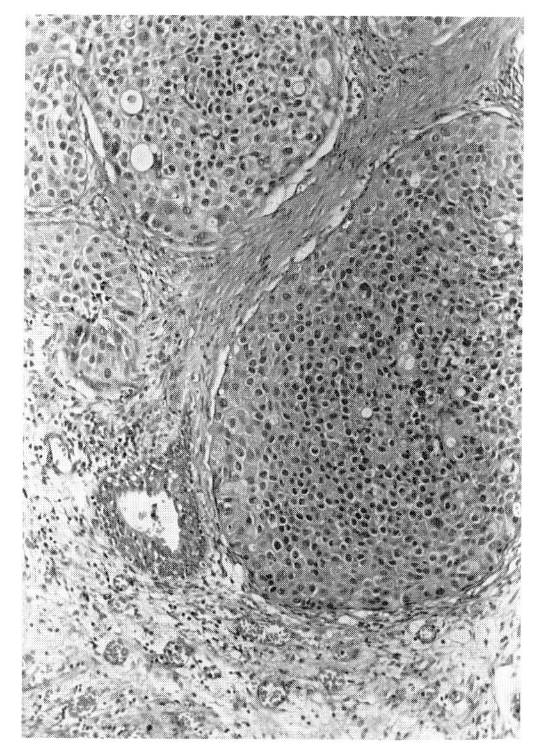

Figure 2.9 TCC in prostatic ducts. CIS filling prostatic ducts, note intact basement membrane and rounded configuration of the expanded ducts.

that up to 1.6% of high-grade prostatic carcinomas are negative for both markers[18].

CARCINOMA IN SITU (CIS)

Carcinoma in situ is now a well-known clinical entity. Arising *de novo* and causing symptoms, or occurring in association with papillary TCC, its treatment with intravesical BCG or chemotherapy has become established. In many institutions the histological diagnosis is regarded as straightforward and recent multiauthor texts would reflect this consensus[124,135]. However, a recent publication from a multicentre trial in the USA[116] reveals a worrying lack of uniformity in diagnosing CIS which suggests that further discussion is needed.

The cytological abnormalities of mild, moderate and severe dysplasia were described by Nagy *et al.*[83] to equate to the epithelial changes seen in Grades 1, 2 and 3 urothelial carcinoma, as if the urothelium covering the papillary fronds were to be flattened out. Hence, mild urothelial dysplasia shows slightly disordered nuclear polarity, slight variation in nuclear size and irregularity, slight nuclear crowding and finely granular nuclear chromatin. Moderate dysplasia shows altered nuclear polarity, moderate variation in nuclear size and prominent irregularity, more marked nuclear crowding and predominantly fine nuclear chromatin, with some cells showing more marked chromatin granularity. Severe dysplasia shows markedly altered polarization, variable nuclear size, prominent nuclear irregularity and coarsely granular, uneven nuclear chromatin. In all grades of dysplasia superficial umbrella cells are preserved and it is the intermediate and basal cells that show the varying degrees of atypia as described above.

In typical CIS (as originally described by Melicow[71]) the full thickness of the urothelium is replaced by cytologically malignant cells (Figure 2.10). Superficial umbrella cells are lost. There is marked nuclear enlargement and pleomorphism, complete loss of nuclear polarity, nuclei show coarse, uneven chromatin and nucleoli are large and irregular. Mitoses are common and may be abnormal, but are not a necessary diagnostic feature. There is loss of cell cohesion with shedding of malignant cells from the surface, which has been known in the past as 'denuding cystitis'[31]. As discussed in Chapter 6, if a biopsy shows only a layer of atypical basal cells, cytological examination of the urine is essential to identify the obviously malignant cells that have been shed from the urothelium of the biopsy. The underlying connective tissue stroma shows capillary proliferation and sometimes a chronic inflammatory infiltrate.

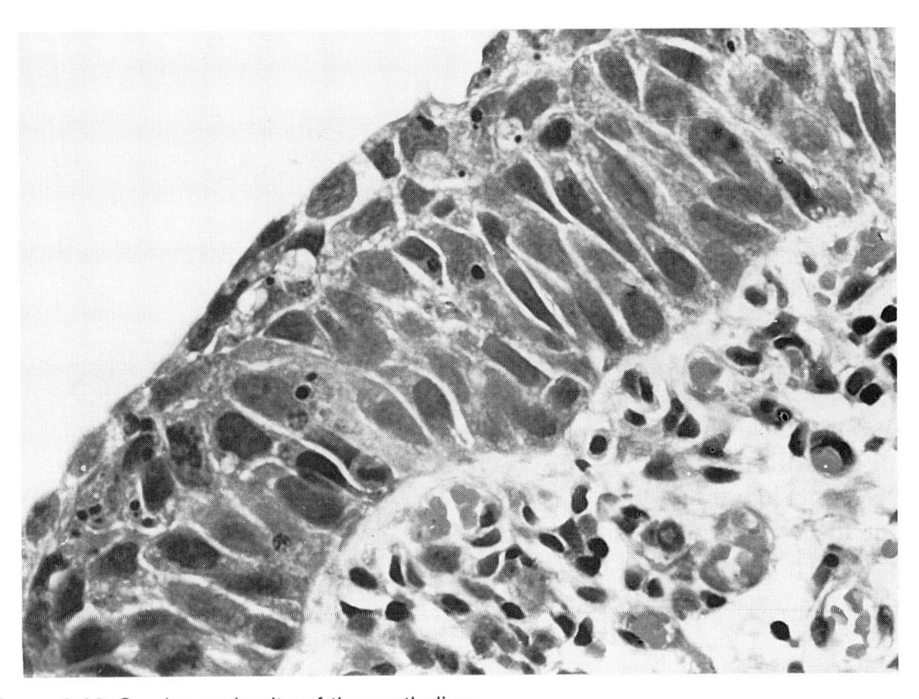

Figure 2.10 Carcinoma in situ of the urothelium.

From the foregoing description it is apparent that cellular abnormalities are similar in CIS and 'severe dysplasia'. The difference (according to the classical description) between CIS and severe dysplasia is the presence of abnormalities throughout the **full thickness** of the urothelium in CIS (Figure 2.10), compared with the retention of surface umbrella cells in severe dysplasia. Many authors now agree that it is the marked degree of cellular abnormality that is important rather than the presence or absence of umbrella cells. Thus, it is suggested that severe dysplasia and CIS can be combined in one category[41,81].

Orozco *et al.*[91] have described a pagetoid variant of carcinoma in situ, in which large cells with abundant pale cytoplasm, large nuclei and prominent nucleoli infiltrate between normal urothelial cells or undermine adjacent urothelium. This variant apparently never occurs as the primary or only manifestation of the disease and the clinical significance is considered to be identical to typical carcinoma in situ[107]. CIS may also show a small cell pattern with small, densely aggregated cells showing nuclear characteristics similar to the usual type of CIS[81]. Von Brunn's nests may be involved by CIS with a spectrum of atypia ranging from a few atypical cells to complete replacement by neoplastic cells. These nests may be preserved even when the surface urothelium has been completedly denuded.

Several studies have demonstrated difficulty in obtaining reproducibility in grading dysplasia, not only between pathologists in different institutions, but also by the same pathologist on different occasions[83,105,108]. Robertson *et al.*[108] reported that agreement on the degree of dysplasia in random bladder mucosal biopsies was 'fair' for high-grade lesions, but 'very poor' for low-grade lesions. Richards *et al.*[105] reported similar inter- and intra-observer variation which, in their opinion, undermined the clinical value of bladder mapping mucosal biopsies. Acknowledging these difficulties, some workers have recommended a reduction in the number of grades of dysplasia to two categories, combining mild and moderate dysplasia as 'dysplasia', and combining severe dysplasia with CIS as 'carcinoma is situ'[82,105]. A workshop of clinicians and pathologists reviewing existing reported information agreed that it was difficult to draw a sharp boundary between marked dysplasia and carcinoma in situ, and that these could be combined in one category for the purposes of clinicopathological correlation[41].

Mostofi and co-workers[75,76] suggested that the terminology of intraepithelial neoplasia used for dysplasia of the cervix, be applied to the bladder, with three grades of urothelial CIS replacing mild, moderate and severe dysplasia. However, Freidell *et al.*[41] recommended that this terminology should not be used for the bladder, although it appears to have been adopted by some pathologists[116] and is a potential source of continuing confusion. Grignon[48] supports the view of Friedell *et al.*[41], stating:

> CIS grades 1, 2 and 3 . . . a terminology we discourage. Several studies have documented the lack of reproducibility in grading dysplasia. Many authors continue to advocate attempting to grade these atypical lesions whereas others, recognising the lack

of reproducible criteria would combine high grade dysplasia with carcinoma in situ, acknowledging this as a clinically significant lesion that requires the same treatment.

Grignon[48] favours a three-grade system:

- **low-grade dysplasia**, which is non-neoplastic and indistinguishable from a variety of reactive atypias;
- **moderate dysplasia**, for truly dysplastic lesions falling short of the criteria for severe dysplasia/carcinoma in situ; and
- **carcinoma in situ**, for cytological changes amounting to severe dysplasia or carcinoma in situ as described above.

From the clinical management point of view a distinction has to be made between those patients that require treatment, and those that do not (Chapter 6). There is certainly controversy over the biological significance of dysplasia[82]. Some regard it as an embryonic form of malignancy which, if left undisturbed, will mature to invasive cancer. A second theory contends that dysplasia is merely a marker of an epithelium responding abnormally to some combination of genetic and environmental factors and that the risk of invasive cancer may increase with a number of abnormal features present. This is reflected in differences in clinical practice[116] although, in general, in Europe only patients with the severest dysplasia ('Grade 3') would usually be treated with BCG. For the future, as clinical practice appears to be variable, it is essential that a common terminology should be used to enable meaningful evaluation of the outcomes of treatment. For this purpose, the above classification by Grignon would be most appropriate.

Primary CIS, without concurrent or pre-existing bladder cancer, and **secondary** CIS show identical histological appearances[90] but opinions vary as to their aggressiveness. Fukui *et al.*[42] suggested that the apparently favourable prognosis of patients with secondary CIS appeared to be attributable to its early detection by close follow-up of previous papillary tumours and the resulting prompt treatment with intravesical therapy. However, Orozco *et al.*[90] found progression or death from bladder cancer to be unusual in patients presenting with primary CIS, but common in individuals with secondary or concurrent CIS, concluding that the appearance of urothelial CIS identified patients with at least localized resistance to the development of invasive bladder cancer. These opposing views cannot be reconciled at the present time and only long-term study of large numbers of patients with the different types of CIS will resolve this apparent discrepancy.

VARIANTS OF TRANSITIONAL CELL CARCINOMA

Transitional cell carcinoma with focal squamous or glandular dedifferentiation are the most common variants of TCC, usually occurring in moderately or poorly

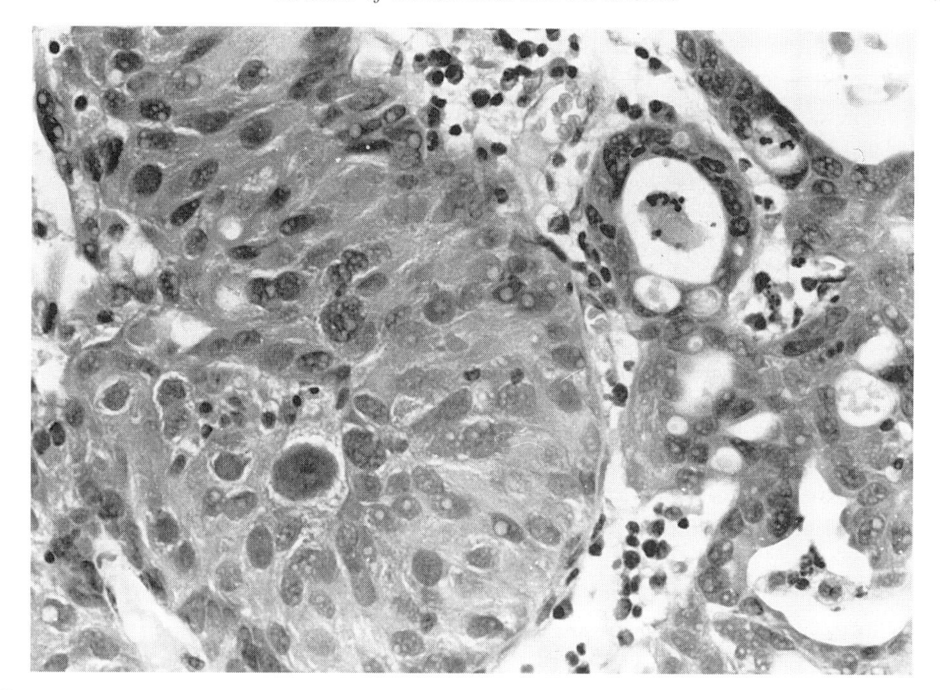

Figure 2.11 Poorly differentiated TCC with focal glandular metaplasia.

differentiated invasive cancers[134]. Up to 20%[110] of TCCs show squamous metaplasia and glandular differentiation is present in 7%[13] of tumours. These tumours with mixed differentiation are classified by the WHO as transitional cell carcinoma with either squamous or glandular metaplasia, or both. Some authors[27] consider that this terminology is satisfactory for a urothelial carcinoma with a small component of glandular or squamous dedifferentiation, but that carcinomas with a truly 'mixed' appearance and with more than one type of differentiation prominent, should be classified separately in order to gain meaningful data on their behaviour.

In order to make the designation **glandular metaplasia** meaningful, it is necessary to restrict the use of the term to unequivocal gland formation, with true gland-like lumina lined by columnar cells[81]. Figure 2.11 shows glandular metaplasia occurring in a poorly differentiated TCC. Transitional cell carcinoma of all grades often show pseudoglandular spaces containing cell debris which appear to arise by individual cell dropout or necrosis, and does not represent true gland formation. Similarly, **squamous metaplasia** requires the presence of keratinization or intercellular bridges[48]. Tumours with glandular and squamous dedifferentiation are reported to have a worse response to therapy than pure TCC[10,65,69,95]. In a review of patients treated with cisplatin-based chemotherapy for unresectable bladder cancer, Logothetis *et al.*[65] found that only 45% of those patients with tumours of mixed differentiation achieved partial or complete remission, compared with 70% of patients with pure TCC. On further evaluation of the mixed tumours, it appeared that the TCC with adenocarcinomatous elements were responsible for the lower response rate, tumours with squamous metaplasia showing a similar response rate to pure TCC. However, Martin *et al.*[69] found that the presence of squamous metaplasia within individual TCC of the bladder was associated with failure of response to radiotherapy in 78% of patients, compared with a 90% complete response in tumours without this change. Squamous metaplasia within TCC usually does not seem to be associated with vesical squamous metaplasia in the adjacent urothelium[110].

Small cell carcinoma (oat cell) has also been described to occur admixed with urothelial carcinoma and is thought to arise from undifferentiated or stem cells within the urothelium. However, TCC with areas of oat cell carcinoma should be reported as a mixed tumour, although the therapeutic implications of this remain unclear[50].

A further rare variant is urothelial carcinoma with **trophoblastic differentiation**. Some of the early examples were reported as pure choriocarcinoma[4,125] but several cases have been associated with TCC or undifferentiated carcinoma[43,85,121]. TCC may contain syncytiotrophoblastic giant cells which secrete ß-HCG (human chorionic gonadotrophin) into both urine and serum[46] and it has been suggested that choriocarcinoma of the bladder may arise by further metaplasia or dedifferentiation of transitional cell carcinoma to the level of embryonal trophoblast[3,46,121].

SQUAMOUS EPITHELIAL TUMOURS

Squamous papilloma is a papillary tumour with a delicate fibrovascular stroma covered by regular squamous epithelium[77]. Such tumours are rare in the bladder, and in this location squamous papilloma usually represents extension of condyloma acuminatum from the vagina or urethra[122]. Condylomas show the typical koilocytotic cells with abundant clear cytoplasm and hyperchromatic wrinkled nuclei seen in condylomas elsewhere and do not show destructive stromal invasion[134].

Squamous carcinoma is defined in the WHO classification as a malignant epithelial tumour, with cells forming keratin or showing intercellular bridges (Figure 2.12). The tumour must be of one cell type[77]. It is variably reported as comprising 3–7% of urological malignancies in Britain and the USA[13,109], but in areas of endemic schistosomiasis this figure is 75%[30]. In some countries, patients may have a long history of irritation of the bladder mucosa due to recurrent urinary tract infection[37], calculi, calcified schistosoma ova[30] and in-dwelling catheters[37].

Squamous cancers are often associated with keratinizing squamous metaplasia elsewhere in the urothelium[14,74,110] which appears macroscopically as white plaques.

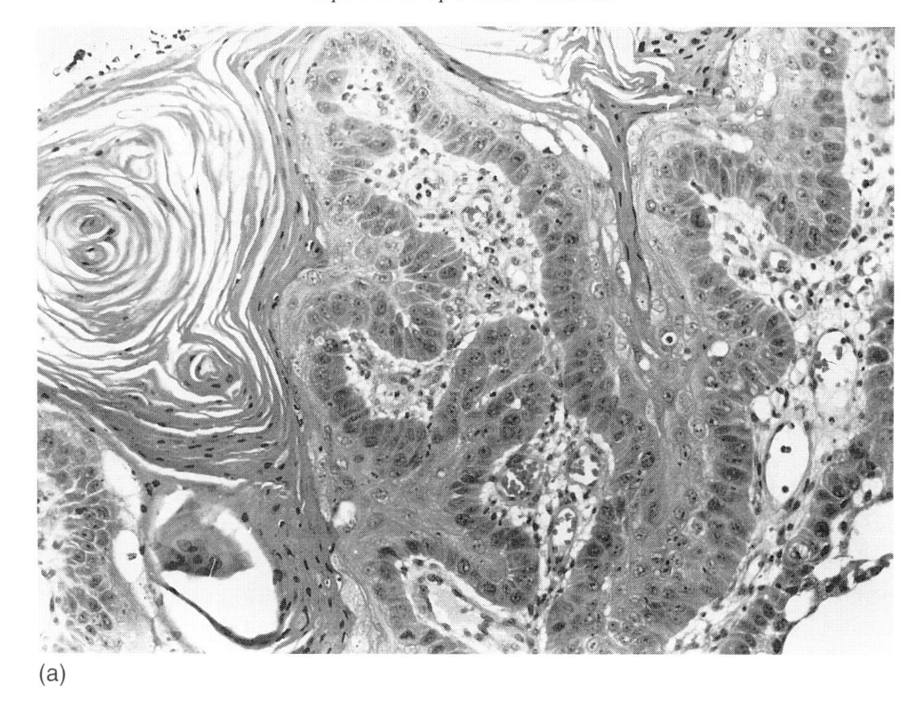

(a)

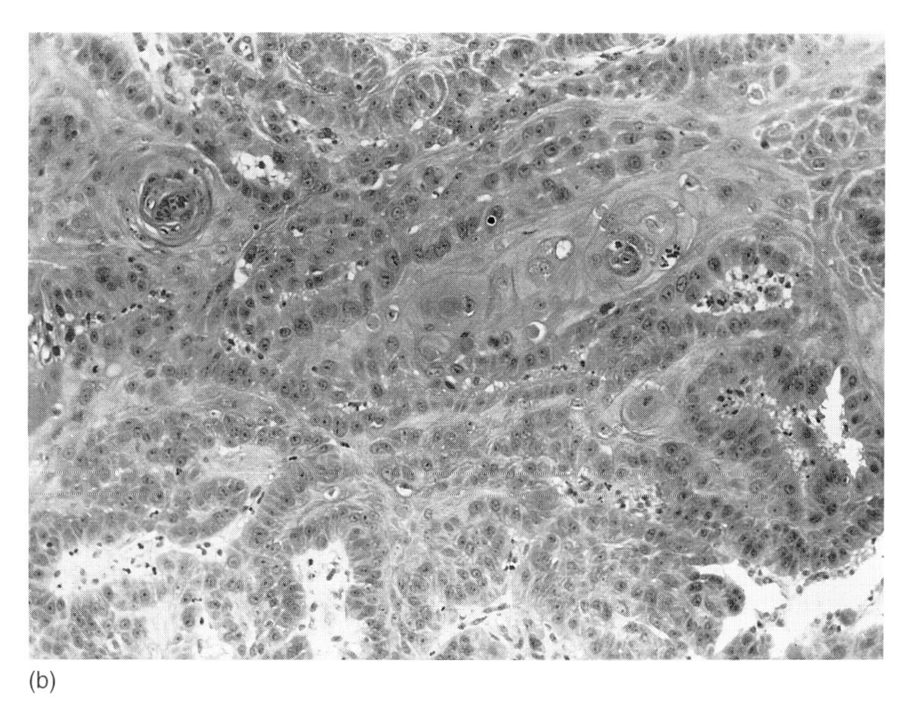

(b)

Figure 2.12 (a) Superficial biopsy of a moderately differentiated keratinizing squamous carcinoma. (b) A deeper biopsy of the same squamous cancer, showing epithelial pearls.

It should be noted that **keratinizing squamous metaplasia** is distinct from the **non-keratinizing vaginal type squamous epithelium** seen at the normal trigone of the female bladder. Squamous carcinomas are often deeply invasive at first diagnosis. Cystoscopically, the 'snowstorm' appearance caused by keratin flakes from the tumour surface, should alert the urologist to the likely diagnosis.

On histological examination, squamous carcinoma invades the lamina propria as sheets and irregular islands of polygonal squamous cells which show well-defined cell borders. Within the islands there are usually concentric aggregates of keratinized cells, forming squamous pearls. The periphery of these tumour islands shows fairly uniform basal cells, which keratinize as they progress to the centre or to the tumour surface[79]. Grading is based on the degree of keratinization within the tumour[79], and most grading schemes recognize well, moderately and poorly differentiated tumours[37,60,109] but it is recommended[27,109] that nuclear features are also taken into account. Well-differentiated tumours show prominent keratinization and intercellular bridges and only slight nuclear abnormalities[60]. Moderately differentiated tumours show moderate nuclear pleomorphism but still show well-developed keratinization. In poorly differentiated tumours, keratinization is less apparent, and only occasional dyskeratotic cells are seen. There is pronounced nuclear pleomorphism and mitotic activity. Non-keratinizing areas cannot be distinguished from high-grade transitional cell carcinoma. The differential diagnosis includes transitional cell carcinoma with squamous dedifferentiation and this diagnosis should be rendered if there is a transitional component or CIS elsewhere in the bladder[110]. The possibility of squamous carcinoma of the cervix and vagina invading the bladder should also be considered.

Verrucous carcinoma

Verrucous carcinoma of the bladder is a rare variant of squamous carcinoma. Only a few cases and one series have been reported[32,55]. Verrucous carcinoma is a papillary or polypoidal, white tumour with a shaggy surface, covered by friable keratinous debris. Tumour invades the bladder wall with a pushing growth margin. Histology shows papillomatous and markedly acanthotic squamous epithelium, with hyperkeratosis at the surface and large bulbous rounded downward outgrowths of bland squamous cells with minimal cytologic atypia, at the deep edge. Superficial biopsies will include only bland squamous cells leading to an erroneous impression of benign squamous epithelium and it is **essential** that diagnostic biopsy includes the full thickness of the tumour, including the pushing margin. Many otherwise typical squamous carcinomas are papillomatous and show focal pushing growth margins, so that the diagnosis of verrucous carcinoma can only be made after thorough examination of the whole tumour[134]. Young[134] recommends use of the term verrucoid carcinoma to distinguish these lesions.

GLANDULAR EPITHELIAL TUMOURS

Villous adenoma

Only a few examples of villous tumours with a histological similarity to villous adenoma of the large bowel have been described in the bladder[12,23,73,118]. Villous adenomas show a tubulovillous configuration, with long papillary fronds covered by intestinal type epithelium, which may show pseudostratification and nuclear crowding consistent with epithelial dysplasia. There should be no evidence of glandular invasion into the stroma at the tumour base. Adjacent bladder mucosa may show cystitis glandularis. Caution is advised in interpretation if the base of a well-differentiated papillary glandular neoplasm is not included in the biopsy, particularly if there is a cribriform glandular pattern or high-grade nuclear atypia, as this may represent the surface of a papillary adenocarcinoma[134].

Villous adenoma is relatively more common in the urachus[28,48] where it forms a cystic mass, with locules filled with mucus, which may rupture. The lining epithelium is of tall columnar type with a population of mucous goblet cells, either lining the cyst wall or arranged on papillary fronds, closely resembling the villous adenoma of the large bowel. Rupture of the cyst into the peritoneal cavity at surgery may result in risk of pseudomyxoma peritonei.

Adenocarcinoma

The frequency of adenocarcinoma of the bladder has ranged in various series from 0.5 to 2%[49]. Primary adenocarcinomas are subdivided into two major groups. About two-thirds arise from the bladder mucosa and include those cases associated with exstrophy[2], schistosomiasis[68] and rare cases associated with endometriosis[5]. A second major group arising from the urachus make up the remaining third. Secondary adenocarcinoma in the bladder represents invasion from adjacent structures, notably prostate, colon and ovary[44]. The clinical presentation is similar to that of transitional cell carcinoma of the bladder. Mucinuria is seen in 25% of patients with urachal adenocarcinomas[117]. Urachal tumours occur at a younger age than vesical adenocarcinoma with an average age at diagnosis of 51 years compared with 62 years for vesical adenocarcinoma[49]. Most adenocarcinomas, urachal or vesical, are already advanced at the time of diagnosis[44]. Histological type was generally not a significant predictor of outcome, although the signet ring variant appeared to behave very aggressively[49].

Both urachal and vesical adenocarcinomas probably have a similar pathogenesis, with metaplasia of transitional epithelium to a glandular type of epithelium, possible development of dysplasia[86] and ultimately progression to adenocarcinoma[9,78]. Widespread intestinal metaplasia with replacement of significant portions of the urothelium by colonic-type mucosa, occurs as a response to chronic irritation and apparently identifies a disturbed urothelium at risk of cancer[20]. The high frequency of adenocarcinoma in exstrophy appears to be a result of the presence of metaplastic glandular epithelium in the exstrophic bladder. Grignon *et al.*[49] reported cystitis glandularis in the adjacent bladder epithelium of 24% of patients with primary vesical (non-urachal) adenocarcinoma, and intestinal metaplasia in 32%. Two patients showed adenocarcinoma in situ. Papillary adenocarcinoma in situ of the bladder has been described in association with urothelial CIS[86] and there has been speculation that a dysplastic papillary adenomatous epithelium may represent an intermediate morphological pattern between cystitis glandularis and adenocarcinoma of the bladder.

Adenocarcinomas are seen more commonly than transitional cell carcinomas in the dome, trigone and bladder neck[59]. Tumours tend to be infiltrating, and local spread in and outside the bladder wall is disproportionately large compared with growth within the bladder cavity[59,79]. Some may appear gelatinous, with obvious mucin production. Signet ring carcinoma may show a diffuse thickening of the bladder wall, with little mucosal abnormality and a linitis plastica-like appearance. Tumours may be subdivided into five histological subtypes[9,49] and may show a single histological pattern or any combination of patterns in both primary and urachal adenocarcinoma, with the exception that clear cell adenocarcinoma has not been reported in the urachus or in exstrophy[134].

1. Intestinal or enteric type: the architecture and cytology of which resemble those of typical colonic adenocarcinoma (Figure 2.13).
2. Mucinous or colloid type: single tumour cells and nests of cells are seen floating in lakes of mucin.
3. Signet ring type: with single tumour cells distended with mucin and a diffuse infiltrative pattern.
4. Clear cell type with abundant, clear, glycogen-rich cytoplasm and hobnail cells in a tubular or papillary pattern.
5. Glandular, not otherwise specified.

The differential diagnosis of primary adenocarcinoma in the bladder includes secondary tumour from the prostate, colon or female genital tract[134]. Epstein *et al.*[34] reported that immunoperoxidase staining for prostate-specific acid phosphatase is positive in some primary adenocarcinomas of the bladder, but that these tumours are negative for prostate-specific antigen. Signet ring carcinoma may be mistaken for nephrogenic metaplasia and metastatic gastric signet ring carcinoma[134]. Clear cell adenocarcinoma may be confused with metastatic renal cell carcinoma, clear cell carcinoma from the female genital tract and nephrogenic adenoma.

Distinction between primary bladder adenocarcinoma and urachal adenocarcinoma may be difficult and cannot be made on the histological type of the tumour. Mucin histochemistry and immunocytochemistry have not been shown to aid this separation[49]. Location in the dome of the bladder, absence of cystitis cystica or cystitis glandularis, predominant involvement of the muscularis rather than the

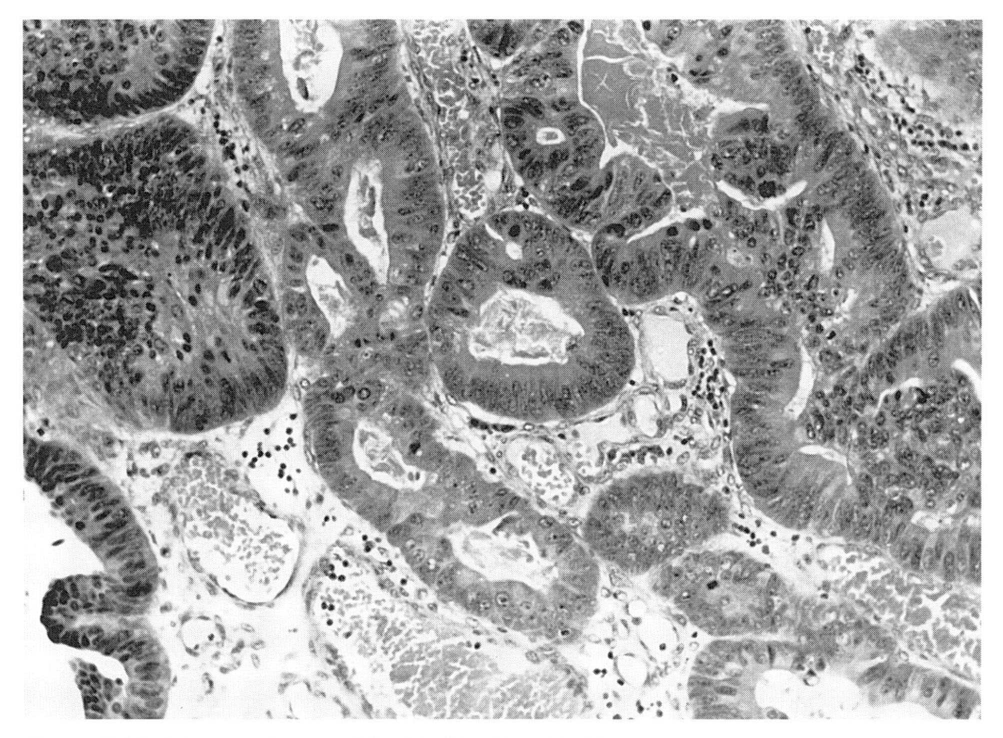

Figure 2.13 Adenocarcinoma of the bladder, intestinal type.

submucosa, with intact or ulcerated bladder epithelium, demonstration of a urachal remnant connected with the neoplasm and presence of a suprapubic neoplastic mass have been proposed by Wheeler and Hill[126] as criteria for the identification of urachal adenocarcinoma. To these criteria Mostofi *et al.*[78] added sharp demarcation between the tumour and surface epithelium and ramification in the bladder wall with extension to the space of Retzius, anterior abdominal wall or umbilicus. Sheldon *et al.*[117] remarked that regrettably, many of the reported cases are described in insufficient detail to ensure that they meet all the criteria.

SMALL (OAT) CELL CARCINOMA (Figure 2.14)

Small cell carcinoma of the bladder is rare, making up 0.5% of a series of 3778 bladder malignancies[17]. It is histologically and morphometrically identical to small cell undifferentiated carcinoma of the lung[17]. Small cell carcinoma is more common in males, the average age is approximately 70 years[98], and tumour is usually of an advanced stage at diagnosis[50]. Clinical presentation is similar to that of TCC but its rapidly lethal course is much less favourable and may justify its recognition as a

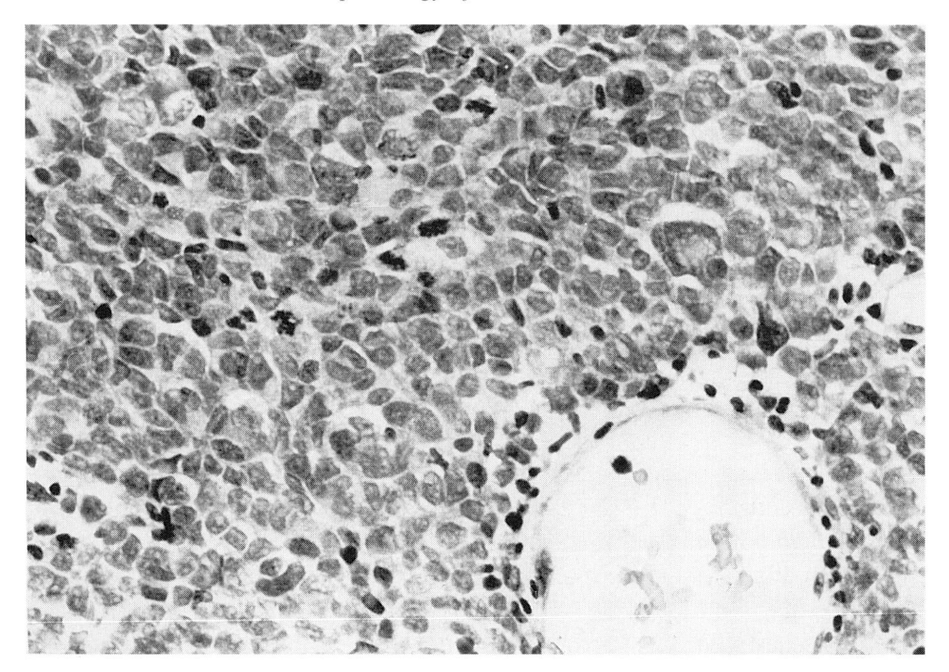

Figure 2.14 Small cell carcinoma of intermediate cell type. Note the nuclear moulding and numerous mitotic figures.

separate entity for treatment purposes[17]. Grignon *et al.*[50] recorded that radical cystectomy with adjuvant chemotherapy appears to be the treatment of choice and that although overall survival was poor, some cases responded well to therapy. It is necessary to exclude a lung primary tumour, but metastasis to the bladder of small cell carcinoma of lung is rare[45].

Tumours are large, polypoid and frequently ulcerated[136]. Histologically, small cell tumours show large nests and sheets of small darkly staining tumour cells with little cytoplasm, separated by fibrous stroma, with sparse lymphocytic infiltration. There are large areas of necrosis and crush artefact is common. Nuclei are small, hyperchromatic and rounded to oval with evenly distributed chromatin and insignificant nucleoli[17]. Nuclear moulding may be present and mitoses are numerous. An occasional multinucleate giant cell may be seen. The tumours may be subdivided into three groups according to the definitions of the World Health Organization[132], as applied to tumours in the lung:

1. oat cell carcinoma type;
2. intermediate cell type; and
3. combined cell type.

In the intermediate cell type, the cells are slightly larger with more abundant cytoplasm and discernible nucleoli. In the combined type, small cell carcinoma is seen in association with another recognizable tumour type, such as TCC or adenocarcinoma or both[50]. There may be carcinoma in situ or dysplasia elsewhere in the bladder mucosa but no continuity with the tumour has been observed[17].

Immunohistochemistry of small cell carcinoma

Blomjous *et al.*[17] reported positive immunoreactivity to at least one epithelial marker, including epithelial membrane antigen, pan-cytokeratin and CAM 5.2, in 18 cases of small cell carcinoma. Neuroendocrine markers neurone-specific enolase (NSE) and synaptophysin showed a diffuse, finely granular cytoplasmic staining pattern in 13/18 cases. Immunoreaction for chromogranin A and Leu 7 was only seen in a minority of cases and was focal in distribution. There was no reaction to neuroendocrine markers in 3/18 cases. Electron microscopy showed rounded, dense-core membrane-bound granules, with diameters 150–250 nm in 44% of the cases. Using a combination of neuroendocrine markers and electron microscopical analysis, the diagnosis was confirmed in 78% of cases.

The histogenesis of small cell carcinoma is unknown but it has been suggested that it may arise by malignant transformation of neuroendocrine cells in the bladder mucosa[17] or from multipotent stem cells that represent the common precursors of various types of bladder cancer[17,98]. Young and Eble[136] have suggested that the frequent association of small cell carcinoma with urothelial carcinoma in the bladder may represent yet another differentiation pathway for urothelial carcinoma.

REVIEW PATHOLOGY

The variation in interpreting biopsies of normal bladder mucosa has been mentioned already and it should be no surprise that similar subjectivity is apparent in the reporting of tumour histology. Ooms *et al.*[89] reported considerable differences in the grading of TCCs by seven pathologists reviewing 57 biopsies. Some disagreed by two grades (Grade 1 versus Grade 3) and two pathologists revised their own grading in 50% of patients on a second examination. Abel *et al.*[1] reported that a review pathologist 'changed' the pT category of 14, and the grade of 13 of 99 tumours, compared with the original pathologist's report. A second 'review' pathologist then disagreed with the pT category of the first review pathologist in half of these cases. This study examined the original slides only, with no additional sections being cut by the review pathologists. Of 29 patients deemed to have muscle invasion by the original pathologist, five were 'downstaged' to pT1 and one to pTa. Of 24 pT1 tumours, seven were redesignated pTa and one of 46 pTa was 'upstaged' to pT1. The second review pathologist then upstaged seven of the 13 downstaged cases, two to pT1 from pTa and five to muscle infiltration again.

Witjes *et al.*[130] reported the outcome of central review pathology of 450 patients in a trial of intravesical therapy for Ta T1 TCC. Conformity between local and review pathology was 79.3% for pT category, 70.2% for tumour grade and the combination of both only 59.7%. Only five patients were upstaged to muscle invasion by the review pathologist, but of 54 pT1 G3 tumours (local pathologist) 16 were considered to be of lower grade or pTa by the review pathologist. Conversely, the review pathologist diagnosed 61 pT1 G3 tumours of which 25 had been described as pTa or G1 or G2 by the local pathologist.

On first reading, these discrepancies appear disturbing, especially as variations of similar magnitude have been found for both grade and stage (pTa, pT1) in Denmark[87] and other large but unpublished reviews of patients in Medical Research Council and EORTC trials. Fortunately for multicentre trials stratification procedures, randomization and large numbers avoid the introduction of bias and, as found by Witjes *et al.*[130], these interpathologist variations do not affect the trial results. Although the diagnosis of muscle invasion was a problem in the study of Abel *et al.*[1], in the recent MRC/EORTC trial of neoadjuvant CMV (cisplatin, methotrexate and vinblastine) the national review pathologists confirmed the muscle invasion reported by local pathologists of 78 institutions in 88% of patients (MRC/EORTC unpublished data).

As discussed in Chapter 4, both grade and T category have long been regarded as important prognostic factors. How does this general view fit with such histological variability? The prognostic factor analysis of Parmar *et al.*[93] confirmed that grade and category are significant prognostic factors, but not as significant as two other factors that are not so prone to variation. The latter were the number of tumours at diagnosis and the presence or absence of tumour at the first follow-up cystoscopy. Not all 'tumours' seen with a cystoscope are cancer and biopsy confirmation of the diagnosis is essential. 'Mistakes' can be made on cystoscopy, but they are less likely when counting the number of tumours (one or more than one) or when deciding with the help of biopsies whether TCC is present or absent. Not surprisingly these proved to be much stronger predictors of recurrence than grade and stage, and may now be used as the principal determinants for cystoscopic follow-up and the need for intravesical chemotherapy.

As discussed elsewhere in this book, the finding of CIS, submucosal invasion, muscle invasion and tumour grade are of daily relevance to the clinical management of bladder cancer. The management problems posed by T1 G3 tumours is a recurring theme. The foregoing discussion emphasizes the need for caution, but especially for regular discussion between urologist and pathologist, and a readiness to seek the help of an experienced uropathologist. Clearly it is not just the T1 G3 tumours that should be discussed but the T1 G2 tumours as well: 15/93 were upgraded to T1 G3 in the study of Witjes *et al.*[130]. This could create considerable extra work in a large unit but for most urology practices this would not be big problem. Only by taking the greatest possible care with both the endoscopic assessment **and** the histological examination of bladder cancers, will the patient be assured of the most appropriate care.

REFERENCES

1. Abel PD, Henderson D, Bennett MK *et al.* (1988) Differing interpretations by pathologists of the pT category and grade of transitional cell cancer of the bladder. *Br J Urol* **62**: 339–342.
2. Abeshouse BS (1943) Exstrophy of the bladder, complicated by adenocarcinoma of the bladder and renal calculi. *J Urol* **49**: 254–289.
3. Abrass RP, Temple-Camp CRE and Pontin AR (1989) Choriocarcinoma and transitional carcinoma of the bladder: a case report and review of the clinical evolution of disease in reported cases. *Eur J Surg Oncol* **15**: 149–153.
4. Ainsworth RW and Cresham GA (1960) Primary choriocarcinoma of the urinary bladder in a male. *J Pathol Bacteriol* **79**: 185–192.
5. Al-Izzi MS, Horton LWL, Kelleher J *et al.* (1989) Malignant transformation of the urinary bladder. *Histopathol* **14**: 191–198.
6. Amin MB, Gómez JA and Young RH (1997) Urothelial transitional cell carcinoma with endophytic growth patterns. *Am J Surg Pathol* **21**: 1057–1068.
7. Anderström C, Johansson S and Nilsson S (1980) The significance of lamina propria invasion on the prognosis of patients with bladder tumors. *J Urol* **124**: 23–26.
8. Anderström C, Johansson S and Petterson S (1982) Inverted papilloma of the urinary tract. *J Urol* **127**: 1132–1134.
9. Anderström C, Johansson SL and von Schultz L (1983) Primary adenocarcinoma of the urinary bladder. A clinicopathologic and prognostic study. *Cancer* **52**: 1273–1280.
10. Angulo JC, Lopez JI, Flores N *et al.* (1993) The value of tumour spread, grading and growth pattern as morphological predictive parameters in bladder carcinoma. A critical revision of the TNM classification. *J Cancer Res Clin Oncol* **119**: 578–593.
11. Angulo JC, Lopez JI, Grignon DJ and Sanchez-Chapado M (1995) Muscularis mucosa differentiates two populations with different prognosis in stage T1 bladder cancer. *Urology* **45**: 47–53.
12. Assor D (1978) A villous tumour of the bladder. *J Urol* **119**: 287–288.
13. Ayala AG and Ro JY (1989) Pre-malignant lesions of the urothelium and transitional cell tumours, in *Pathology of the Urinary Bladder*, (ed. RH Young), Churchill Livingstone, New York, pp. 65–101.
14. Benson RC, Swanson SK and Farrow GM (1984) Relationship of leukoplakia to urothelial malignancy. *J Urol* **131**: 507.
15. Berqvist A, Ljungqvist A and Moberger G (1965) Classification of bladder tumours based on the cellular pattern: preliminary report of a clinicopathological study of 300 cases with a minimum follow up of 8 years. *Acta Chir Scand* **130**: 371–378.
16. Birch BRP and Harland SJ (1989) The pT1 G3 bladder tumour. *Br J Urol* **64**: 109–116.

17. Blomjous CEM, Vos W, De Voogt HJ *et al.* (1989) Small cell carcinoma of the urinary bladder. *Cancer* **64**: 1347–1357.
18. Bostwick DG (1997) *Urologic Surgical Pathology*, (eds DG Bostwick and JB Eble), Mosby, St Louis, pp. 342–421.
19. Brodsky GL (1992) Pathology of bladder cancer. *Hematol Oncol Clin N Am* **6**: 59–80.
20. Bullock PS, Thoni DE and Murphy WM (1987) The significance of colonic mucosa (intestinal metaplasia) involving the urinary tract. *Cancer* **59**: 2086–2090.
21. Carbin BE, Ekman P, Gustafson H *et al.* (1991) The grading of human urothelial carcinoma based on nuclear atypia and mitotic frequency. 1. Histological description. *J Urol* **145**: 968–971.
22. Carbin BE, Eckman P, Gustafson H *et al.* (1991) The grading of human urothelial carcinoma based on nuclear atypia and mitotic frequency. 2. Prognostic importance. *J Urol* **145**: 972–976.
23. Channer JL, Williams JL and Henry L (1993) Villous adenoma of the bladder. *J Clin Path* **46**: 450–452.
24. Chibber PJ, McIntyre MA, Hindmarsh JR *et al.* (1981) Transitional cell carcinoma involving the prostate. *Br J Urol* **53**: 605–609.
25. Deen S and Ball RY (1994) Basement membrane and extracellular interstitial matrix components in bladder neoplasia – evidence of angiogenesis. *Histopathol* **25**: 475–481.
26. De Meester LJ, Farrow GM and Utz DC (1975) Inverted papillomas of the urinary bladder. *Cancer* **36**: 505–513.
27. Eagan JW (1989) Urothelial neoplasms; pathologic anatomy, in *Uropathology*, (ed. GS Hill), Churchill Livingstone, New York, Vol. 2, pp. 719–792.
28. Eble JN (1989) Abnormalities of the urachus, in *Pathology of the Urinary Bladder*, (ed. RH Young), Churchill Livingstone, New York, pp. 213–243.
29. Eble JN and Young RH (1989) Benign and low-grade papillary lesions of the urinary bladder. A review of the papilloma–papillary carcinoma controversy and a report of five typical papillomas. *Sem Diagnos Pathol* **6**: 351–371.
30. El Bolkainy MN, Mokhtar NM, Ghoneim MA *et al.* (1981) The impact of schistosomiasis on the pathology of bladder carcinomas. *Cancer* **48**: 2643–2648.
31. Elliott GB, Moloney PJ and Anderson GH (1973) 'Denuding cystitis' and in situ urothelial carcinoma. *Arch Pathol* **96**: 91–94.
32. El-Sebai, Sherif M, El Bolkainy MN *et al.* (1974) Verrucose squamous carcinoma of bladder. *Urology* **4**: 407.
33. Epstein JI (1995) Transitional cell carcinoma, in *Prostate Biopsy Interpretation*, 2nd edn, Lippincott-Raven, Philadelphia pp. 221–234.
34. Epstein JI, Kuhajda FP and Lieberman PH (1986) Prostate-specific acid phosphatase immunoreactivity in adenocarcinomas of the urinary bladder. *Human Pathol* **17**: 939–942.

35. Esrig D, Freeman JA, Elmajian DA *et al.* (1996) Transitional cell carcinoma involving the prostate with a proposed staging classification for stromal invasion. *J Urol* **156**: 1071–1076 .
36. Farrow GM (1979) A pathologist's role in bladder cancer. *Sem Oncol* **6**: 198–206.
37. Faysal MH (1981) Squamous cell carcinoma of the bladder. *J Urol* **126**: 598–599.
38. Fossa SD, Reitan JB, Ous S *et al.* (1985) Prediction of tumour progression in superficial bladder carcinoma. *Eur Urol* **11**: 1–5.
39. Francis RR (1979) Inverted papilloma in a 14 year old male. *Br J Urol* **51**: 327.
40. Friedell GH, Bell JR, Burney SW *et al.* (1976) Histopathology and classification of urinary bladder carcinomas. *Urol Clin N Am* **3**: 53–70.
41. Friedell GM, Soloway MS, Hilgar AG and Farrow GM (1986) Summary of workshop on carcinoma in situ of the bladder. *J Urol* **136**: 1047–1048.
42. Fukui I, Yokokawa M, Sekine M *et al.* (1987) Carcinoma in situ of the urinary bladder. *Cancer* **59**: 164–173.
43. Gallagher L, Lind R and Onasu R (1984) Primary choriocarcinoma of the urinary bladder in association with undifferentiated carcinoma. *Human Pathol* **15**: 793–795.
44. Gill HS, Dhillon HK and Woodhouse CRJ (1989) Adenocarcinoma of the urinary bladder. *Br J Urol* **64**: 138–142.
45. Goldstein AG (1967) Metastatic carcinoma to the bladder. *J Urol* **98**: 209–211.
46. Grammatico D, Grignon DJ, Eberwein P *et al.* (1993) Transitional cell carcinoma of the renal pelvis with choriocarcinomatous differentiation: immunohistochemical and immunoelectron microscopic assessment of human chorionic gonadotrophin production by transitional cell carcinoma of the urinary bladder. *Cancer* **71**: 1835–1841.
47. Greene LF, Hanash KA and Farrow GM (1973) Benign papilloma or papillary carcinoma of the bladder? *J Urol* **110**: 205–207.
48. Grignon DJ (1997) Neoplasms of the urinary bladder, in *Urologic Surgical Pathology,* (eds DG Bostwick and JN Eble), Mosby, St Louis, pp. 214–305.
49. Grignon DJ, Ro JY, Ayala AG *et al.* (1991) Primary adenocarcinoma of the urinary bladder. *Cancer* **67**: 2165–2172.
50. Grignon DJ, Ro JY, Ayala AG *et al.* (1992) Small cell carcinoma of the urinary bladder. *Cancer* **69**: 527–536.
51. Hardeman SW and Soloway MS (1988) Transitional cell carcinoma of the prostate, diagnosis, staging and management. *World J Urol* **6**: 170–174.
52. Hasui Y, Osada Y, Kitada S and Nishi S (1994) Significance of invasion to the muscularis mucosae on the progression of superficial bladder cancer. *Urology* **43**: 782–786.

53. Heney NM, Ahmed S, Flanagan MJ *et al.* (1983) Superficial bladder cancer: progression and recurrence. *J Urol* **130**: 1083–1086.
54. Heney NM, Proppe K, Prout GR *et al.* (1983) Invasive bladder cancer: tumor configuration, lymphatic invasion and survival. *J Urol* **130**: 895–897.
55. Horner SA, Fischer HA, Barada JH *et al.* (1991) Verrucous carcinoma of the bladder. *J Urol* **145**: 1261–1263.
56. Jakse G, Loidl W, Seeber G and Hofstadter G (1987) Stage T1 grade 3 transitional cell carcinoma of the bladder: an unfavourable tumour? *J Urol* **137**: 39–43.
57. Jewett HJ, King LR and Shelley WJ (1964) A study of 365 cases of infiltrating bladder cancer: relation of certain pathological characteristics to prognosis after extirpation. *J Urol* **92**: 668–678.
58. Jordan AM, Weingarten J and Murphy WM (1987) Transitional cell neoplasms of the urinary bladder. Can biological potential be predicted from histologic grading? *Cancer* **60**: 2766–2774.
59. Kamat MR, Kulkarni JN and Tongaonkar HB (1991) Adenocarcinoma of the bladder: study of 14 cases and review of the literature. *Br J Urol* **68**: 254–257.
60. Koss LG (1974) Tumours of the urinary bladder. *Atlas of Tumour Pathology*, (2nd series, Fascicle 11), Armed Forces Institute of Pathology, Washington, DC.
61. Kunze E, Schauer A and Schmitt M (1983) Histology and histogenesis of two different types of inverted urothelial papillomas. *Cancer* **51**: 348–358.
62. Larsen NP, Steinberg GD, Brendler CB *et al.* (1990) Use of the *Ulex europaeus* agglutinin 1 (UEA1) to distinguish vascular and pseudovascular invasion in transitional cell carcinoma of bladder with lamina propria invasion. *Mod Pathol* **3**: 83–88.
63. Lazarevic B and Garret R (1978) Inverted papilloma and papillary transitional cell carcinoma of the urinary bladder. *Cancer* **42**: 1904–1911.
64. Lerman RI, Hutter RV and Whitmore WF (1970) Papilloma of the urinary bladder. *Cancer* **25**: 333–342.
65. Logothetis CJ, Dexeus FH, Chong C *et al.* (1989) Cisplatin, cyclophosphamide and doxorubicin chemotherapy for unresectable urothelial tumours: The MD Anderson experience. *J Urol* **141**: 33–37.
66. Lopez JI and Angulo JC (1995) The prognostic significance of vascular invasion in stage T1 bladder cancer. *Histopathol* **27**: 27–33.
67. Mahdadevia PS, Koss LG and Tar IJ (1986) Prostatic involvement in bladder carcinoma. *Cancer* **58**: 2096–2102.
68. Makar N (1962) Some observations on pseudoglandular proliferation in the bilharzial bladder. *Acta Unio Inter Contra Cancrum* **18**: 599–607.
69. Martin JE, Jenkins BJ and Zuk RJ (1989) Clinical imprtance of squamous metaplasia in invasive transitional cell carcinoma of the bladder. *J Clin Path* **42**: 250–253.

70. Matzkin H, Soloway MS and Hardeman S (1991) Transitional cell carcinoma of the prostate. *J Urol* **146**: 1207–1212 .
71. Melicow MM (1952) Histological study of vesical urothelium intervening between gross neoplasms in total cystectomy. *J Urol* **68**: 261–279.
72. Miller A, Mitchell JP and Brown WJ (1969) The Bristol Bladder Tumour Registry. *Br J Urol* **41** (Suppl.): 1–64.
73. Miller DC, Gang DG, Gavris VE *et al.* (1983) Villous adenoma of the urinary bladder; a morphologic or biologic entity. *Am J Clin Pathol* **79**: 728–731.
74. Morgan RJ and Cameron KM (1980) Vesical leukoplakia. *Br J Urol* **52**: 96–100.
75. Mostofi FK and Sesterhenn IA (1984) Pathology of epithelial tumours and carcinoma in situ of the bladder, in *Bladder Cancer. Part A: Pathology, Diagnosis and Surgery,* (eds R Kus, S Khoury, LJ Denis, GP Murphy and JP Karr), Alan R Liss, New York, pp. 55–74.
76. Mostofi FK, Sesterhenn IA and Davis CJ (1988) Dysplasia versus atypia versus carcinoma in situ of bladder, in *Difficult Diagnoses in Urology*, (ed. DL McCullough), Churchill Livingstone, New York.
77. Mostofi FK, Sobin LF and Torloni H (1973) *Histological Typing of Urinary Bladder Tumours,* International Histological Classification of Tumours, No. 10, World Health Organization, Geneva.
78. Mostofi FK, Thomason RV and Dean AL (1955) Mucous adenocarcinoma of the urinary bladder. *Cancer* **8**: 741–758.
79. Murphy WM (1989) Diseases of the urinary bladder, urethra, ureters and renal pelvis, in *Urological Pathology,* (ed. WM Murphy), WB Saunders, Philadelphia, pp. 34–146.
80. Murphy WM (1994) ASCP survey on anatomic pathology examination of the urinary bladder. *Am J Clin Pathol* **102**: 715–723.
81. Murphy WM, Beckwith JB and Farrow GM (1994) Tumours of the kidney, bladder and related urinary structures. *Atlas of Tumour Pathology,* (3rd series, fascicle 11) Armed Forces Institute of Pathology, Washington, DC.
82. Murphy WM and Soloway MS (1982) Urothelial dysplasia. *J Urol* **127**: 849–854.
83. Nagy GK, Frable WJ and Murphy WM (1982) Classification of pre-malignant urothelial abnormalities, in *Pathology Annual*, (ed. SC Sommers and PP Rosen), Appleton-Century, New York, Vol. 17, Part 1, pp. 219–235.
84. National Bladder Cancer Collaborative Group A (1997) Surveillance, initial assessment and subsequent progress of patients with superficial bladder cancer in a prospective longitudinal study. *Cancer Res* **37**: 2907–2910.
85. Norton KD and Burnett RA (1988) Choriocarcinoma arising in transitional cell carcinoma of the bladder. *Histopathol* **12**: 325–328.
86. O'Brien AME and Urbanski SJ (1985) Papillary adenocarcinoma in situ of the bladder. *J Urol* **134**: 544–546.

87. Olsen LH, Overgaard S, Frederikesen P *et al.* (1993) The reliability of staging and grading of bladder tumours. *Scand J Urol Nephrol* **27**: 349–353.
88. Olsen S (1984) Urothelium and urothelial neoplasia, in *Tumours of the Kidney and Urinary Tract*, Munksgaard, Copenhagen, pp. 125–135.
89. Ooms ECM, Anderson WAD, Allons CL *et al.* (1983) Analysis of the performance of pathologists in the grading of bladder tumours. *Human Pathol* **14**: 140–143.
90. Orozco RE, Martin AA and Murphy WM (1994) Carcinoma in situ of the urinary bladder. *Cancer* **74**: 115–122.
91. Orozco RE, van der Zwaag R and Murphy WM (1993) The pagetoid variant of urothelial carcinoma in situ. *Human Pathol* **24**: 1199–1202.
92. Pagano F, Bassi P, Galetti T *et al.* (1991) Results of contemporary radical cystectomy for invasive bladder cancer: a clinicopathological study. With an emphasis on the inadequacy of the tumor, nodes and metastases classification. *J Urol* **145**: L45–50.
93. Parmar MKB, Friedman LS, Hargreave TB and Tolley DA (1989) Prognostic factors for recurrence and follow up policies in the treatment of superficial bladder cancer: a report from the British Medical Research Council Subgroup on superficial bladder cancer (Urological Cancer Working Party). *J Urol* **142**: 284–288.
94. Paschkis R (1927) Über adenome der Harnblase. *J Urol Chir* **21**: 315.
95. Paulson DF (1993) Critical review of radical cystectomy and indications of prognosis. *Sem Urol* **11**: 205–213.
96. Pauwels RPE, Schapers RFM, Smeets AWGB *et al.* (1988) Grading in superficial bladder cancer. (1) Morphological criteria. *Br J Urol* **61**: 129–134.
97. Petersen RO (1986) Urinary bladder, in *Urologic Pathology*, JP Lippincott, Philadelphia, pp. 279–416.
98. Podesta AH and True LD (1989) Small cell carcinoma of the bladder. *Cancer* **64**: 710–714.
99. Potts AF and Hirst E (1963) Inverted papillomas of the bladder. *J Urol* **90**: 175–179.
100. Prout GR, Barton BA, Griffin PP *et al.* (1992) Treated history of non-invasive Grade 1 transitional cell carcinoma. *J Urol* **148**: 1413–1419.
101. Pugh RCB (1973) The pathology of cancer of the bladder. *Cancer* **32**: 1267–1274.
102. Ramani P, Birch BRP, Harland SJ *et al.* (1991) Evaluation of endothelial markers in detecting blood and lymphatic channel invasion in pT1 transitional carcinoma of the bladder. *Histopathol* **19**: 551–554.
103. Reese JH, Freiha FS, Geb AB *et al.* (1992) Transitional cell carcinoma of the prostate in patients undergoing radical cystoprostatectomy. *J Urol* **147**: 92–95.

104. Renfer LG, Kelley J and Belville WD (1988) Inverted papilloma of the urinary tract; histogenesis, recurrence and associated malignancy. *J Urol* **140**: 832–834.
105. Richards B, Parmar MKB, Anderson CK *et al.* (1991) Interpretations of biopsies of 'normal' urothelium in patients with superficial bladder cancer. *Br J Urol* **67**: 369–375.
106. Ro JY, Ayala AG and El-Naggar AK (1987) Muscularis mucosae of the urinary bladder: importance of staging and treatment. *Am J Surg Pathol* **11**: 668–673.
107. Ro JY, Staerkel GA and Ayala AG (1992) Cytologic and histologic features of superficial bladder cancer. *Urol Clin N Am* **19**: 435–453.
108. Robertson AJ, Swanson Beck J, Burnett RA *et al.* (1990) Observer variability in histopathological reporting of transitional cell carcinoma and epithelial dysplasia in bladders. *J Clin Path* **43**: 17–21.
109. Rundle JSH, Hart JL, McGeorge A *et al.* (1982) Squamous cell carcinoma of the bladder. A review of 114 patients. *Br J Urol* **54**: 522–526.
110. Sakamoto N, Tsuneyoshi M and Enjoji M (1992) Urinary bladder carcinoma with a neoplastic squamous component: a mapping study of 31 cases. *Histopathol* **21**: 135–141.
111. Sakamoto N, Tsuneyoshi M, Naito S *et al.* (1993) An adequate sampling of the prostate to identify prostatic invovement by urothelial carcinoma in bladder cancer patients. *J Urol* **149**: 318–321.
112. Schapers RF, Pauwels RP, Havenith MG *et al.* (1990) Prognostic significance of type IV collagen and laminin immunoreactivity in urothelial carcinoma of the bladder. *Cancer* **66**: 2583–2588.
113. Schapers RFM, Pauwels RPE, Wijnen JThM *et al.* (1994) A simplified grading method of transitional cell carcinoma of the urinary bladder: reproducibility, clinical significance and comparison with other prognostic parameters. *Br J Urol* **73**: 625–631.
114. Schellhammer PF, Bean MA and Whitmore WF (1977) Prostatic involvement by transitional cell carcinoma; pathogenesis, patterns and prognosis. *J Urol* **118**: 399–403.
115. Schellhammer PF, Ladega LE and Moriarty RP (1995) Intravesical Bacillus Calmette–Guérin for the treatment of superficial transitional cell carcinoma of the prostatic urethra in association with carcinoma of the bladder. *J Urol* **153**: 53–56.
116. Sharkey FE and Sarosdy MF (1997) The significance of central pathology review in clinical studies of transitional cell carcinoma in situ. *J Urol* **157**: 68–71.
117. Sheldon CA, Clayman RV, Gonzalez R *et al.* (1984) Malignant urachal lesions. *J Urol* **131**: 1–8.
118. Soli M, Bercovitch E, Botteghi B *et al.* (1987) A rare case of mucous-secreting villous adenoma of the bladder. *It J Surg* **3**: 261–264.

119. Solsona E, Iborra I, Ricós JL *et al.* (1995) The prostate involvement as prognostic factor in patients with superficial bladder tumours. *J Urol* **154**: 1710–1713 .
120. Stein BS, Rosen S and Kendall R (1984) The association of inverted papilloma and transitional cell carcinoma of the urothelium. *J Urol* **131**: 751–752.
121. Tinkler SD, Roberts JT, Robinson MC *et al.* (1996) Primary choriocarcinoma of the urinary bladder; a case report. *Clin Oncol* **8**: 59–61.
122. Walther M, O'Brien DP and Birch HW (1986) Condylomata acuminata and verrucous carcinoma of the bladder: case report and literature review. *J Urol* **135**: 362–365.
123. Webb JN (1992) Aspects of tumours of the urinary bladder and prostate gland, in *Recent Advances in Histopathology,* (eds PP Anthony and RNM MacSween), Churchill Livingstone, Edinburgh, pp. 157–176.
124. Webb KN (1990) Histopathology of bladder cancer, in *Scientific Foundations of Urology,* (eds GD Chisholm and WR Fair), Heinemann Medical Books, Oxford, pp. 549–561.
125. Weinberg T (1939) Primary chorionepithelioma of the urinary bladder in a male patient. *Am J Pathol* **15**: 783–795.
126. Wheeler JD and Hill WT (1954) Adenocarcinoma involving the urinary bladder. *Cancer* **7**: 119.
127. Whitesel JA (1982) Inverted papilloma of the urinary tract: malignant potential. *J Urol* **127**: 539–540.
128. Whitmore WF (1979) Surgical managment of low stage bladder cancer. *Sem Oncol* **6**: 207–216.
129. Wishnow KI and Ro JY (1988) Importance of early treatment of transitional cell carcinoma of the prostatic ducts. *Urology* **32**: 11–12.
130. Witjes JA, Kiemeney LALM, Schaasfa HE *et al.* (1994) The influence of review pathology on study outcome of a randomised multicentre superficial bladder cancer trial. *Br J Urol* **73**: 172–176.
131. Wood DP, Montie JE, Pontes JE *et al.* (1989) Transitional cell carcinomaof the prostate in cystoprostatectomy specimens removed for bladder cancer. *J Urol* **141**: 346–349.
132. World Health Organization (1982) Histological typing of lung tumors. *Am J Clin Pathol* **77**: 123–136.
133. Younes M, Sussman J and True LD (1990) The usefulness of the level of the muscularis mucosae in the staging of invasive transitional cell carcinoma of the urinary bladder. *Cancer* **66**: 543–548.
134. Young RH (1989) Unusual variants of primary bladder carcinoma and secondary tumours of the bladder, in *Pathology of the Urinary Bladder*, (ed. RH Young), Churchill Livingstone, New York, pp. 103–138.

135. Young RH (1996) Pathology of bladder cancer, in *Comprehensive Textbook of Genito-urinary Oncology*, (eds NJ Vogelzang, PT Scardino, WU Shipley and DS Coffey), Williams and Wilkins, Baltimore, pp. 326–337.
136. Young RH and Eble JN (1991) Unusual forms of carcinoma of the urinary bladder. *Human Pathol* **22**: 948–964.
137. Young RH and Oliva E (1996) Transitional cell carcinomas of the urinary bladder that may be underdiagnosed. *Am J Surg Pathol* **20**: 1448–1454.

3

Transurethral resection and staging of bladder cancer

Reginald R. Hall

Transurethral resection, histology and the staging of bladder cancer are so closely linked that it is difficult to discuss one without the others. The cystoscopic appearance of the tumour and the rest of the bladder, the findings of bimanual palpation, the visual appearance of the tumour and bladder wall as the resection proceeds and the details of histological examination are the fundamental pieces of information that determine the tumour stage and guide treatment. As most new bladder tumours are diagnosed by flexible cystoscopy as part of an outpatient consultation, the patient will attend for endoscopic assessment knowing that they probably have bladder cancer but uncertain about its prognosis. For cancers not invading the detrusor muscle the transurethral resection (TUR) combines the definitive treatment and the provision of biopsy material for histology and staging all in one process. For muscle invasive bladder cancers thorough endoscopic resection of the tumour mass is the initial step in bladder conserving curative treatment.

ENDOSCOPIC ASSESSMENT AND TRANSURETHRAL RESECTION

Anaesthesia

The initial TUR should be performed under general anaesthetic. In some patients this may not be feasible and spinal (epidural) anaesthesia may be unavoidable, but is not ideal. The endoscopic procedure and bimanual palpation must be thorough and this requires good relaxation of the abdominal wall. If anaesthesia proves difficult and the patient is straining or coughing, the urologist should wait until the situation has stabilized to permit safe resection wherever this is necessary in the bladder. The secret of success for bimanual palpation is an empty bladder in a fully relaxed patient.

Transurethral resection (TUR)

Endoscopic resection of bladder tumours (TURBT) has been a standard procedure throughout the working lifetime of most urologists in current practice, but in terms of the history of urology it is a relatively recent development. The resectoscope was developed originally for the prostate. Beer in 1910[4] was the first to describe endoscopic methods to treat bladder cancers, followed by Gibson in 1935[18] and McDonald and co-workers in 1947[51]. By the late 1940s both the diathermy 'hot loop' and cold punch resectoscopes were in regular use in North America and some European institutions but their use for bladder tumours appears to have been limited. Reynolds *et al.*[67] reported in 1949 the superiority of the diathermy resectoscope over the cold punch for bladder cancer, and in 1955 Thompson and Kaplan[80] described detailed technique for TUR of bladder cancer, reporting generally favourable results in 300 patients, including poorly differentiated cancers. However, as recently as 1962 Jones and Swinney[42] observed that 'transurethral resection of bladder tumours has not received general acceptance' and recommended the technique on the basis of its suitability and efficacy in 148 patients out of a total referral population of 750 with bladder cancer.

To this day, the endoscopic training of most urologists concentrates on TUR of the prostate with less attention to the resection of bladder tumours. With the advent of alternative medical and surgical treatments for benign prostatic hyperplasia, TUR has already become a less frequent operation. For bladder cancer in the foreseeable future there is no prospect of any alternative to conventional TUR, perhaps combined with electrode vaporization. This means that those urologists who intend to care for patients with all stages of bladder cancer (as distinct from tertiary referral urologists who limit themselves to radical surgery) need to acquire and maintain these skills by regular, thoughtful endoscopic practice. **Using the right equipment is important**. Although it is possible to resect bladder cancers with a standard resectoscope, even with the two-handed instruments based on the original Stern–McCarthy design[5], there is no excuse not to use the single-handed, continuous suction, double sheath 'Iglesias' resectoscope[34,35]. Used properly with low pressure inflow and variable, continuous suction attached to the outlet, the resection is made as safe as is possible for the patient and easy for the surgeon[21]. Attention should also be paid to the quality of the diathermy current in order to minimize electrocautery artefact that impedes histological examination. Too high a cutting current will cause arcing that will destroy precious tissue. A current heavily blended with coagulation will cause unnecessary thermal coagulation. A pure cutting current, or a blend with minimal coagulation, is best for resecting material for biopsy. Unblended coagulation is fine for haemostasis.

Laser vaporization, as an alternative to TUR[31,41,75], and diathermy **fulguration**[58] have become common practice for the destruction of some bladder tumours. Both methods may be appropriate for small tumours (less than 5 mm) detected during follow-up, when they can be combined with cold cup biopsy using the flexible cystoscope without anaesthesia. However, they are not adequate for the initial assessment of a first-time bladder cancer. The new range of grooved, cylindrical electrodes for **vaporization** rather than resection have become popular for TUR of the prostate because they appear to be more haemostatic than the loop electrode[50]. They may prove useful in removing the main bulk of large bladder tumours[53] but can only be an aid to resection, which remains the only means of obtaining generous, good-quality material for histological examination.

It should be remembered always that the purpose of resecting a bladder cancer is threefold:

1. To remove the cancer, as a (curative) surgical procedure.
2. To assess endoscopically the depth of tumour invasion.
3. To obtain samples of tumour and bladder wall for histological examination to determine the type, grade and extent of the tumour.

To achieve the first objective, all tumour visible in the bladder should be removed or destroyed. For the majority of Ta and T1 tumours this is a straightforward procedure. Some patients with numerous multiple tumours are described as 'unresectable', particularly if there are large areas of confluent papillary tumour or numerous tumours on the anterior bladder wall or dome. In such circumstances some urologists give up on TUR and hope that intravesical chemotherapy or BCG will destroy what remains. This is not good practice. Phase II trials of all the intravesical agents currently in use have demonstrated that 6–8-week courses of treatment, as 'chemoresection', destroy residual marker tumours in only 50% of patients[7,15]. The most effective means of clearing a bladder of multiple Ta or T1 bladder cancer is transurethral resection, completed if necessary on a second occasion 2–4 weeks later. For papillary tumours the role of intravesical therapy is as adjuvant or prophylactic treatment to reduce the risk of recurrence after a complete TUR. It is not a substitute for skilled, patient resection. Given reasonable time and care, very few bladders cannot be cleared of Ta or T1 tumours by endoscopic resection. If I am prepared to spend 4 or 5 hours performing cystectomy and bladder reconstruction for a patient, I should be prepared to spend 2 or 3 hours completing the TUR, in two or three sessions if necessary, in the justified hope of avoiding cystectomy.

If the tumour extends outside the bladder, complete removal may not be possible, but most muscle invasive bladder cancers are 5 cm or less in diameter and relatively few occur at the bladder dome where there would be a risk of intraperitoneal perforation. In the international trial of neoadjuvant CMV (cisplatin, methotrexate and vinblastine) chemotherapy for muscle invasive bladder cancer[22], 80% of patients recruited to the study had tumours 5 cm or less in diameter[36]. Any surgeon involved in the management of bladder cancer patients should be able to resect tumours of this size without undue difficulty. Furthermore, several reviews of cystectomy for muscle invasive bladder cancer have demonstrated that there was no residual tumour in the cystectomy specimen in up to 17% of patients who were planned for cystectomy and for whom there would have been no reason to attempt complete endoscopic removal[28]. Thus, with a little more time and effort most T2 and T3 tumours, including some T3b cancers can be resected if the urologist has the desire to do so.

Tissue for histology

The second objective is achieved by careful observation during the course of the resection, and keeping a record of the findings. The surgeon can obtain a very good idea of the depth of tumour invasion as the TUR proceeds. It is this visual and tactile information that determines the depth and lateral extent of the operation. It should also guide the separation of the resected tissue into sequential, numbered specimens for the pathologist, to maximize the third objective and the histological confirmation of depth of invasion. Small tumours may be removed entirely as a single or only a few pieces of tissue. Provided these include some superficial detrusor muscle there is no need to take further cold cup biopsies from the floor, or to separate the pieces for the pathologist. However, for larger and especially muscle invasive cancers, separation as three or four labelled samples is advised, as follows.

- Sample 1. All exophytic tumour down to the level of the detrusor.
- Sample 2. The tumour base, including underlying and surrounding normal-looking tissue.
- Sample 3. Further normal-looking tissue from the resection floor to confirm the completeness of the depth of resection. If the tumour penetrates the full thickness of the bladder wall, this third specimen should be taken with caution and may be difficult to obtain as normal perivesical connective tissue and fat does not cut easily with a diathermy resectoscope.
- Sample 4. If there is any doubt about the lateral extent of the tumour, the presence of adjacent CIS or the clearance of resection margins, a further 'loopful' of tissue all round the margin should be removed and examined separately.

If **multiple tumours** are present there is little to be gained by separate histological examination of each tumour unless one is thought to be invading muscle, or areas of CIS are to be distinguished from papillary tumour. The identification of CIS is important as the choice of adjuvant intravesical treatment may be different from that for Ta or T1 tumours. If there are numerous small tumours in the bladder, resection of all may remove unnecessarily large amounts of bladder wall in total. Similarly, the resection of large areas of CIS may predispose to later bladder shrinkage. For both situations, provided the largest tumours or most suspect representative areas of CIS have been resected, complete ablation of normal mucosa can be achieved with minimal damage to underlying tissue using a ball electrode or 'Vaportrode' with a high cutting diathermy current. For tumours invading muscle, if bladder conserving treatment is planned, knowledge of the site of the tumour within the bladder is important for radiotherapy, and to guide follow-up biopsies if no residual tumour is visible at subsequent cystoscopy.

Completeness of endoscopic resection

Studies of the **completeness of TUR** have been sparse. Oosterlink *et al.*[59] reported the experience of the EORTC Genito-urinary Group in a trial of single-dose epirubicin in solitary Ta T1 bladder tumours. To ensure the quality of the surgery and the inclusion of only solitary tumours in the trial, cystoscopy was performed 4 weeks after the initial TUR. A surprising number of patients had further, presumably residual tumour. A few may have been rapidly growing new tumours but the findings emphasize the need to check the thoroughness of tumour TUR very carefully. For may years some European colleagues have performed a second TUR routinely. This was commendable practice in the days when telescopes and light sources were of poor quality, but should not be necessary with modern equipment.

Another study of the adequacy of TUR was reported by Kolozsy[46] who examined tissue resected from the margins and floor of bladder tumour resections after they were considered optically to be complete. Overall 35% of 462 TURs were found to be incomplete; 12.7% of pTa tumours, 36.2% of pT1, 55.9% of pT2 and 83% of pT3. The most common site of residual tumour was the floor of the resection (26%). In 16% of patients, perivesical adipose tissue was demonstrated on histological examination. Without knowing the circumstances of the institution concerned and the quality of the endoscopic equipment available, it is difficult to judge the applicability of these findings. However, the lesson appears to be clear: when the TUR of a bladder cancer appears to be complete, it probably is not, and the removal of a generous amount of further, apparently normal-looking tissue from the margins and floor is advisable.

New techniques have become available that may help to ensure that the endoscopic resection of bladder tumours and any concomitant CIS is complete. **Fluorescence cystoscopy**[43,48] uses a xenon arc lamp to detect the light-induced fluorescence of the photosensitizer protoporphyrin 1X (Pp 1X) induced by the topical application of 5-amino-levulinic acid (ALA). Compared with the normal urothelium areas of dysplasia, CIS and papillary tumour fluoresce and are thus made visible, having been invisible to conventional white-light cystoscopy. Overall sensitivity and specificity of the technique are reportedly high[39,49] but Jichlinski *et al.*[40] observed false-positive fluorescence in 35% (71/205) of biopsies. Of the 'lesions' detected by this method, 27/41 were areas of mild and moderate dysplasia that would not usually be of concern. The technique therefore tended to overdiagnose abnormalities and if used in routine practice would probably result in unnecessary TUR and overtreatment of some patients. However, for patients with CIS it could enhance assessment and follow-up by enabling mucosal biopsies to be 'targetted' morc appropriately. Koenig *et al.*[45] used a nitrogen laser to induce fluorescence in untreated bladder mucosa (without photosensitizer) and reported that the technique could distinguish between transitional cell carcinoma and dysplasia/ inflammation. Further study with larger patient numbers is required.

PT1 G3 tumours

The high recurrence and local progression rates of pT1 G3 tumours are well recognized[37]. The fact that so many recur at the same site as the first TUR suggests that the latter has been incomplete, leaving behind adjacent foci of T1 cancer, or failing to diagnose the T2 tumour actually present. In either circumstance a repeat, wide, full-thickness TUR no more than 2–3 weeks after the first TUR should address both problems, although the contribution of this second TUR to local control has not been proved. pT1 G3 tumours are also associated frequently with CIS elsewhere in the bladder. It may be this, as much as residual deeper tumour cells, that is responsible for later progression. The lack of benefit to be gained by taking multiple random biopsies of normal-looking urothelium for Ta and T1, G1 or G2 tumours has been discussed. However, T1 G3 tumours warrant full endoscopic assessment before making any treatment decision and random biopsies to search for concomitant CIS should be taken at the same time as the repeat TUR.

Random biopsies and invasive bladder cancer

Random biopsies of the bladder would not appear to provide any information of clinical relevance in patients with muscle invasive bladder cancer if they are to undergo radical cystectomy. However, if bladder conservation is being considered, mucosal biopsies could be helpful, but data to support this view are limited. CIS in the mucosa adjacent to muscle invasive cancers is a common finding. When concomitant CIS has been documented elsewhere in the bladder it seems to predict for a poor outcome following radical radiotherapy (Chapter 8). Patient selection criteria for brachytherapy and partial cystectomy are strictly defined and for both of these treatment modalities the absence of concomitant CIS is a prerequisite for successful treatments. Patients treated by TUR alone[76] or combined with systemic therapy[23,79], TUR and chemoradiation or standard radical radiotherapy[73,74] all run the risk of developing new superficial tumours during long-term follow-up. A few develop second muscle invasive bladder cancers at a different site in the bladder[79]. Whether this could be predicted by random biopsies, and prevented by adjuvant intravesical treatment, is pertinent to speculate. In view of the growing interest in bladder-preserving treatments for muscle invasive bladder cancer, further study of the role of random biopsies in this particular group of patients would appear to be worthwhile.

Biopsies of the prostatic urethra

Transitional cell carcinoma in the prostatic urethra, prostatic ducts and invading the prostate stroma is not a common occurrence, but when it does exist it has a significant impact on the management of the patient[25]. Most of our information comes from the retrospective review of cystoprostatectomy specimens, made available by the radical treatment of muscle invasive bladder cancer, CIS or patients with multiple recurrent T1 tumours[13,26,60]. As a result these reports probably overestimate the frequency of the problem, but several features have practical implications for the pretreatment endoscopic evaluation of bladder cancer[25].

Dr Robinson in Chapter 2 has noted that the prognosis of non-invasive TCC in the prostatic urethra or prostate ducts is probably similar to that of CIS or T1 tumours in the bladder and may be treated by TUR of the prostate and intravesical BCG[8,10,25,77]. Any breach of the basement membrane of the urothelium of the prostatic urethra leads to a much higher risk of disease progression and radical cysto-prostato-urethrectomy is the current treatment of choice[13,26]. Invasion of the prostate stroma from the urethra will be more likely to occur than muscle invasion of the bladder wall because the lamina propria is very thin in the prostatic urethra and non-existent around prostatic ducts[25]. It is therefore just as important to biopsy the prostatic urethra as it is to take random biopsies of the bladder in all patients with CIS and pT1 G3 cancers before deciding treatment. The diagnosis of TCC in the prostate is achieved best by TUR biopsies of the floor of the prostatic urethra at 5 and 7 o'clock that extend from the bladder neck to the distal end of the verumontanum and include underlying prostate stroma[69,88], together with other more anterior biopsies of the prostatic urethra. In a review of patients treated by cystectomy without synchronous urethrectomy, Hardeman and Soloway[26] found that the extent of TCC in the prostate predicted for urethral recurrence. The latter occurred in 0/8 of patients when tumour was confined to the prostatic urethra, 2/8 in prostate ducts and 9/14 with prostate stromal infiltration. They therefore recommended synchronous cysto-prostato-urethrectomy for the third group of patients and that the prostate should be biopsied preoperatively in all patients planned for cystectomy. In his commentary on Chapter 1, Dr Herr clearly supports the view that some patients with TCC in the prostatic urethra may be suitable for orthotopic bladder substitution, emphasizing the need for careful TUR biopsy of the prostate when cystectomy is a treatment option.

SIGNIFICANCE OF COMPLETE TUR FOR T2 AND T3 CANCERS

Prout[64] found that survival was greater in patients whose biopsy TUR had been complete compared with other patients undergoing cystectomy in whom tumour remained in the operative specimen. He suggested that the difference in survival might be due to the dissemination of cancer cells during the cystectomy from the residual tumour in the bladder[65]. Miller and Johnson[56] had reported higher 5-year survival for patients treated by full-dose radiotherapy if their T2 and T3 tumours had been resected completely. Similarly, Shipley *et al.*[74] reported the long-term outcome to be better for those patients in whom the TUR biopsy had removed all tumour. However, Prout's explanation would not have applied to the findings of Miller and Johnson[56] and Shipley *et al.*[74] as cystectomy was not part of the definitive treatment.

The most probable explanation is that smaller invasive bladder cancers have a better prognosis than large tumours[12]; small tumours are easier to resect, and therefore are more likely to be resected completely. Thus, the completely resected tumours observed by these authors are likely to have been smaller and to have had an inherently better prognosis. The TUR itself would have had no impact on cure. This view was supported by Thrasher *et al.*[81] who compared survival of patients whose cystectomy specimen was pT0 with those whose residual pT category was the same as their pre-cystectomy category. Within each category survival was equivalent and the authors concluded that 'a pT0 cystectomy functions, by survival analysis, in a manner similar to one with the stated clinical stage'.

While the above applies to the development of metastases and survival, completeness of the TUR may contribute to local tumour control and bladder preservation. Dr Roberts' review in Chapter 10 and the previous review by Herr[28] illustrate that TUR alone may be curative for some patients with T2 and T3 cancers. The completeness of the surgical removal of the tumours is a determinant of success for these patients. As a general principle, external beam radiotherapy is more effective if the amount of tumour is small. As suggested by Miller and Johnson[56] and Shipley *et al.*[74] for patients who opt for radiotherapy instead of cystectomy, a determined effort to remove as much tumour as possible by TUR before commencing radiotherapy may improve local tumour control. The multivariate analysis of Shipley[73] had shown complete TUR to be an independent prognostic factor for local tumour control, albeit based on a small number of patients. All too often urologists who intend to refer their patient for radiotherapy rather than cystectomy, take no more than a biopsy, and overlook the possibility that they could facilitate the irradiation and maybe improve the outcome by half an hour's endoscopic work. Concern that full-thickness bladder resection will cause problems with increased radiotherapy toxicity has not been confirmed in practice. Endoscopically complete full-thickness TUR of T3 bladder cancers can be followed by radiotherapy in 3–4 weeks without problems.

Systemic chemotherapy is also believed to be most effective when confronted with minimal tumour volume. Studies have shown that selected patients can be cured by thorough TUR combined with systemic chemotherapy, or by consolidating a good chemotherapy response in the bladder with TUR (Chapter 10), but as discussed by Herr[28] how much these patients owe to the chemotherapy and how much to TUR is unknown.

No randomized trial has examined the role of TUR in achieving successful bladder preservation for muscle invasive bladder cancer[29]. However, in view of the knowledge that some patients can be cured by TUR alone, others by TUR combined with chemotherapy, and that debulking TUR may improve the likelihood of success with radiotherapy, if a patient wishes to avoid cystectomy, as thorough a TUR as possible should be performed. If after the wishes of the patient become clear it is realized that the biopsy TUR was an incomplete surgical procedure, a second TUR is both logical and appropriate.

FULL-THICKNESS TUR

Full-thickness resection of the bladder wall, exposing perivesical fat and connective tissue, has been mentioned several times in this review. I am aware that some colleagues are opposed to the operation, having reservations about its safety, and others have had unfavourable experience with its outcome.

Full-thickness TUR is not a new operation. It was described by Barnes *et al.*[3] 50 years ago and has been performed regularly by many surgeons since. Its opponents are concerned by three potential hazards:

- The operation may be dangerous as a result of haemorrhage or the TUR syndrome.
- Perforation of the bladder wall will carry malignant cells unnecessarily into the extravesical tissues and increase the risk of loco-regional recurrence.
- Attempts at full-thickness resection of the posterior wall and dome of the bladder will lead to intraperitoneal perforation and wide dissemination of cancer cells.

Operative risks

Resecting deep in the detrusor, especially in the vicinity of large cancers, and outside the bladder may encounter large vessels that retract out of sight and are difficult to control. Opening large veins runs the risk of intravenous absorption of large quantities of irrigant which causes hyponatraemia, hypokalaemia or the TUR syndrome. Even more major blood vessels, such as the superior and inferior vesicle arteries and large branches of the internal iliac vein, may be only a few millimetres away from the extravesical part of a T3 cancer. The obturator nerve is a notorious cause of inconvenience to resectors of bladder tumours and is even more likely to cause trouble during full-thickness resection. All of these factors are true but, just like any other major operation, full-thickness TUR is a safe and effective procedure provided that the patients are selected carefully, the appropriate equipment is used and the surgeon has the necessary skills. The most serious risks arise from fibrosis outside the bladder fixing the outer surface of the bladder to vascular and nervous structures in the deep pelvis, or adhesions of bowel to the dome of the bladder. To avoid such risk, full-thickness resection should not be performed if there has been previous pelvic sepsis, genito-urinary tuberculosis, rectal or gynaecological surgery or previous pelvic radiotherapy.

Extraperitoneal perforation

Anxiety about tumour dissemination outside the bladder is a very valid concern but one for which there is little documented evidence. Hetherington *et al.*[30] reported the development of deeply invasive cancer at the site of TUR in two out of five patients who had undergone resection of the intramural ureter as part of nephroureterectomy, by the method described by McDonald *et al.*[52] and modified by Abercrombie[1]. This was attributed to the dissemination of cancer cells outside the bladder from the transected ureter. The opposite experience was reported by Carr *et al.*[9] who suggested that careful surgical technique avoided the problem, these authors having observed no invasive recurrence in 19 patients treated by the Abercrombie operation. Solsona *et al.*[76] used radical TUR without any other treatment for 59 patients with T2 or T3 bladder cancer followed for a median of 58 months. Their report did not mention specifically the number of patients progressing outside the bladder but of the 11 patients (out of 59) with persistent or recurrent invasive tumour, 10 were amenable for successful salvage surgery or radiotherapy. By implication therefore their resection which 'in most cases [included] perivesical fat' did not lead to significant perivesical extension of the cancer.

Grosse *et al.*[20] reviewed the outcome of 150 full-thickness resections in 95 patients in our institution, with a median follow-up of 45 months. This included 126 full-thickness TURs or TUR biopsies all for transitional cell carcinoma of the bladder and all exposing perivesical fat, extending up to 2 cm outside the bladder in some patients. Sixty-seven were in the presence of muscle invasive cancer (T2 = 10, T3 = 57) of which 59% were poorly differentiated. During follow-up 12.7% of patients developed T4b cancers, of which two had been T2 and seven were T3 (G2 = 3, G3 = 6) at the time of surgery. All patients had received systemic chemotherapy or radiotherapy following TUR. In a recent review of TUR combined with systemic chemotherapy as definitive treatment for 50 patients with muscle invasive bladder cancer, Thomas *et al.*[79] found no patient to have progressed to T4b, some of these patients having been included in the previous review by Grosse *et al.*[20]. Martinez-Pineiro *et al.*[55] treated 13 patients in a similar manner with radical TUR extending into perivesical fat, plus systemic chemotherapy. Three patients had persistent invasive tumour and were treated by cystectomy or radiotherapy; none progressed to T4b. In summary, the risk of local progression following full-thickness resection is not negligible but, on the basis of the foregoing limited data, appears to be no greater than the loco-regional relapse rate that follows both radiotherapy and cystectomy.

Perforation to the peritoneal cavity

The depth of the peritoneal vesico-rectal pouch is very variable and cannot be predicted prior to TUR. Some bladders are unexpectedly thin and it is inevitable that the peritoneum will be opened in the occasional patient, despite careful selection. Dissemination of cancer cells in the peritoneum poses at least a theoretical hazard and should be avoided if at all possible. TUR biopsy of any tumour at the dome or upper posterior wall should proceed cautiously and if the 'wispy' deep detrusor is exposed at the margin of the resection, attempts at complete TUR should be abandoned. If conservative surgery is the desired treatment option, partial cystectomy is then preferable.

If intraperitoneal perforation is suspected, it should not be ignored. The TUR site should be inspected carefully and, if necessary, probed gently with the resectoscope loop so that a definite decision is made endoscopically. Inserting a catheter and awaiting the outcome of a subsequent cystogram is not appropriate. If perforation is confirmed, a lower midline laparotomy is performed, paying particular attention to protect the wound from contamination with malignant cells. The perforation is oversewn, or a wide partial cystectomy performed if TUR was incomplete. Thorough washing of the peritoneum and pelvis with sterile water probably minimizes the risk of seeding or abdominal wall metastasis[79].

Points of technique for full-thickness TUR

A general problem of TUR of bladder tumours is that many urologists regard it as a minor procedure of relatively little interest compared with the surgical prowess associated with cystectomy and other radical cancer operations. For many Ta or T1 tumours it is very straightforward but if there are 20 or 30 papillary tumours, or if a 5 cm palpable T3 mass is present, considerable endoscopic skill and patience is required. It cannot be emphasized too strongly that in these cases the continuous suction Iglesias double resectoscope sheath transforms the operation. It is also essential to guarantee maximum safety. Before any resection is started the degree of bladder distension should be stabilized by balancing the inflow and the outflow suction. For tumours on the posterior wall or dome, a 'bladder wall' loop electrode should be used instead of thc standard 90° electrode. A pure, unblended cutting high-frequency current is best for resection and minimizes coagulation damage to biopsy material that hinders histopathological interpretation.

Deep resection of the detrusor should not commence until all exophytic tumour has been removed. Thereafter, tumour deep in the bladder wall should be resected one layer at a time, avoiding deep excavation in one place, to maintain a clear view of the whole resection area. Every blood vessel should be coagulated individually as it is transected. If large veins are opened, the inflow should be stopped or reduced to obtain a clear view of the extent of the opening and to minimize absorption of irrigant until the vein has been sealed. This is particularly important outside the bladder where veins may retract and be lost from view. The full extravesical extent of a tumour can usually be assessed when normal, soft, perivesical fat is seen. At this point the feasibility of completing the resection may be judged by repeating the bimanual palpation and assessing the extent of the residual mass. Invasive cancer is usually firm and easy to resect; normal fat and extravesical connective tissue are difficult to cut with the resectoscope and tend to be pushed out of harm's way by the irrigating fluid. If overdistension of the bladder and inadvertent fluid extravasation occurs, the latter can be removed by the resectoscope suction with the inflow turned off. Thorough haemostasis is essential. Nursing staff responsible for postoperative care must be aware of what has been done, must supervise postoperative saline irrigation of the bladder carefully and maintain continuous free catheter drainage for 1 week. Problems reported by colleagues have usually arisen because the catheter has been removed too soon, haemostasis has not been thorough or postoperative care has been less than optimal. The surgical alternative to TUR is cystectomy, the morbidity and mortality of which is far greater. The pros and cons of bladder sparing treatment versus cystectomy for muscle invasive bladder cancer are debated in Chapters 8 and 9. For most patients bladder sparing is certainly an option. Its success may depend in part on the endoscopic skill of the urologist.

Laparoscopy has been used to only a limited extent for the lymph node staging of bladder cancer and for this purpose anecdotal experience suggests that it offers little benefit. However, as an aid to full-thickness resections it could reduce the risk of intraperitoneal perforation by permitting an assistant to watch the peritoneal surface of the bladder throughout the resection. For staging tumours of the dome and upper posterior wall, it could detect full-thickness invasion and improve the safety of deep TUR biopsies. So far as I am aware this technique has never been tried. Some may consider the standard 5 mm umbilical trocar and telescope rather invasive; finer instruments would be an advantage. Nevertheless, if used selectively for the assessment of palpable tumours at the bladder dome of upper two-thirds of the posterior wall, the advantages seem obvious provided there has been no previous pelvic surgery, radiotherapy or sepsis.

Postoperative care

It is wise to leave a urethral catheter draining the bladder at the conclusion of every TUR for bladder cancer. The occasional patient will bleed postoperatively despite the best endoscopic technique. If the resection is neither extensive nor deep, and the drainage clear, the catheter may be removed after a few hours and the patient returned home. The second reason for leaving a catheter is that all new Ta or T1 tumours will almost certainly merit an instillation of intravesical chemotherapy within 24 hours (Chapter 4). There is also the possibility that irrigation of the bladder following TUR may reduce subsequent tumour recurrence; this, too, is discussed in

Chapter 4. Patients undergoing tumour resections larger than about 2 cm that include detrusor muscle usually benefit from catheter drainage for 1–2 days, with continuous irrigation for a few hours as a precaution against clot retention if bleeding should occur. For any full-thickness resection of the bladder wall that extends into the perivesical connective tissue, catheter drainage for 1 week is essential[6]. If urinary extravasation occurs the consequences are serious and there is no justification for taking unnecessary risks.

TUMOUR STAGING

The T category is a basic determinant of treatment for bladder cancer, particularly the recognition of muscle invasion. If all muscle invasive cancers are to be treated by cystectomy, subdivisions of muscle invasion are not needed: CIS, Ta, T1, T2 and 'inoperable' would suffice for the T part of TNM. However, many patients prefer to avoid cystectomy if possible. Also, we know that cystectomy is not necessary for every patient with muscle invasive bladder cancer. Clinical studies of the past three decades have shown that bladder-preserving treatment is an option for selected good-risk subsets of patients, and some form of adjuvant treatment is needed to improve the outcome for bad-risk patients. Thus there is an even greater need for a tumour classification of muscle invasive bladder cancer that reflects both prognosis and the different forms of treatment available.

For most patients muscle invasion is the critical point at which endoscopic and intravesical treatments are abandoned and radical therapy is employed. For many years invasion beyond the middle of the detrusor muscle has been regarded as the next prognostic event[38] and has been acknowledged again in the 1997 edition of the TNM classification[83]. Finally, invasion beyond the outer surface of the bladder carries the worst prognosis for survival[60] and has been the stimulus in recent years for some urologists to seek the assistance of adjuvant chemotherapy to try and improve the outcome of cystectomy. In terms of treatment choices that have serious implications for patients, the assessment of the depth of tumour invasion is a crucial factor. These three critical stages are recognized in the 1997 TNM classification[83] (Table 3.1) as:

- muscle invasion: T2 versus T1;
- halfway point of detrusor: T2b versus T2a (previously T3 versus T2)[82]; and
- extravesical extension: T3 versus T2 (previously T3b versus T3a)[82].

The need for pretreatment clinical staging

Now that bladder-preserving treatments have to be considered, it is self-evident that treatment decisions have to be made before treatment is commenced. From the patient's perspective, his or her doctor needs to have all the necessary information

Table 3.1 1997 TNM classification of bladder cancer

T-primary tumour

Rules for classification

The classification applies only to carcinomas. Papilloma is excluded. There should be histological or cytological confirmation of the disease. The following are the procedures for assessing T, N and M categories:

T categories – physical examination, imaging and endoscopy
N categories – physical examination and imaging
M categories – physical examination and imaging

T-primary tumours

The suffix (m) should be added to the appropriate T category to indicate multiple tumours
The suffix (is) may be added to any T to indicate presence of associated carcinoma in situ

TX	Primary tumour cannot be assessed	
T0	No evidence of primary tumour	
Ta	Non-invasive papillary carcinoma	
Tis	Carcinoma in situ	
T1	Tumour invades subepithelial connective tissue	
T2	Tumour invades muscle	
	T2a	Tumour invades superficial muscle (inner half)
	T2b	Tumour invades deep muscle (outer half)
T3	Tumour invades perivesical tissue	
	T3a	Microspically
	T3b	Macroscopically (extravesical mass)
T4	Tumour invades any of the following: prostate, uterus, vagina, pelvic wall, abdominal wall	
	T4a	Tumour invades prostate, uterus, vagina
	T4b	Tumour invades pelvic wall, abdominal wall

available **before** any treatment is started, but herein lies a problem. The significance of the critical prognostic events has been discovered by the examination of cystectomy specimens. With the whole bladder available for step sectioning and meticulous inspection, measurement of the depth of invasion through the detrusor muscle or detection of microscopic infiltration of perivesical fat is a relatively simple process; but how are these findings to be translated to the assessment of the T category **before** the bladder has been removed? So far as biopsies are concerned 'the pathologist can function optimally only when enough tissue is available for examination. If deep tissue is not available in the biopsy specimen, this fact must be stated in the pathology report'[14]. Inclusion of some muscle in the TUR specimen should not be a problem so that the diagnosis of any muscle invasion, compared with none (T2 versus T1) should be straightforward, providing the muscularis mucosae is not forgotten: but how can the histologist determine the midpoint of the detrusor from a jumble of TUR biopsies? How often do TUR biopsies contain perivesical fat? How else can the pathologist know that the tissue is 'deep'[71]?

Histology cannot provide the whole answer and this was recognized by the early editions of the TNM classification which stated very clearly the minimum requirements for assessing the depth of invasion of bladder tumours[82]. These were 'clinical examination, urography, cystoscopy, bimanual examination under anaesthesia, biopsy and transurethral resection of the tumour prior to definitive treatment'. If these requirements could not be met, the description 'TX' should be used, indicating that assessment was incomplete or unknown. Whitmore[86] was quite clear that these procedures were the **minimum requirements** for the clinical staging of bladder cancer. However, some urologists do not perform bimanual palpation because they consider it inaccurate – or, more precisely, because they 'do not believe in it'. Instead, CT scanning or magnetic resonance imaging (MRI) is the preferred staging technique. More recent editions of TNM have sacrificed the mimimum criteria of the old editions for medico-legal or other reasons. The latest TNM classification[83] (Table 3.1) states only that 'physical examination, imaging and endoscopy . . . are the procedures for assessing T category'. The possibility that the assessment or biopsy may be incomplete and inadequate has also been ignored. The prudence of such imprecise terminology may have its advantages but the lack of agreed procedures undermines the very unanimity and credibility that a universal classification is supposed to ensure. The decision to perform cystectomy, or to recommend radiotherapy or neoadjuvant chemotherapy has an enormous impact on a patient's life. The information on which it is based needs to be precise and as reliably predictive as possible.

Imaging

Ultrasound

Bladder cancers can be examined ultrasonically by abdominal, transrectal (TRUS) or transurethral probes. Of the three, transurethral ultrasound is the most accurate but is not generally available[72,89]. The more readily available TRUS is not worthwhile for bladder cancer[85,89]. Overall transurethral ultrasound has been reported to show a 90% sensitivity for separating muscle invasive from superficial bladder tumours, but with a specificity of only 76%; 24% of Ta, T1 tumours were overstaged and 10% of T2 tumours were understaged[72]. Detection of extravesical invasion was less encouraging, with 71% sensitivity, transurethral ultrasound failing to detect extravesical cancer in 29% of patients[47].

Computed tomography (CT)

CT scanning suffers from the inherent disadvantage that it relies upon the disruption of normal radiographic tissue planes which are not distinct at the boundary between the detrusor muscle and adjacent organs or tissues. CT has proved to be particularly limited for tumours of the bladder base and dome. Furthermore, the inflammatory change and fluid extravasation that follows TUR is not distinguishable from tumour by CT. To avoid the latter, CT would have to be performed after flexible cystoscopy but before any biopsy. Although some colleagues follow this sequence of investigation, as it would involve most patients with newly diagnosed bladder cancer, the demands for CT scanning and the cost of the procedure make this an impractical and cost-ineffective proposition. If used only for those tumours found to be invasive on biopsy, an interval of 3–4 weeks is essential before scanning to avoid TUR artefacts.

Not surprisingly, CT scanning has not proved accurate for bladder staging[11,17,57,84,89]. Voges *et al.*[84] found 67% of T1 bladder tumours to be incorrectly overstaged, 30% of T2 tumours understaged and 20% overstaged. Even when CT has been performed before any biopsy, its overall accuracy for staging has been only 35–60%[11,17,89]. Specifically for the detection of extravesical tumour results are better, but overall sensitivity and specificity are both about 80%[72,89]. Sager *et al.*[68] were able to achieve remarkable CT definition of invasive bladder cancers by using rapid sequence imaging, controlled bladder distension and multiple contrast materials. Unfortunately, such commendable imaging techniques have not proved practicable in general use.

Magnetic resonance imaging (MRI)

MRI has several advantages over CT scanning. Images are possible in multiple planes, intravenous contrast is not required and the separation of bladder and tumour from extravesical tissues and adjacent organs is more distinct. T_1-weighted images separate tumour from urine and perivesical fat; T_2-weighted images depict invasion of muscle, prostate and vagina. These theorectical advantages notwithstanding, the results of early comparisons of MRI with CT have not been particulary impressive[33,63,72]. Gadolinium-enhanced MRI may prove more successful and MRI technology continues to evolve[33,78]. However, in a study of 48 patients (25 Ta, T1; 23 muscle invasive) Scattoni *et al.*[70] found only the following diagnostic accuracy: T_1-weighted MRI, 58%; T_2-weighted MRI, 71%; dynamic gadolinium-enhanced MRI, 81%; and late gadolinium-enhanced MRI, 56%. MRI detected extravesical tumour successfully but tended to overstage lesser degrees of invasion. Reviewing the published literature and the underlying technical principles of different forms of MR imaging, these authors gave the impression that currently available MRI is unlikely to be any more useful than CT or transurethral ultrasonography.

Positron emission tomography (PET)

PET uses an external camera to detect the uptake of cyclotron generated nuclear isotopes by tumours after their intravenous injection. Experience with PET is very limited. [^{18}F]fluorodeoxyglucose is considered to be a tumour-specific agent but it is excreted in urine and is therefore unsuitable for examining bladder cancers[11]. C-choline is not so excreted and has been shown to visualize bladder cancers clearly, but so, too, was prostate cancer and benign prostatic hyperplasia[44].

CT-guided percutaneous needle biopsy

CT-guided needle biopsy has improved the specificity of CT scanning alone for the detection of pelvic lymph node metastases and other solid tumours. In the hope that the technique would improve the staging of bladder cancer, two groups[32,54] explored the possibility of obtaining full-thickness core biopsies of invasive bladder cancers by passing a needle across the distended bladder and through the tumour to its extravesical margin, perpendicular with the bladder wall. An alternative technique was to take core biopsies of tumour within and outside the detrusor, tangential to the bladder wall. Preliminary data from small numbers of patients confirmed the feasibility of the technique, but further experience is required to assess its value.

Bimanual palpation

At the Fourth International Consensus Meeting on Bladder Cancer the group reviewing imaging for bladder cancer staging 'felt that the identification of local extension, by either ultrasound or CT was more accurate than bimanual palpation'[62]. They recommended that imaging studies should be obtained prior to loco-regional manipulation (presumably TUR biopsy) and that IVP and CT were the optimal imaging studies. The feelings of this group do not appear to have been sensitive to the limitations of CT imaging described in the published literature, summarized above, nor to the problem of performing CT before taking biopsies in all patients suspected of having muscle invasive cancer.

More recently, Wijkstrom *et al.*[87] took the opposite view: 'Modern imaging methods have failed to improve the pre-treatment assessment of tumour stage. Bimanual palpation is crucially important in clinical staging and there is a need for further standardisation and refinement of this procedure.'[87] These authors confirmed the prognostic importance of distinguishing extravesical invasion (50% 5-year survival) from tumours confined to the bladder wall (85% survival) on the basis of operative pathological stage. However, more importantly they demonstrated that this distinction could be anticipated accurately, preoperatively by bimanual palpation. Patients with no palpable mass after TUR had 83% 5-year survival compared with 50% for those with a residual mass. These authors performed the bimanual palpation **after** the TUR, although they did not state if TUR attempted to remove all the visible cancer or only exophytic tumour, as was stipulated by the original TNM classification.

Following the removal of bimanual palpation from the TNM classification in 1987, Hendry *et al.*[27] demonstrated in multicentre analysis that bimanual examination was, none the less, the most prognostic variable in assessing bladder tumour stage and could replace the T category. Of the two factors that made up the T category (histological depth of invasion and palpable mass) it was the bimanual palpation that made the most important contribution to the prediction of survival. They therefore concluded that the omission of bimanual palpation from TNM was a mistake. Gospodarowicz[19] reported in 1994 that patients with a clearly palpable extravesical mass had a significantly lower complete response rate to radiotherapy and a significantly lower survival than patients with no extravesical mass. These authors defined a tumour mass as extravesical if after TUR the palpable mass was larger than the size of any visible tumour remaining in the bladder. Finally, Fossa *et al.*[16] reported that in patients with biopsy-proven muscle invasive bladder cancer treated with preoperative radiotherapy and cystectomy, bimanual palpation provided greater prognostic discrimination than biopsy assessment of the histological depth of invasion. Patients could be 'divided into two prognostic groups: those with and those without palpable tumour after TURB-T'.

Three of the four series reviewed above stated that the bimanual palpation was performed after TUR but the thoroughness, or completeness of the TUR in each case is not known. Gospodarowicz[19] commented correctly that most reports did not separate T3a and T3b patients (stage B2 and C tumours). The reason was that endoscopic practice was so variable. Some urologists would take only a generous biopsy, others would resect flat with the bladder wall and others would attempt to resect the whole palpable tumour mass. Clearly the definition of extravesical tumour invasion based on residual palpable tumour 'after TURBT' cannot be applied uniformly in all institutions[29]. Furthermore, some urologists prefer not to manipulate the bladder when they have achieved haemostasis at the end of a resection. For those who favour full-thickness resection it is inconvenient to stop half way to assess what other urologists would consider to be a 'residual' tumour. For all these reasons the bimanual examination has come to be regarded as an unreliable basis for tumour staging, is not performed at all by some colleagues and is not a stated requirement for TNM.

Despite these problems of practice and attitude, the above review confirms that bimanual palpation is the most reliable way to separate good prognosis from bad prognosis muscle invasive bladder cancers. It is more accurate than the histological depth of invasion with biopsies and, in the absence of good-quality comparative data for ultrasound, CT and MRI scanning, appears to be more accurate and reliable than any form of imaging currently available. As suggested by Wijkstrom *et al.*[87] there is a need to standardize this important procedure.

Thus, in the interests of uniformity and to standardize 'good practice' for staging muscle invasive bladder cancers it seems most logical to agree the following.

- Resect all exophytic tumour, to the level of adjacent normal mucosa.
- Take a few more TUR biopsies sufficient to prove muscle invasion. Stop any bleeding.
- Empty the bladder completely and remove the resectoscope.
- Perform a careful bimanual palpation to stage the tumour.
- If bladder preservation is planned, reintroduce the resectoscope and complete the removal of all visible and palpable tumour, if possible.
- If cystectomy is planned, pass a catheter and check haemostasis.

This simple technique may require minor changes in endoscopic practice for some urologists (some may have to learn how to do bimanual palpation), but it provides a standardized staging procedure that is easy to understand, is universally applicable, requires no special equipment, applies equally to all forms of radical or conservative treatment and is based on prognostic factors of demonstrated significance. For clinical staging purposes, whether tumour is histologically extravesical or not is immaterial; **it is the presence of the mass** that is of prognostic significance. For surgeons who are accustomed to pathological, post-cystectomy staging this may be a difficult concept to grasp and probably explains why all editions of TNM have, unhelpfully, qualified the simple clinical staging by the addition of unrealistic histological criteria. By excluding biopsy assessment of depth of invasion from the T2 and T3 categories, the confusion caused by trying to apply the findings of surgical staging to pretreatment clinical staging is avoided. If in the future imaging techniques are shown by direct comparison to be superior to bimanual palpation, the latter procedure could be made redundant.

Bladder cancer invading adjacent organs

TCC of the prostate

A number of publications have drawn attention to the anomalous situation of TCC involving the prostatic urethra, prostate ducts and invading prostate stroma-[25] (Chapter 2). For none of these is there a place in the T category as bladder cancer; they are currently classified as carcinoma of the urethra or prostate[2,83]. From the clinical point of view this is unsatisfactory because treatment of TCC in the prostate is always considered to be integral with the treatment of urothelial cancer arising in the bladder. Carcinoma in situ and non-invasive papillary urothelial tumours arising from the prostatic urethra or prostatic ducts can be controlled by TUR and respond to intravesical BCG[8,25,77]. When it comes to radical surgery for TCC of the lower urinary tract, bitter experience has taught that the bladder **and** prostate must be regarded as a single pathological entity. Cystectomy, when performed for bladder cancer, does not remove only the bladder but is a cystoprostatectomy.

The current T4a category includes 'tumour that invades the prostate, uterus, vagina'[83]. This is usually taken to mean muscle invasive bladder cancer (T3) that invades the prostate from the bladder, but some urologists extend this to include TCC invading the prostate stroma from the urethra, or any concomitant prostatic TCC, whether an invasive cancer is present in the bladder or not[60,61]. Pugh[66] suggested in 1981 that 'there is a good case for modifying the UICC scheme [TNM] by being more precise with regard to prostatic involvement'. Many subsequent authors have supported this view that the staging of TCC arising in the prostatic urethra and ducts, or invading the prostate stroma from these sites, should be separated from tumours invading the prostate by direct extension from the bladder[25].

Review of the published literature suggests that the progression-free survival for non-invasive TCC in the prostatic urethra (75%) is the same as for non-invasive TCC involving prostate ducts and acini (71%), but much better than urothelial cancer invading the prostate stroma from the urethra (26%)[25]. In turn, the prognosis for tumours invading from the prostatic urethra is considerably better than for those invading from a contiguous cancer in the bladder[25,60,61]. For this reason, Pagano *et al.*[60] suggested that the two types of invasive TCC involving the prostate should be separated as 'true T4a' and 'false T4a'. Several alternative classifications have been proposed for the staging of prostatic TCC. Hardeman and Soloway[26] suggested three groups based on the likelihood of urethral recurrence after cystectomy:

1. CIS confined to the prostatic urethra;
2. ductal or acinar involvement limited by the basement membrane; and
3. prostate stroma invasion.

Chibber *et al.*[10] recognized the same three groups but included bladder and prostatic non-invasive tumours together and did not distinguish non-contiguous from contiguous stromal invasion. Pugh[66] proposed a subdivision of the TNM pT4a category: P4aa for tumour in ducts an acini only and P4ab for stromal invasion. Most recently Esrig *et al.*[13] agreed that the inclusion of all urothelial tumours involving the prostate in a single category is inappropriate. They proposed 'P1str and P2/P3a. str' to describe prostate stromal invasion when it is found in association with bladder tumours in a cystectomy specimen.

All of these proposals address part of the problem but none meets the need to 'stretch' the TNM and AJCC systems to accommodate **all** aspects of these infrequent but important tumours without adding new categories that are too numerous or too complex for everyday use, and without requiring cystectomy. For this reason the following modification to the TNM classification has been proposed[25].

Proposal

The simplest solution would be to reorganize the TNM T4 category as follows.

- T4a: non-invasive TCC of prostatic urethra, ducts or acini.
- T4b: invasive TCC of the prostatic stroma, arising from prostatic urethra or ducts but not as contiguous extension of an invasive cancer in the bladder.
- T4c: invasion of the prostate by a tumour of the bladder (T4a in the current classification).

It should be remembered that the histopathologist will be unable, in most biopsies, to distinguish between contiguous TCC invading the bladder and non-contiguous TCC arising within the prostatic urethra. The proposed T classification is, as always, based upon the combination of cystoscopic findings, bimanual palpation and histology of TUR biopsies. Redefinition of a category in this way has proved very popular and practical for early stage prostate cancer, with the introduction of the 'T1c' category in 1992. It would, of course, leave no place for the current T4b tumours that extend to the pelvic side wall. Rather than complicate the T4 category with a fourth subdivision, current T4b tumours should be renamed 'T5' as proposed previously by Hall and Prout[24].

Bladder cancers fixed to the pelvic side wall

The T4 category has always included two very different groups of patients: those with tumours that invade the prostate, vagina and uterus but are still operable (T4a), and those with inoperable tumours fixed to the pelvis (T4b). Separating the latter group of fixed tumours into another T category would resolve a recurring source of confusion. How often do publications refer to T4 tumours without clarifying whether they were T4a or T4b? These are entirely different tumours with different prognoses suitable for very different treatment, but the reader has no idea what is being reported. By calling inoperable, fixed tumours 'T5' any possibility of confusion is removed.

This is contrary to the rules of TNM which heretofore has only used four T categories to describe primary tumours at any site. However, these criteria are of arbitrary, historical origin and the need to improve the staging of bladder cancer is more important. A new T5 category would acknowledge the inoperability and extreme prognosis of bladder cancers fixed to the pelvic side wall, separate them very clearly from less advanced tumours that have a better prognosis, and permit the inclusion of non-invasive TCC of the prostatic urethra and ducts in a T category of bladder cancer.

SUMMARY

The TNM classification has been in use for bladder cancer for more than 23 years. In general it has proved very successful and its combination with the American

Table 3.2 Proposed T classification for bladder cancer

TX	Primary tumour cannot be assessed	
T0	No evidence of primary tumour	
Ta	Non-invasive papillary carcinoma	
Tis	Carcinoma in situ	
T1	Tumour invades subepithelial connective tissue	
T2	Tumour invades detrusor muscle but no mass palpable on bimanual examination after TUR of exophytic tumour	
T3	Tumour invades muscle, with mass palpable after TUR of exophytic tumour; the mass is freely mobile within the pelvis	
T4	Tumour involves prostate or adjacent organs	
	T4a	Non-invasive TCC in prostatic urethra, ducts or acini
	T4b	TCC invades prostate stroma, arising from prostatic urethra or ducts, but not as contiguous extension of any invasive cancer in the bladder
	T4c	Invasion of adjacent organs, e.g. prostate, vagina, by a tumour arising in the bladder
T5	Invasive cancer arising in bladder or prostatic urethra invading pelvic wall or abdominal wall	

TCC, transitional cell carcinoma; TUR, transurethral resection

(AJCC) classification was a welcome step forward. The introduction of the Ta category in 1979 caused considerable debate at the time, but has proved useful and is accepted universally. Discussions in journals and at a series of international consensus conferences have illustrated continuing problems with the staging of muscle invasive bladder cancers that have yet to be resolved. It is hoped that the proposal set out in this chapter (summarized in Table 3.2) will be accepted as a basis for consensus that will reflect published evidence, such as it is, differences in clinical practice and individual preference for different treatments, and the need to compare outcomes of treatment in such a way that survival and quality of life for patients with bladder cancer will be improved in the future.

REFERENCES

1. Abercrombie GF. (1972) Nephro-ureterectomy. *Proc R Soc Med* **65:** 1021–1022.
2. American Joint Committee on Cancer (1992) *Manual for Staging of Cancer*, 4th edn, JB Lippincott, Philadelphia.
3. Barnes RW, Dick AL, Hadley HL and Johnston OL (1977) Survival following transurethral resection of bladder carcinoma. *Cancer Res* **37**: 2895–2897.
4. Beer E (1910) Removal of neoplasms of the urinary bladder. A new method, employing high-frequency (Oudin) currents through a catheterising cystoscope. *J Am Med Ass* **54**: 1768–1769.

5. Blandy JP and Knotley RG (1993) *Transurethral Resection*, 3rd edn, Butterworth–Heinemann, Oxford, p. 6.
6. Blandy JP and Knotley RG (1993) Technique of resecting tumours of the bladder, in *Transurethral Resection*, 3rd edn, (eds JP Blandy and RG Knotley), Butterworth–Heinemann, Oxford, pp. 105–116.
7. Bono AV, Hall R, Denis L *et al.* (1996) Chemoresection in Ta T1 bladder cancer. *Eur Urol* **29**: 385–390.
8. Bretton PR, Herr HW, Whitmore WF *et al.* (1989) Intravesical Bacillus Calmette–Guerin therapy for in situ transitional cell carcinoma involving the prostatic urethra. *J Urol* **141**: 853–856.
9. Carr T, Powell PH, Ramsden PD and Hall RR (1987) The risk of tumour implantation following 'Abercrombie' modified nephroureterectomy. *Br J Urol* **59**: 99–100.
10. Chibber PJ, McIntyre MA, Hindmarsh JR *et al.* (1981) Transitional cell carcinoma involving the prostate. *Br J Urol* **53**: 605–609.
11. Colleen S, Ekelund L, Henrikson H *et al.* (1981) Staging of bladder carcinoma with computed tomography. *Scand J Urol Nephrol.* **15**: 109–113.
12. Dubben HH and Beck-Borholdt HP (1997) An obvious and underestimated predictive assay: Precise, cheap and easy prediction of radiotherapy outcome using tumour volume. *Eur J Cancer* **33** (Suppl. 8): S97, Abstract 428.
13. Esrig D, Freeman JA, Elmajian DA *et al.* (1996) Transitional cell carcinoma involving the prostate with a proposed staging classification for stromal invasion. *J Urol* **156**: 1071–1076.
14. Farrow GM (1979) A pathologist's role in bladder cancer. *Sem Oncol* **6**: 198–206.
15. Fellows GJ, Palmar MKB, Grigor KM *et al.* (1994) Marker tumour response to Evans and Pasteur BCG in multiple recurrent pTa/pT1 bladder tumours. Report from the Medical Research Council Subgroup on Superficial Bladder Cancer (Urological Working Party). *Br J Urol* **73**: 639–644.
16. Fossa SD, Ous S and Berner A (1991) Clinical significance of the 'palpable mass' in patients with muscle-infiltrating bladder cancer undergoing cystectomy after pre-operative radiotherapy. *Br J Urol* **67**: 54–60.
17. Fryjordet A and Skatun J (1983) Staging of urinary bladder cancer by computerised tomography compared to clinical staging and post-operative pathologic staging. *J Oslo City Hosp* **33**: 77–81.
18. Gibson TE (1935) Treatment of bladder tumours with the McCarthy resectosocope. *J Urol* **34**: 8–9.
19. Gospodarowicz MK (1994) Staging of bladder cancer. *Sem Surg Oncol* **10**: 51–59.
20. Grosse JO, Whiteway J, Hall RR and Essenhigh DM (1992) Full thickness resection (FTR) for bladder cancer. *Proc Br Ass Urol Surg 1992*, p. 6.
21. Hall RR (1992) Transurethral resection for bladder carcinoma. *Probl Urol* **6**: 460–470.

22. Hall RR (1996) Neoadjuvant CMV chemotherapy and cystectomy or radiotherapy in muscle invasive bladder cancer. First analysis of MRC/EORTC international trial. *Proc Am Soc Clin Oncol* **15**: 244, Abstract 612.
23. Hall RR, Newling DW, Ramsden PD *et al.* (1984) Treatment of invasive bladder cancer by local resection and high dose methotrexate. *Br J Urol* **56**: 668–672.
24. Hall RR and Prout GR (1990) Staging of bladder cancer: is the tumour, node, metastasis system adequate? *Sem Oncol* **17**: 517–523.
25. Hall RR and Robinson MC (1998) Transitional cell carcinoma in the prostate. *Eur Urol Update Series* **7**: 1–7.
26. Hardeman SW and Soloway MS (1990) Urethral recurrence following radical cystectomy. *J Urol* **144**: 666–669.
27. Hendry WF, Rawson NSB, Turney L *et al.* (1990) Computerisation of urothelial carcinoma records: 16 years' experience with the TNM system, *Br J Urol* **65**: 583–588.
28. Herr HW (1992) Transurethral resection in regional advanced bladder cancer. *Urol Clin N Am* **19**: 695–700.
29. Herr HW and Scher HI (1990) Surgery of invasive bladder cancer: is pathologic staging necessary? *Sem Oncol* **17**: 590–597.
30. Hetherington JW, Ewing R and Philp NH (1986) Modified nephroureterectomy: a risk of tumour implantation. *Br J Urol* **58**: 368–370.
31. Hofstetter A, Frank F and Keiditsch E (1981) Endoscopic Neodymium-YAG laser application for destroying bladder tumours. *Eur Urol* **7**: 278–282.
32. Hoshi S, Orikasa S, Yoshikawa K *et al.* (1991) Whole layer needle biopsy for evaluation of neoadjuvant therapy to invasive bladder cancer. *Nippon Hinyokika Gakki Zasshi* **82**: 1649–1655.
33. Husband JE (1992) Staging bladder cancer. *Clin Radiol* **46**: 153–159.
34. Iglesias JJ, Ellendt EP, Madduri SD *et al.* (1977) Hydraulic haemostasis in the transurethral resection of the prostate using the Iglesias continuous suction resectoscope. *J Urol* **117**: 306–308.
35. Iglesias JJ, Navio-Nino S, de Boisgisson PH and Seebode JJ (1974) Nuevo resectoscopio de Iglesias con irrigacion continna y succion simultanea. *Arch Esp Urol* **27**: 657.
36. International Collaboration of Trialists on behalf of the Medical Research Council Advanced Bladder Cancer Working Party, the EORTC Gentio-urinary Group, the Australian Bladder Cancer Study Group, the National Cancer Institute Canada, Finnbladder, Norwegian Bladder Cancer Study Group and Club Urologico Espanol du Tratamiento Group (1998) Neoadjuvant CMV chemotherapy for muscle invasive bladder cancer: result of the international trial BAO6 (MRC) 30894 (EORTC), to be published.
37. Jakse G, Loidl W, Seeber G and Hofstadter F (1987) Stage T1, grade 3 transitional cell carcinoma of the bladder: an unfavourable tumour? *J Urol* **137**: 39–43.

38. Jewett HJ, King LR and Shelley WM (1964) A study of 365 cases of infiltrating bladder cancer: relation of certain pathological characteristics after extirpation. *J Urol* **92**: 668–678.
39. Jichlinski P, Wagnieres G, Forrer M *et al.* (1997) Clinical assessment of fluorescence cystoscopy during transurethral bladder resection in superficial bladder cancer. *Urol Res* **25** (Suppl. 1): 53–56.
40. Jichlinski P, Wagnieres G, Forrer M *et al.* (1997) Interet clinique de la cystoscopie a fluorescence dans la detection des carcinomes a epithelium de transition superficiels de la vessie. *An d'Urol* **31**: 43–48.
41. Johnson DE (1994) Use of the Holmium:YAG laser for treatment of superficial bladder carcinoma. *Lasers Surg Med* **14**: 213–218.
42. Jones HC and Swinney J (1962) The treatment of tumours of the bladder by transurethral resection. *Br J Urol* **34**: 215–220.
43. Kennedy JC, Pottier RH and Pross DC (1990) Photodynamic therapy with endogenous protoporphyrin 1X: basic principles and present clinical experience. *J Photochem Photobiol B: Biol.* **6**: 143–148.
44. Kishi H, Hirano Y, Kosaka N and Hara T. (1997) Imaging of carcinoma of bladder and prostate using PET with C-choline. *Br J Urol* **80** (Suppl. 2): 262, Abstract 1030.
45. Koenig F, McGovern FJ, Althausen AF *et al.* (1996) Laser induced auto fluorescence diagnosis of bladder cancer. *J Urol* **156**: 1597–1601.
46. Kolozsy Z (1991) Histopathological 'self control' in transurethral resection of bladder tumours. *Br J Urol* **67**: 162–164.
47. Koraitim M, Kamal B, Metwalli N and Zaky Y (1995) Transurethral ultrasonographic assessment of bladder carcinoma: its value and limitation. *J Urol* **154**: 375–378.
48. Kriegmair M, Baumgartner R, Knechel R *et al.* (1994) Fluorescence photodetection of neoplastic urothelial lesions following intravesical instillation of 5-aminolevulinic acid. *Urology* **44**: 836–841.
49. Kriegmair M, Baumgartner R, Kneckel R *et al.* (1996) Detection of early bladder cancer by 5-aminolevulinic acid induced porphyrin fluorescence. *J Urol* **155**: 105–109.
50. Lim LM, Patel A, Ryan TP *et al.* (1997) Quantitative assessment of variables that influence soft-tissue electrovapourisation in a fluid environment. *Urology* **49**: 851–856.
51. McDonald HP, Philip AJ and Williams DC (1947) Endoscopic treatment of tumors of the bladder. *J Am Med Ass* **134**: 500–501.
52. McDonald HP, Upchurch WE and Sturdevant CE (1952) Nephroureterectomy; a new technique. *J Urol* **67**: 804–809.
53. McKiernan JM, Kaplan SA, Santarosa RP *et al.* (1996) Transurethral electrovapourisation of bladder cancer. *Urology* **48**: 207–210.
54. Malmstrom PU, Lonnemark M, Busch C and Magnusson A (1992) Staging of bladder carcinoma by computer tomography-guided transmural core biopsy.

Scand J Urol Nephrol **27**: 193–198.

55. Martinez-Pineiro JA, Leon JJ and Martinez-Pineiro L (1991) Aggressive TURB combined with systemic chemotherapy for locally invasive TCC of the urinary bladder. A second report. *Eur J Cancer* **27** (Suppl. 2): S104, Abstract 605.
56. Miller LS and Johnson DE (1973) Megavoltage radiation for bladder carcinoma: alone, post-operative, or pre-operative, in *Seventh National Cancer Conference Proceedings,* pp. 771–782 (quoted by Shipley [1988])[73].
57. Nurmi M, Kateveno K and Punala P (1988) Readability of CT in pre-operative evaluation of bladder cancer. *Scand J Urol Nephrol* **22**: 125–128.
58. Oosterlink W, Kurth JH, Schroder F *et al.* (1993) A prospective European Organisation for Research and Treatment of Cancer Genito-urinary Group randomised trial comparing transurethral resection followed by a single intravesical instillation of epirubicin or water in single stage Ta, T1 papillary carcinoma of the bladder. *J Urol* **149**: 749–752.
59. Oosterlink W, Kurth KH, Schroder F *et al.* (1993) A plea for cold biopsy, fulguration and immediate bladder instillation with epirubicin in small superficial bladder tumours. Data from the EORTC GU Group Study 30863. *Eur Urol* **23**: 457–459.
60. Pagano F, Bassi P, Dragoferrante GL and Carbeglio A (1994) Transitional cell carcinoma of the prostate in patients with bladder cancer: the need for modifying the TNM classification, in *23rd Congress of Sociéte Internationale d'Urologie, Sydney*, pp. 139, Abstract 243.
61. Pagano F, Bassi P, Galetti TP *et al.* (1991) Results of contemporary radical cystectomy for invasive bladder cancer: a clinicopathological study with an emphasis on the inadequacy of the tumour, nodes and metastases classification. *J Urol* **145**: 45–50.
62. Paulson D, Denis L, Orikasa S *et al.* (1995) Optimal staging procedures, including imaging, to define prognosis of bladder cancer. *Int J Urol* **2** (Suppl. 2): 1–7.
63. Persad R, Kabala J, Gillatt D *et al.* (1993) Magnetic resonance imaging in the staging of bladder cancer. *Br J Urol* **71**: 566–573.
64. Prout GR, Jr (1976) The surgical management of bladder carcinoma. *Urol Clin N Am* **3**: 149–175.
65. Prout GR, Jr, Griffin PP and Shipley WU (1979) Bladder carcinoma as a systemic disease. *Cancer* **43**: 2532–2539.
66. Pugh RCB (1981) Histological staging and grading of bladder tumours, in *Bladder Cancer, Principles of Combination Therapy*, (eds RTE Oliver, WF Hendry and J Bloom), Butterworths, Oxford, pp. 3–8.
67. Reynolds LR, Schulte TL and Hammer HJ (1949) Bladder tumors – a clinical evaluation of radical transurethal management. *J Urol* **61**: 912–916.
68. Sager EM, Talle D, Fossa SD *et al.* (1987) Contrast enhanced computed tomography to show perivesical extension. *Acta Radiol.* **28**: 307–311.

69. Sakamoto N, Tsuneyoshi M, Naito S and Kumazawa J (1993) An adequate sampling of the prostate to identify prostatic involvement by urothelial carcinoma in bladder cancer. *J Urol* **149**: 318–321.
70. Scattoni V, da Pozzo LF, Colombo R *et al.* (1996) Dynamic Gadolinium-enhanced magnetic resonance imaging in staging of superficial bladder cancer. *J Urol* **155**: 1594–1599.
71. Schroeder FH, Cooper EH, Debruyne FM *et al.* (1988) TNM classification of genitourinary tumours 1987 – position of the EORTC Genitourinary Group. *Br J Urol* **62**: 502–510.
72. See WAL and Fuller JR (1992) Staging of advanced bladder cancer. Current concepts and pitfalls. *Urol Clin N Am* **19**: 663–683.
73. Shipley WU (1988) Review of factors predicting improved tumour control and survival from invasive bladder carcinoma following external beam radiation therapy, in *Management of Advanced Cancer of Prostate and Bladder*, (eds PH Smith and M Pavone-Macaluso), Alan R Liss, New York, pp. 437–446.
74. Shipley WU, Prout GR, Kaufman D *et al.* (1987) Invasive bladder carcinoma. The importance of initial transurethral surgery for improved survival with full dose irradiation. *Cancer* **60** (Suppl. 3): 514–520.
75. Smith JA and Dixon JA. (1984) Argon laser phototherapy of superficial transitional cell carcinoma of the bladder. *J Urol* **131**: 655–656.
76. Solsona E, Iborra I, Ricos JV *et al.* (1992) Feasibility of transurethral resection for muscle-infiltrating carcinoma of the bladder: prospective study. *J Urol* **147**: 1513–1515.
77. Solsona E, Iborra I, Ricos JV *et al.* (1995) The prostate involvement as prognostic factor in patients with superficial bladder tumours. *J Urol* **154**: 1710–1713.
78. Tachibana M, Baba S, Deguchi N *et al.* (1991) Efficacy of gadolinium-diethylenetriaminepentaacetic acid-enhanced magnetic resonance imaging for differentiation between superficial and muscle-invasive tumour of the bladder: a comparative study with computed tomography and transurethral ultrasonography. *J Urol* **145**: 1169–1173.
79. Thomas DJ, Roberts JT, Hall RR and Reading J (1998) Radical TUR and chemotherapy in the treatment of muscle invasive bladder cancer. Long term follow up, in press.
80. Thompson GJ and Kaplan JH (1955) Advantages of transurethral remvoal of certain bladder tumors. *J Urol* **73**: 270–279.
81. Thrasher JB, Frazier HA, Robertson JE and Poulson DF (1994) Does a stage pT0 cystectomy specimen confer a survival advantage in patients with minimally invasive bladder cancer? *J Urol* **152**: 393–396.
82. Union Internationale Contre le Cancer (UICC) (1978) *TNM Classification of Malignant Tumours*, 3rd edn, (ed. MH Harmer), UICC, Geneva, pp. 113–117.

83. Union Internationale Contre le Cancer (UICC) (1997) *TNM Classification of Malignant Tumours*, 5th edn, (eds LH Sobin and Ch Wittekind), Wiley-Liss, New York, pp. 187–190.
84. Voges GE, Tauschke E, Stockle M *et al.* (1989) Computerised tomography: an unreliable method for accurate staging of bladder tumours in patients who are candidates for radical cystectomy. *J Urol* **142**: 972–974.
85. Watanbe H, Neshina T and Ohe H (1983) Staging of bladder tumours by transrectal ultrasonography and U.I. octason. *Urol Radiol* **5**: 11–15.
86. Whitmore WF (1979) Surgical management of low stage bladder cancer. *Sem Oncol* **6**: 207–216.
87. Wijkstrom H, Norming U, Lagerqvist M *et al.* (1997) Survival as a function of clinical and surgical tumour stage at cystectomy in transitional cell bladder carcinoma. A long-term follow up of 276 consecutive patients. *Br J Urol* **79** (Suppl. 4): 45–46, Abstract 182.
88. Wood DP, Montie JE, Pontes JE and Levin HS (1989) Idenitification of transitional cell carcinoma of the prosate in bladder cancer patients: a prospective study. *J Urol* **142**: 83–85.
89. Yaman O, Baltachi S, Arikan N *et al.* (1996) Staging with computed tomography, transrectal ultrasonography and transurethral resection of bladder tumour: comparison with final pathological stage in invasive bladder carcinoma. *Br J Urol* **78**: 197–200.

Commentary

Harry W. Herr

This review of TUR and staging of bladder cancer is an excellent example of what an experienced urologist can accomplish in evaluating and treating bladder cancer. I agree with virtually everything that has been said. Consequently, I can add very little of substance other than a few caveats.

COMPLETE TUR

Despite the number or extent of tumours present within the bladder, as complete a TUR as possible should be accomplished. This may take multiple procedures, especially in cases with multiple papillary and *in situ* lesions, but it is necessary before considering intravesical therapy. It is unrealistic to expect topical therapy to eradicate unresected superficial tumours. Intravesical therapy, especially BCG, is most effective when the bladder is free of tumour or when tumour burden is minimal (microscopic carcinoma in situ). Similarly, in cases of invasive bladder cancer, response to chemotherapy or radiation may be enhanced when the bulk of tumour is resected as completely as possible.

How does the surgeon know when he or she has reached the depth of tumour invasion? Deep normal detrusor muscle and fat cut poorly with the resectoscope loop, as opposed to the tumour. The experienced endoscopist knows, in most cases, when the tumour has been completely resected.

Videoendoscopy, in my view, is a major advance in the evaluation and conservative treatment of bladder cancer. It magnifies the view and facilitates a safer and surer resection. I have used video transurethral resection for the past 5 years in over 2000 cases and wonder how I ever resected without it.

TUR BIOPSY

A great deal of weight is placed on the histological evaluation of TUR specimens, but very little attention is given to this aspect of the TUR procedure. A pathologist's interpretation of tumour characteristics is only as good as the material provided by the urologist. Improvements in loop design (such as the Olympus A23 186) may help to maintain three-dimensional architecture of the tumour specimen and reduce cautery artefact[3]. The technical aspects of using a blended current as stressed by Professor Hall cannot be overemphasized in this regard.

STAGING

In cases of invasive tumours confined endoscopically to the bladder (no palpable mass under anaesthesia), a CT scan is of little use in management. I use CT scans sparingly[2].

REPEAT TUR

A second TUR is advised in most, if not all, cases of bladder cancer, especially superficial tumour[1]. A TUR is a stochastic process that is both diagnostic and therapeutic for superficial and invasive tumours, especially if one favours bladder-sparing when possible. A repeat TUR may reduce the uncertainty inherent in staging bladder tumours by biopsy and resection. I perform a second TUR 4–6 weeks after the first in all cases and use the information provided by the second procedure to recommend further therapy. This is especially true in cases of Ta bladder tumours, even in cases where I did the original resection. Recurrent T1 tumours or presence of unresected submucosal invasion and adjacent carcinoma in situ are present in between 20 and 40% of patients who undergo a repeat TUR at the site of the original tumour[4,5]. This fact alone, plus the improved outcome with conservative therapy of such patients having a repeat TUR, should persuade one of the value of this strategy. In cases of confirmed muscle infiltration, a repeat TUR shows no evidence of tumour in up to 20% of patients. Such patients are treated by TUR alone and careful follow-up, or in some cases, by systemic chemotherapy and conservative surgery or radiation rather than cystectomy. The 5- and 10-year results of bladder-sparing approaches selected on the basis of a repeat TUR are comparable with cystectomy and justify the rationale of this strategy.

COMMENTARY REFERENCES

1. Herr HW (1996) Uncertainty and outcome of invasive bladder tumors. *Urol Oncol* **2**: 92–95.
2. Herr HW and Hilton S (1996) Routine CT scans in cystectomy patients: does it change management? *Urology* **47**: 324–325.
3. Herr HW and Reuter VE (1998) Evaluation of a new resectoscope loop for transurethral resection of bladder tumors. *J Urol,* in press.
4. Klan R, Lou V and Huland H (1991) Residual tumor discovered in routine second transurethral resection in patients with stage T1 transitional cell carcinoma of the bladder. *J Urol* **146**: 316–318.
5. Wolf H, Iversen H, Rosenkilde R *et al.* (1987) Transurethral surgery in the treatment of invasive bladder cancer (T1 and T2). *Scand J Urol Nephrol* **104**: 127–132.

4

Prognostic factors, treatment options and follow-up of Ta and T1 bladder tumours

Philip H. Smith and Reginald R. Hall

INTRODUCTION

The management of so-called 'non-invasive' bladder cancer (categories Ta and T1) involves transurethral resection of the primary tumour or tumours, with or without adjuvant intravesical chemo- or immunotherapy, followed by regular surveillance (currently by flexible cystoscopy in the majority of institutions) with or without intermittent cytological examination and imaging of the upper tracts.

At the time of the initial transurethral resection (TUR) it is recommended that all visible tumour be removed and that separate biopsies of the tumour base to include muscle be obtained. Traditionally, imaging of the upper tract has been regarded as vital initially to exclude tumours of the upper tracts. The infrequency of concomitant tumours may not justify this practice, and to detect other upper-tract abnormalities, sonography is more cost effective than intravenous urography[16]. Many urologists also believe that upper-tract imaging is important at intervals during follow-up. The evidence for regular review of the upper tract is, however, less compelling if no tumour is found at the initial examination[66], and the definitive statement upon the need for and frequency of such follow-up has yet to be made.

The number of tumours, tumour category (Ta or T1), grade and size are all indicative of the probability of recurrence. Tumours are recurrent in approximately 70% of patients, but invasive disease is seen in fewer than 10% entered at first diagnosis into randomized clinical trials, despite long-term follow-up.

In this discussion it is assumed that surgeons in all communities have facilities for endoscopic resection of bladder tumours, but it is recognized that the availability of chemotherapy and immunotherapy varies considerably. The prevalence of smoking and exposure to industrial carcinogens, both significant aetiological factors, show marked variations between and within differing communities. It is against this general background that prognostic factors, treatment options and follow-up must be considered.

PROGNOSTIC FACTORS AT DIAGNOSIS

The key prognostic factors in determining treatment of those tumours which do not invade muscle include the T category, the histological grade, the size of the lesion, the number of tumours, the presence or absence of recurrence at the time of the first cystoscopy and the frequency with which recurrence occurs[27,44,51]. The control of known aetiological factors may also be of importance and many urologists believe that all patients with bladder tumours should be advised to stop smoking (Chapter 13).

T category

Although the majority of urologists call both Ta and T1 tumours 'non-invasive', the purist would object on the grounds that the T1 tumour has already shown invasive characteristics by breaching the basement membrane. In fact, the patient with a category T1 tumour is known to have a worse prognosis since the risk of developing invasive disease is higher in these patients. Kurth *et al.*[27], reporting on 576 patients from two EORTC trials followed for a median of 4 years, showed that the T category was not related to the recurrence rate, but that invasive disease was twice as common in those with T1 disease. In a prognostic factor analysis of 417 newly diagnosed Ta T1 tumours, Parmar *et al.*[44] found that T category alone was a poor predictor of superficial recurrence: there was no difference between categories Ta and T1 in terms of recurrence-free rate.

Grade

As in all tumours, the more differentiated lesions have a better prognosis. In patients with superficial bladder cancer, the majority of lesions are well or moderately differentiated and there tend to be two clear subgroups in prognostic terms since most of the Ta lesions are G1 (well differentiated), while a small percentage of the T1 lesions are G3 and have a very adverse prognosis. In keeping with many previous authors, Kurth *et al.*[27] observed marked differences in rates of invasion and in death from bladder cancer in relation to grade. In this analysis, 10% of patients had poorly differentiated (G3) disease, whereas G1 and G2 disease was almost evenly divided in the remaining 90%. The rates of invasion and death from bladder cancer were highest in those with G3 disease (30% and 16% respectively), and these authors suggested that the important separation lay between G1 and the remainder. However, the larger overview conducted by Pawinski *et al.*[45] showed the opposite; invasive disease developing in 7.2%, 12% and 23.6% for G1, G2 and G3 tumours, respectively, as graded by the local pathologist and 5.4%, 8.6% and 29.2% by referee pathologists (Pawinski, unpublished data).

The problems presented by the patient with **pT1 G3** disease have caused increasing concern over the past 15 years. They were reviewed by Birch and Harland[4] and emphasized by the Medical Research Council trial[39] that showed T1 G3 patients had a chance of invasive disease or death from bladder cancer approximately 10 times that of patients with other Ta T1 tumours. After 8 years 50% of those with G3 disease had either died of bladder cancer or developed invasive disease.

Number of tumours

The prognostic importance of the number of tumours has been emphasized by both the EORTC Group analyses and that of the MRC. As an indicator of progression it was found that patients with one to six tumours had a lower rate of subsequent invasive disease (12% versus 18%) and of death from bladder cancer (9% versus 17%), than those with seven tumours or more[27]. The EORTC overview showed a similar difference in outcome for patients presenting with one tumour (progression in 7.6%) compared with multiple tumours (progression in 15.6%; Pawinski, unpublished data). As a predictor of long-term superficial tumour recurrence the number of tumours at first diagnosis is one of the two strongest prognostic factors[44], the second being the presence or absence of tumour at the first 3-month follow-up cystoscopy.

Three-month follow-up cystoscopy

The prognostic factor analysis of Parmar *et al.*[44] was based on 379 patients with newly diagnosed Ta T1 bladder cancer entered in a trial of prophylactic thiotepa, the results being validated in a second group of 502 patients in a subsequent trial of intravesical mitomycin. Multivariate analysis showed that the single most important prognostic factor for subsequent superficial tumour recurrence was the result of the 3-month cystoscopy, i.e. whether tumour was present or not. The number of tumours at presentation was the next most important factor. Thereafter the inclusion of tumour grade, category, maximum diameter and site within the bladder did not greatly improve the predictive ability (Table 4.1). The authors concluded 'since the result of the three month cystoscopy is such an important prognostic factor, it seems advantageous to obtain this information before determining the appropriate follow up schedule'[44]. This significance of the first follow-up cystoscopy has been confirmed by Kurth *et al.*[27].

Table 4.1 Strength of prognostic factors for freedom from tumour recurrence for newly diagnosed Ta T1 bladder cancer (from Parmar *et al.*[44])

	Chi-square on 1 degree of freedom (logrank)	*P* (logrank)
Result of 3-month cystoscopy (tumour present or absent)	26.0	<0.00005
Number of tumours at diagnosis (single or more than one)	17.6	<0.00005
Size of largest tumour	10.4	0.001
Histological grade		
Reference pathology	6.9	0.009
Local pathology	0.08	0.4
Site of tumour involving posterior wall	5.7	0.02
T category: pTa versus pT1		
Reference pathology	2.6	0.11
Local pathology	5.2	0.02

Other prognostic factors

The analyses of these[27,44] and other authors[20,23,36] have also explored the importance of other factors, including age, sex, size of largest tumour, prior recurrence rate, the position of the tumour(s) and the findings of random mucosal biopsies, as prognostic factors. Kiemeney *et al.*[23] used all the available predictive information to generate a prognostic index score. While this discriminated between the risk of recurrence and progression of low- and high-risk groups of patients, the predictability for individuals was highly inaccurate. Parmar *et al.*[44] derived prognostic groupings that are eminently practicable and could guide future patient management. Their prognostic strength and clinical attraction lie in the simplicity and reliability of the two cystoscopic and histological observations required (Table 4.2, Figure 4.1).

Table 4.2 Prognostic subgroups of patients with Ta T1 bladder cancer at time of first-check cystoscopy, 3 months after initial TUR (from Parmar *et al*[44].)

	Diagnosis	3/12 Cystoscopy
Group 1	Solitary tumour	No recurrence
Group 2	Solitary tumour	Recurrence
	or multiple tumours	No recurrence
Group 3	Multiple tumours	Recurrence

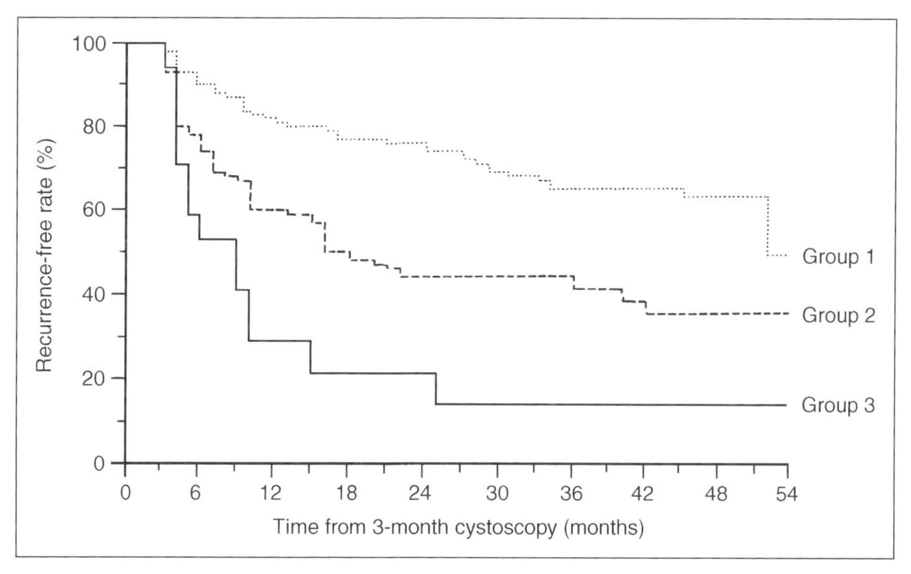

Figure 4.1 Survival-free recurrence for prognostic groups described in Table 4.2. Analysis of MRC trial in newly diagnosed Ta T1 bladder cancer in which adjuvant thiotepa has no effect on recurrence rates. (From Parmar *et al.* (1989)[44] Prognostic factors for recurrence and follow-up policies in the treatment of superficial bladder cancer: report from the British Medical Research Council Subgroup on Superficial Bladder Cancer (Urological Cancer Working Party). *J Urol* **142**: 284–288, with permission from Williams & Wilkins.)

PROGNOSTIC FACTORS IN THE PATIENT WITH RECURRENT TUMOUR

Approximately 70% of patients develop recurrences. If these are single, small and occasional both patient and urologist may be reassured that there is little risk of the development of invasive disease. Biopsy and cystodiathermy are almost certainly adequate for the majority of these lesions. At the other end of the spectrum, the patient who has frequent and multiple recurrences is at a much higher risk of the development of invasive disease and needs more careful management. The problem in this small group of patients is the time at which to advise cystectomy.

Overall the prognostic factors which are of importance to the patient at the time of first diagnosis of bladder cancer are those which are of importance for both repeated recurrence and the development of invasive disease. As time goes by the frequency of recurrences (the recurrence rate) also becomes a dominant factor in determining outcome[27,53].

Though frequent recurrences and the presence of multiple tumours at recurrence are inconvenient for both the patient and the surgeon, the presence of recurrent multiple tumours at the first-check cystoscopy and a finding that three or more of subsequent check cystocopies reveal tumour are probably the factors upon which a clinician should concentrate in determining whether subsequent disease invasion is likely[27]. If such a pattern is seen, especially in a patient with poorly differentiated (G3) disease, alternative therapy is likely to be required, either cystectomy or radiotherapy.

TREATMENT OPTIONS

Until the 1980s the vast majority of Ta and T1 tumours were treated by TUR or cystodiathermy without additional therapy. Within the past 20 years there has been increasing interest in the potential value of intravesical instillations of cytotoxic drugs, BCG or biological response modifiers such as interleukin 2 or α-interferon, in the hope that such agents would prevent recurrence or reduce its incidence, and that this type of treatment might also prevent the development of invasive disease. Recurrent tumours may develop because of a field change or because of implantation of cells liberated at the time of the initial TUR. Many surgeons irrigate the bladder for 18–24 hours after operation, in part to keep the bladder clear of clot and in part to minimize the risk of implantation. Similarly, many urologists prefer to give the first instillation of any intravesical agent within 24 hours of TUR, since this must minimize the capacity of any cells to implant and will also inhibit the growth of any subclinical lesions at the earliest opportunity. The evidence in favour of early irrigation and of early instillation is considered below.

TUR alone

The impact of TUR alone upon subsequent recurrence rate and the development of invasive disease can be assessed by comparing the outcome in randomized studies in which TUR alone is one of the arms to which patients have been randomized. From the meta-analysis undertaken by Pawinski *et al.*[45] it is clear that intravesical cytotoxic chemotherapy reduces the recurrence rate by approximately 50%. Previously Oosterlink *et al.*[43] and Tolley *et al.*[61,62] reported that single doses of intravesical epirubicin or mitomycin C reduced recurrence rates significantly, even in good-risk patients with the lowest likelihood of recurrence. From these trials it can be concluded that the benefit of intravesical chemotherapy lies primarily with the first instillation, since the percentage reduction in recurrence rate was as good as those of other trials in which multiple instillations were given.

The implications for the patient and the surgeon are considerable and it was for this reason that Hall *et al.*[19] advocated a single instillation of cytotoxic chemotherapy for **all** patients at the time of their first TUR of bladder tumour.

Unfortunately there is still no evidence that single or multiple instillations of cytotoxic agents reduce the incidence of invasive disease or the death rate from bladder cancer in those initially diagnosed as having 'superficial' disease[45]. None the less, if the recurrence rate following TUR alone in good-risk patients can be reduced by nearly a half by a single instillation of intravesical chemotherapy, TUR should no longer be considered optimal treatment for Ta T1 bladder cancer at the time of initial diagnosis and treatment.

Patients with recurrent tumours have an equal or higher chance of further recurrence[27]. It could be argued therefore that every treatment of a Ta or T1 tumour, whether primary or recurrent and whether by cystodiathermy, laser or TUR, should be accompanied by a dose of intravesical chemotherapy in the immediate postoperative period.

Bladder irrigation

While some urologists are happy to leave a catheter without irrigation following TUR of a bladder tumour, relying upon the diuretic response of an intravenous infusion to keep the urine clear of blood clot, others believe firmly that bladder irrigation is preferable in controlling both clotting and in washing out any tumour debris, thus minimizing the risk of tumour cell implantation.

In the hope of resolving this issue the Medical Research Council organized a trial of postoperative irrigation following initial TUR of superficial bladder tumours with intravesical chemotherapy at the time of first recurrence. Following complete resection of clinical Ta or T1 bladder tumours, patients were randomized to be irrigated with isotonic or hypotonic fluid or to receive no such irrigation over the first 18 hours following TUR. Nine hundred and twenty-nine patients have been randomized and the results are awaited.

Intravesical instillation of cytotoxic agents

For therapy

At least 30 different agents have been recommended for intravesical instillation within the past 40 years, often in an uncontrolled way and usually in an attempt to treat lesions (chemoresection) rather than to prevent recurrence[57]. Most information is available from studies involving Adriamycin® (doxorubicin), epirubicin (4-epiadriamycin), mitomycin C, epodyl (ethoglucid) and thiotepa, initially for therapy and later for prophylaxis[56,57,58]. All the agents have been shown to produce complete or partial remissions in approximately 60% of patients, the drugs being effective particularly against the smaller tumours, less than 1 cm in diameter.

Table 4.3 EORTC and MRC trials comparing adjuvant intravesical chemotherapy with TUR alone[26,39,43,52,62]

Trial no.	Treatments compared	Number of patients
EORTC		
30751	Thiotepa versus VM-26 versus TUR alone	370
30790	Adriamycin® versus Epodyl versus TUR alone	443
30863*	Epirubicin versus sterile water following complete TUR	512
MRC		
BS01	Thiotepa one instillation versus five instillations versus TUR alone	417
BS03	Mitomycin C one instillation versus five instillations versus TUR alone	502

*In this trial a *single* instillation only was given

Prophylaxis

Within the past 20 years there have been co-ordinated attempts to investigate the potential of intravesical instillations of these cytotoxic agents for prophylaxis in randomized studies, particularly those of the urological group of the EORTC (European Organization for Research and Treatment of Cancer), the British Medical Research Council (MRC) and of the South-West Oncology Group (SWOG) in the United States. The chief studies of these groups that compared adjuvant chemotherapy with TUR alone and were considered mature enough for meta-analysis are listed in Table 4.3[45]. These studies are concerned with prophylaxis against recurrence rather than with the treatment of existing tumours. Long-term information on the effect of the intravesical chemotherapy is now becoming available from many of these studies, but as far as we are aware, there is no universally agreed choice of drug, dose, timing, frequency or duration of therapy.

Present understanding of intravesical chemotherapy

Number of instillations

Whereas the MRC has investigated the outcome following a single instillation or five instillations at 3–4-monthly intervals, the earlier studies of the EORTC used weekly instillations for 4 weeks and monthly thereafter for a further 11 months. The more intensive regimen of 15 instillations rather than five, or even one, is more disturbing for the patient and more expensive. The differences in outcome need to be evaluated to ensure that the most effective but least intrusive and most cost-effective treatment is offered.

Oosterlink *et al.*[43] have shown that a single instillation of epirubicin, 80 mg in 40 ml of normal saline given within 24 hours of the first TUR, halved the rate of tumour recurrence in low-risk patients from 30% to 15%, compared with a similar instillation of sterile water. Similarly, in the MRC trial comparing one and five instillations of mitomycin C (40 mg in 40 ml of normal saline) with TUR only, a single instillation showed a reduction in the recurrence rate of 40% when compared with a control group for up to 5 years[61,62]. These results justify the use of a single intravesical instillation within 24–48 hours of the initial TUR in good-risk patients, i.e. those with solitary tumours at the time of diagnosis.

For those with multiple tumours at diagnosis or with recurrence at the first cystoscopy following the initial TUR, the MRC trial of mitomycin C suggests five instillations at 3-monthly intervals (at the time of attendance for check cystoscopy) may be adequate[61,62] as this reduces the recurrence rate by 50% compared with the control. This result is equivalent to that obtained by the more conventional regimens of weekly instillation for 8 weeks and monthly instillation thereafter adopted many years ago by the EORTC[7,26,52]. The more demanding regimens involving weekly instillation may now be reserved for those patients with multiple and frequently recurrent tumours, many of whom have poorly differentiated lesions. In such patients an intensive course of intravesical chemotherapy (or BCG) may be of assistance or may be only a prelude to cystectomy.

Duration of instillations

For the majority of patients a single instillation at the time of the initial TUR is likely to be adequate[43,62] as 60% of Ta T1 tumours are solitary at presentation and of good prognosis[48]. A further instillation may then be given at the time of any subsequent recurrence necessitating transurethral resection.

Though the MRC study of mitomycin C showed no significant difference between the results of one and five instillations, it must be accepted that there was a difference in favour of five rather than one instillation and it is almost certainly preferable to advise a course of therapy rather than a single instillation for those in whom recurrence is frequent. In this situation one may adopt the regimen of the MRC or the more intensive therapy used by the EORTC and SWOG. Whichever approach is thought preferable, the duration of treatment will be of importance to the patient and the health economist.

In EORTC trials 30831 and 30832, in which immediate or deferred instillations of mitomycin C or Adriamycin® were followed by treatment limited to 6 months or continuing for 1 year, there was a marginal benefit for early instillation but none for treatment continued longer than 6 months[8]. The reduction in recurrence rate was of the order of 5–6% and was not significant. The poorest results were seen in those with delayed treatment without maintenance therapy. Again the differences were not significant.

Following the work of Lamm *et al.*[31] with intravesical BCG, it has become common practice to give booster instillations every 6 months for up to 3 years. The rationale for this (repeated immunostimulation) does not apply to cytotoxic drugs, but for the purposes of comparability some recent trials have used similar repeated instillations of cytotoxic agents. Schwaibold *et al.*[53] compared three intensive and prolonged regimens for mitomycin C and Adriamycin® and concluded that 45 instillations over 3 years constitute the most effective regimen. These authors made no comment on toxicity or cost; such a regimen may be beneficial but further comparative data will be required to justify its routine use.

Choice of drug

Adriamycin®, epirubicin, epodyl and mitomycin C have all been shown to be of similar value in therapy and prophylaxis and to have similar levels of local toxicity with minimal chance of significant absorption. Epodyl is no longer available and cisplatin is no longer recommended[14]. The choice of treatment currently lies between Adriamycin®, epirubicin and mitomycin C in Europe. Thiotepa is still in common use in the United States but instillations of 30 mg in 50 ml of normal saline either once or at 3-monthly intervals have been shown to be ineffective[39]. Also, higher doses can occasionally cause significant toxicity[10,67].

Dose and concentration

The doses recommended have varied as have the concentrations. In general a dose of 40–50 mg in 40–50 ml of normal saline is likely to be effective without significant hazard. Higher doses may be appropriate for single instillation but tend to be associated with a higher incidence of chemical cystitis in those patients requiring frequent and multiple instillations.

CHEMORESECTION OF ‘MARKER’ TUMOURS

As intravesical instillation has become more generally accepted, more thought has been given to the fact that up to 40% of tumours appear to be resistant to any given drug. This is of limited importance when the drugs are being used therapeutically for chemoresection, but has great significance in prophylaxis, especially if long-term therapy is being considered. As a consequence there has been increasing discussion of the possible value of using drugs for prophylaxis only when their effect in a given patient has been demonstrated by their effectiveness on a ‘marker lesion’. Those in favour argue that this targets the use of these agents more effectively[65]. Those against express concerns about the possibility of progression of any such lesion during the period of trial therapy.

The matter has been considered recently by Bono *et al.*[6] who reported an analysis of two EORTC studies in which a well-defined marker lesion was left *in situ* at the time of TUR in patients with multiple primary or recurrent Ta T1 bladder cancer. In one study mitomycin C was used and in the other epirubicin. In each study eight weekly instillations of mitomycin C or epirubicin were given, starting 7–15 days after TUR of all tumours but the marker lesion. Check cystoscopies were carried out 2 weeks after completion of the instillations. At this time any lesion still present was resected. In the absence of any lesion, a biopsy of the marker site was performed to provide histological proof of tumour regression. In these studies a complete response was seen in 50% of the patients treated with mitomycin C and 55.6% of those receiving epirubicin[6]. Of 108 patients treated with mitomycin C two cases of muscle invasive disease were observed at the time of the first-check cystoscopy. Of these, one was at the site of the marker lesion and one at a site where no tumour had been seen initially. In this latter patient a T2 tumour was seen on the anterior wall of the bladder whereas the marker lesion on the posterior wall had undergone a complete response. Of the 40 eligible patients treated with epirubicin, 39 completed the full cycle of eight administrations. No case of muscle invasion at the marker lesion site or elsewhere was observed. These authors[6,65] conclude that marker lesion trials are feasible and reliable for the objective evaluation of anti-tumoural activity of a given drug and suggest that this model should be used in every phase II study designed to assess the efficacy of new agents for intravesical use.

It might also be thought that such studies should now also be undertaken in all situations in which long-term prophylaxis is considered, in order that the clinician can be certain that an agent to which the tumour is sensitive is being used. The advantage of such an approach, given a 30–40% no response rate to any of these agents, clearly outweighs the very small risk of any marker lesion undergoing progression by invasion over the 8–10 weeks of any such preliminary study.

In our view the importance of the work reported by Bono *et al.*[6] lies in its confirmation of the previous report of Popert *et al.*[46]. In this study a complete response of a marker lesion was observed in 46% of 81 patients following a **single** instillation of epirubicin in a dose of 50 or 100 mg (1 mg/ml or 2 mg/ml) in 50 ml of saline retained for 1 hour. In view of the small number of patients involved, no significant difference was found between the two dose regimens. This study does, however, emphasize the possibility of testing the sensitivity of the tumour(s) in a given patient by a single instillation of a drug.

It is interesting to speculate on the time scale of this tumourcidal effect. Absorption of drug must take place during the hour of instillation and intracellular action probably follows over a few hours or days. Tumour necrosis may proceed at a variable rate but should be visible, if not complete, within a few weeks. This time scale would need investigation but, if confirmed, could permit the re-examination of the association, if any, between the response of a marker lesion and the long-term prophylactic benefit of intravesical drugs. EORTC trials 30864 and 30869[6] were designed initially as phase III trials to test this hypothesis. Both failed to recruit because of anxieties about the possible progression of the marker tumour during the 10-week treatment and observation period. Despite the conclusion by Van der Meijden *et al.*[65] that such a delay is not dangerous, many ethical committees, colleagues and patients remain reluctant to participate in marker lesion studies designed in this manner.

If the foregoing speculation of the time scale of tumour necrosis is correct, Popert's findings[46] suggest a more acceptable treatment schedule that could be practicable and would be worth phase III trial. Patients with multiple Ta T1 tumours deemed in need of prophylactic therapy would undergo TUR save a small marker tumour. Following a single dose of postoperative intravesical chemotherapy, the outcome would be determined by flexible cystoscopy 3–4 weeks later. For those patients with visible necrosis of the marker tumour, arrangements would be made for further prophylactic treatment with the same drug. For those without visible benefit, an alternative agent or BCG would be appropriate. The implication for any long-term study involving prophylactic instillation is clear.

OTHER INTRAVESICAL THERAPY

Intravesical **bacille Calmette–Guérin (BCG)** is discussed in Chapter 5. **α-Interferon**[11,15,47,62,70] **and keyhole limpet haemocyanin**[22,33] are alternative forms of intravesical immunotherapy. Experience is limited but responses in CIS and Ta T1 papillary tumours have been reported, some after BCG failure[68]. Serious systemic side-effects do not occur. Their use is still experimental. Intravesical **heparin**[5] has no discernible effect on short-term recurrence rates. **Suramin**[17] has been shown to be active in an animal transitional cell carcinoma model. Intravesical administration could avoid suramin's well-known systemic toxicity but no clinical trials have been reported. Intravesical **methotrexate**[55] was considered insufficiently active to be worth further trial.

ORAL TREATMENT FOR TA T1 TUMOURS AND CIS OF THE BLADDER

Oral prophylactic treatment for bladder cancer could provide a welcome alternative to repeated intravesical instillations, if it were equally effective. **Oral methotrexate**, 50 mg every week for 18 months[18], was reported to be effective, but Nogueira March *et al.*[42] concluded that systemic toxicity could be hazardous. **Etretinate** (a synthetic vitamin A retinoid) has also been shown to reduce the rate of superficial recurrence[1,60] but unwanted side-effects and possible cardiovascular morbidity have prevented its wider application. **Pyridoxine** (vitamin B_6) has been tested and found to have no advantage over placebo in a phase III trial[41]. High doses of **multiple vitamins** have been reported to improve the benefit of BCG. In a double-blind study Lamm *et al.*[34] observed a 40% reduction in the recurrence rate following the addition to standard BCG therapy of daily oral vitamin A, 40 000 u; vitamin B_6, 100 mg; vitamin C, 2 g; vitamin E, 400 IU; and zinc, 90 mg. However, the study was of inadequate power and has not been confirmed. ***Lactobacillus casei***, given three times daily as a powder, prolonged the recurrence-free interval in a pilot study of 58 patients with Ta T1 bladder tumours[2]. Daily oral **UFT** (a mixture of uracil and *N*-[2-tetrahydrofusyl]-5-fluorouracil) also reduced the recurrence rate when given for 2 years and compared with placebo[25]. **Bropiramine** is an oral immunomodulator that has shown activity in urothelial carcinoma in situ[50]. It has been the subject of two international phase III clinical trials until recently, when both trials were closed prematurely.

Whether any of these or other oral agents will find a role in high-risk Ta T1 bladder tumours or carcinoma in situ remains to be seen. It is unlikely that long-term oral therapy would replace single intravesical instillations for good-prognosis patients.

CYTOTOXIC CHEMOTHERAPY OR BCG?

Some urologists, especially in the United States, are convinced that BCG is preferable to the instillation of cytotoxic agents[9,28,29,37,40]. However, several large studies have failed to show any benefit when BCG was compared with mitomycin C[13,24,69], Adriamycin®[3] or epirubicin[38]. Comparison of BCG with chemotherapy in the published literature has been made particularly difficult by the many varied, uncontrolled and sometimes unknown variables, such as tumour and patient prognostic factors, drug dosage and treatment regimens. The efficacy and prophylactic benefit of BCG and chemotherapy may vary according to the grade and category of papillary tumour and CIS, but most trials did not separate low- and high-risk patients. In early studies there was no consensus about the timing and duration of BCG treatment. Ignorance of the underlying mechanism of the systemic side-effects and infection with BCG, its avoidance and its treatment, probably resulted in more serious BCG toxicity than should now occur[59]. Of six trials comparing BCG with mitomycin C, three concluded that BCG was superior[13,24,32,35,49,69]. Of three comparisons with Adriamycin® or epirubicin, one was in favour of BCG[3,30,38].

One of the main reasons for considering BCG to be preferable is the belief that it reduces the rate of tumour progression and the possibility of death from bladder cancer while chemotherapy does not[29]. The latter certainly appears to be true given the result of the overview of Pawinski *et al.*[45]. The evidence that BCG prevents progression is based on two studies that compared TUR alone with TUR and BCG in high-risk patients[12,21]. No randomized study comparing BCG with mitomycin C or Adriamycin® has shown this benefit. The problems of using tumour progression

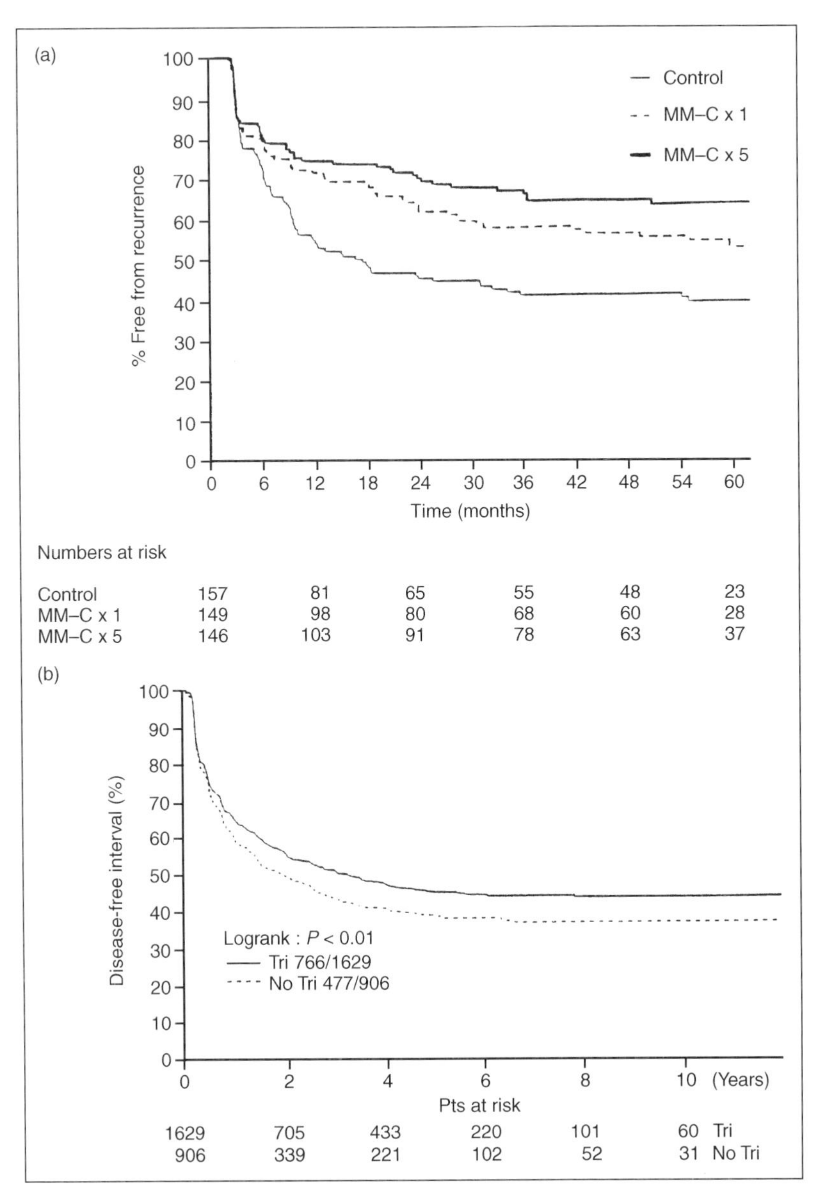

Figure 4.2 Long-term benefit of intravesical chemotherapy. (a) The effect of a single instillation of mitomycin C on recurrence up to 5 years. (From Tolley *et al.* (1996)[62] The effect of intravesical mitomycin C on recurrence of newly diagnosed superficial bladder cancer: a further report with 7 years follow up. *J Urol* **155**: 1233–1238, with permission from Williams & Wilkins.) (b) A similar effect from a variety of intravesical chemotherapies. (From Pawinski *et al.* (1996)[45] A combined analysis of European Organization for Research and Treatment of cancer, and Medical Reserach Council ramdomized clinincal trials for the prophylactic treatment of Ta, T1 bladder cancer. *J Urol* **156**: 1934–1941, with permission from Williams & Wilkins.)

or cancer death as end points in trials of superficial bladder cancer are explained by Sylvester in Chapter 12, from which it is clear that much larger trials will be required to resolve this issue.

The durability of many complete remissions achieved by BCG in carcinoma in situ of the bladder (CIS) and the long-term effectiveness of BCG in Ta T1 prophylaxis lend added encouragement to the proponents of BCG. However, contrary to the conclusion of Lamm in 1992[29], it has now been shown that single doses of mitomycin C and epirubicin have reduced recurrence rates for up to 7 years[43,62] (Figure 4.2), and Pawinski *et al.*[45] have confirmed that the adjuvant effect of chemotherapy is maintained for up to 10 years. How a single instillation can have such long-term impact is not known but its clinical usefulness should not be ignored.

The '6 + 3' BCG regimen of Lamm *et al.*[31] has become popular and may be the most effective. However its toxicity is acknowledged[31]. The inconvenience and toxicity of single instillations of epirubicin and mitomycin C are negligible[43,46,62]. Single instillations of BCG are almost certainly of no benefit. BCG cystitis and mild systemic disturbance are common[54], serious systemic toxicity is rare and treatable. Courses of intravesical chemotherapy cause chemical cystitis as frequently as standard-dose BCG. BCG is available universally, thiotepa, Adriamycin®, epirubicin and mitomycin C are not. Indigenous strains of BCG are cheaper than imported cytotoxics. Future adequately powered trials should help to optimize both BCG and chemotherapy regimens. In the meanwhile an assessment of the patient's risk of relapse balanced by the toxicity, inconvenience of multiple instillations, cost and availability of competing intravesical treatments will determine urologists' choice.

FREQUENCY OF FOLLOW-UP

There is considerable variation in urological practice as to the frequency with which patients undergo check cystoscopies. In many European countries it is the practice to advise cystoscopy 3-monthly for at least 2 years, whereas in the UK some surgeons believe in extending the period between cystoscopies if no recurrence is found. As already discussed, the prognostic groups derived by Parmar *et al.*[44] provide a practical basis for the selection of intravesical chemotherapy[19]. These authors also concluded that one may differentiate between those who need frequent cystoscopic follow-up and those who may be allocated to an annual examination on the basis of the initial tumour(s) and the findings at the first-check cystoscopy (Table 4.2). The reproducibility of the three prognostic groups has been confirmed by Reading *et al.*[48], as illustrated in Figure 4.3. However, their application to the planning of follow-up and their eventual introduction to routine clinical practice will depend upon their safety in terms of tumour progression. The number of patients with disease progression in the MRC prognostic factor data set was small[44]. It remains to be confirmed that no patient in the best prognosis group develops a

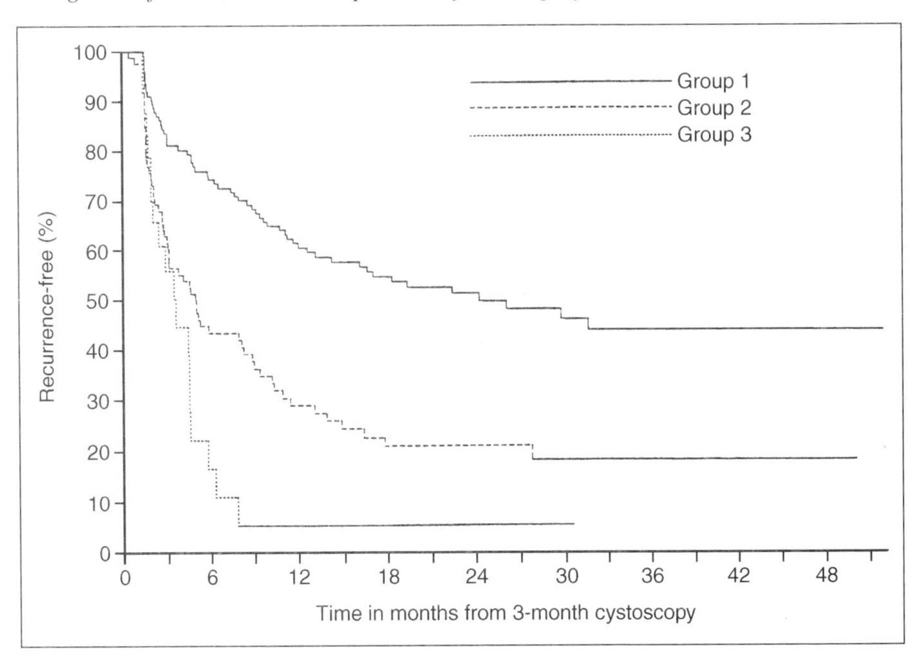

Figure 4.3 Tumour recurrence rates for 232 consecutive newly diagnosed Ta T1 bladder cancer patients, analysed according to the three prognostic groups proposed by Parmar *et al.*[44]. The result is similar to that shown in Figure 4.1, demonstrating that the prognostic groupings apply in routine urology practice as well as in randomized trials. (From Reading *et al.* (1995)[48] The application of a prognostic factor analysis for Ta, T1 bladder cancer in routine urological practice. *Br J Urol* **75**: 604–607, with permission from Blackwell Science, Ltd.)

muscle invasive cancer as a result of the less intensive cystoscopy schedule recommended. Similarly, the number of patients with pT1 G3 tumours was small. Given the high risk of progression in these patients, prolonged 3-monthly follow-up is indicated. Also of importance in some patients with bladder tumours is development of lesions of the upper tract. In general, one may expect to find a tumour of the upper tract at some stage in up to 10%. Of these the vast majority will be in the renal pelvis. The cautious surgeon will certainly undertake an evaluation of the upper tracts by ultrasound or by intravenous urography every 2–3 years.

Cystoscopy, preferably with the flexible cystoscope, remains the standard method for diagnosing recurrent bladder tumours during long-term follow-up. As for the diagnosis of bladder cancer, a number of alternative diagnostic tests have been developed and are currently undergoing evaluation. As yet, none has been shown to be sufficiently reliable to be able to replace cystoscopy (Chapter 2). In the future it is to be hoped that new tests will become available that are sufficiently sensitive and specific to reduce the need for inconvenient, uncomfortable, expensive but currently unavoidable cystoscopy.

REFERENCES

1. Alfthan O, Tarkkanen J, Crohn P *et al.* (1983) Tigsason (etretinate) in prevention of recurrence of superficial bladder tumours. *Eur Urol* **9**: 6–9.
2. Aso Y, Akazan H *et al.* (1992) Prophylactic effect of *Lactobacillus casei* preparation on the recurrence of superficial bladder cancer. *Urol Int* **49**: 125–129.
3. Bassi P and Pagano F (1996) Epirubicin versus BCG as prophylaxis in superficial bladder cancer: a prospective randomised multicentre phase III trial. *J Urol* **155**: 663A, Abstract 1407.
4. Birch BRP and Harland SJ (1989) The pT1 G3 bladder tumour. *Br J Urol* **64**: 109–116.
5. Bitsch M, Hermann GG, Andersen JP and Steven K (1990) Low dose intravesical heparin as prophylaxis against recurrent non-invasive (stage Ta) bladder cancer. *J Urol* **144**: 635–636.
6. Bono AV, Hall RR, Denis L and members of the EORTC Genito-urinary Group (1996) Chemo-resection in Ta-T1 bladder cancers. *Eur Urol* **29**: 385–390.
7. Bouffioux Ch (1991) Intravesical adjuvant chemotherapy treatment in superficial bladder cancer. *Scand J Urol* (Suppl.) **138**: 167–177.
8. Bouffioux C, Kurth KH, Bono A *et al.* (1995) Intravesical adjuvant chemotherapy for superficial transitional cell bladder carcinoma: results of 2 European Organization for Research and Treatment of Cancer radomized trials with mitomycin C and doxorubicin comparing early versus delayed instillations and short-term versus long-term treatment. *J Urol* **153**: 934–941.
9. Brosman SA (1982) Experience in Bacillus-Calmette–Guerin in patients with superficial bladder cancer. *J Urol* **128**: 27–30.
10. Bruce DW and Edgcombe JH (1967) Pancytopenia and generalised sepsis following treatment of cancer of the bladder with instillation of triethyline thiophosphoramide. *J Urol* **97**: 482–485.
11. Calais da Silva F, Furtado L, Reis M *et al.* (1995) Comparison of 2 doses of interferon-alpha-2b in intravesical prophylaxis of superficial bladder tumours. *Eur Urol* **28**: 291–296.
12. De Barardinis E, Giulianelli R, van Heland M, Zarrelli G and Di Silverio F (1997) Management of superficial bladder cancer stage T1 with intravesical Bacillus-Calmette–Guerin therapy: a 10 year study. *J Urol* **157**: 213, Abstract 830.
13. Debruyne FMJ, van der Meijden APM, Witjes JA *et al.* (1992) Bacillus Calmette–Guérin versus Mitomycin C intravesical therapy in superficial bladder cancer. *Urology* (Suppl.) **40**: 11–15.
14. Denis L (1983) Anaphylactic reaction to repeated intravesical instillation of cisplatin. *Lancet* **1**: 1378.
15. Glashan RW (1990) A randomised controlled study of intravesical α-2b-interferon in carcinoma in situ of the bladder. *J Urol* **144**: 658–661.

16. Goessl C, Knispel HH, Miller R and Kaln R (1997) Is routine excretory urography necessary at first diagnosis of bladder cancer? *J Urol* **157**: 480–481.
17. Graham S, Petros JH, Napalkov P *et al.* (1995) Intravesical Suramin in the prevention of transitional cell carcinoma. *Urology* **45**: 59–63.
18. Hall RR, Herring DW, Gibb I and Heath AB (1981) Prophylactic oral methotrexate therapy for multiple superficial bladder cancer. *Br J Urol* **53**: 582–584.
19. Hall RR, Parmar MKB, Richards AB and Smith PH (1994) Proposals for change in cystoscopic follow-up of patients with bladder cancer and adjuvant intravesical chemotherapy. *BMJ* **308**: 257–260.
20. Heney NM, Ahmed S, Flannagan MJ *et al.* (1983) Superficial bladder cancer: progression and recurrence. *J Urol* **130**: 1083–1086.
21. Herr HW, Laudone VP, Badalment RA *et al.* (1988) Bacillus-Calmette–Guerin therapy alters the progression of superficial bladder cancer. *J Clin Oncol* **6**: 1450–1455.
22. Jurincic-Winkler C, Metz KS, Beuth H, Engelmann U and Klippel KF (1995) Immunohistological findings in patients with superficial bladder carcinoma after intravesical instillation of keyhole limpet haemocyanin. *Br J Urol* **76**: 702–707.
23. Kiemeney LAL, Witjes JA, Heijbroek RP, Verbeek ALM and De bruyne FMJ (1993) Predictability of recurrent and progressive disease in individual patients with primary superficial bladder cancer. *J Urol* **150**: 60–64.
24. Krege S, Giani G, Meyer R *et al.* (1996) A randomised multicentre trial of adjuvant therapy in superficial bladder cancer: transurethral resection only versus transurethral resection plus mitomycin C versus transurethral resection plus Bacillus-Calmette–Guerin. *J Urol* **156**: 962–966.
25. Kuboto Y, Hosaka M, Fukushima S and Kondo I (1993) Prophylactic oral UFT therapy for superficial bladder cancer. *Cancer* **71**: 1842–1845.
26. Kurth KH, Tunn A, Ay R *et al.* (1984) Adjuvant chemotherapy in superficial transitional bladder carcinoma: an EORTC (30791) randomised trial comparing Doxorubicin hydrochloride, ethoglucid and TUR alone. *J Urol* **132**: 258–262.
27. Kurth KH, Denis L, Bouffioux C *et al.* (1995) Factors affecting recurrence and progression in superficial bladder tumours. *Eur J Cancer* **31A**: 1840–1846.
28. Lamm DL (1985) Bacillus-Calmette–Guerin immunotherapy for bladder cancer. *J Urol* **134**: 40–47.
29. Lamm DL (1992) Long term results of intravesical therapy for superficial bladder cancer. *Urol Clin N Am* **19**: 573–579.
30. Lamm DL, Blumenstein BA, Crawford ED *et al.* (1991) A randomized trial of intravesical Doxorubicin and immunotherapy with Bacillus-Calmette–Guerin for transitional cell carcinoma of the bladder. *N Engl J Med* **325**: 1205–1209.

31. Lamm DL, Blumenstein BA, Sarosdi M, Grossman HB and Crawford DE (1997) Significant long-term patient benefit with BCG maintenance therapy: a South West Oncology Group study. *J Urol* **157**: 213, Abstract 831.
32. Lamm DL, Crawford ED, Blumenstein BA *et al.* (1993) SWOG 8795: a randomized comparison of Bacillus-Calmette–Guerin and mitomycin C prophylaxis in stage Ta and T1 transitional cell carcinoma of the bladder. *J Urol* **149**: 282A, Abstract 275.
33. Lamm DL, Morales A, Grossman HB *et al.* (1996) Keyhole Limpet Haemocyanin (KLH) immunotherapy of papillary and in situ transitional cell carcinoma of the bladder: a multicentre phase II clinical trial. *J Urol* **155**: 662A, Abstract 1405.
34. Lamm DL, Riggs DR, Schriver JS *et al.* (1994) Megadose vitamins in bladder cancer: a double-blind clinical trial. *J Urol* **151**: 21–26.
35. Lundholm C, Norlen BJ, Ekman P *et al.* (1996) A randomised prospective study comparing long term intravesical instillations of mitomycin C and Bacille Calmette–Guerin in patients with superficial bladder cancer. *J Urol* **156**: 372–376.
36. Lutzeyer W, Rubben H and Dahm H (1992) Prognostic parameters in superficial bladder cancer: an interim analysis of 315 cases. *J Urol* **127**: 250–252.
37. Martinez-Pineiro JH, Jimenez Leon J, Martinez-Pineiro L, Jr *et al.* (1989) Intravesical therapy comparing BCG Adriamycin and Thiotepa in 200 patients with superficial bladder cancer: a randomized prospective study, in *EORTC Genito-urinary Group Monograph No. 6. BCG in Superficial Bladder Cancer*, (eds FMJ Debruyne, L Denis and APM van der Meijden), Alan R Liss, New York, pp. 311–323.
38. Melekos MD, Heracles S, Chionis MD, Paranychianakes MD and Dauaher HH (1993) Intravesical 4'-Epi-Doxorubicin (Epirubicin) versus Bacillus Calmette–Guérin. *Cancer* **72**: 1749–1755.
39. MRC Working Party on Urological Cancer (1994) The effect of intravesical Thiotepa on tumour recurrence after endoscopic treatment of newly diagnosed superficial bladder cancer: a further report with long term follow-up of a MRC randomized trial. *Br J Urol* **73**: 632–638.
40. Netto NR, Jr and Lemos GL (1983) A comparison of treatment methods for the prophylaxis of recurrent superficial bladder tumours. *J Urol* **129**: 33–34.
41. Newling DWW, Robinson MRG, Smith PH *et al.* (1995) Tryptophan metabolites, pyridoxine (vitamin B6) and their influence on the recurrence rate of superficial bladder cancer. *Eur Urol* **27**: 110–116.
42. Nogueira March JL, Ojea A, Figueredo L *et al.* (1985) Evaluation of the efficacy of oral methotrexate in the prevention of recurrence of superficial bladder tumours. *Br J Urol* **57**: 306–307.
43. Oosterlink W, Kurth KH, Schroder F and members of the EORTC GU Group (1993) A prospective EORTC GU Group randomised trial (30863) comparing

transurethral resection followed by a single intravesical instillation of Epirubicin or water in single stage Ta, T1 papillary carcinoma of the bladder. *J Urol* **149**: 749–752.

44. Parmar MKB, Freedman LS, Hargreave TB and Tolley DA (1989) Prognostic factors for recurrence and follow-up policies in the treatment of superficial bladder cancer: report from the British Medical Research Council Subgroup on Superficial Bladder Cancer (Urological Cancer Working Party). *J Urol* **142**: 284–288.
45. Pawinski A, Sylvester R, Kurth KH *et al.* and the members of the EORTC GU Group and the British MRC (1996) A combined analysis of European Organization for Research and Treatment of cancer, and Medical Reserach Council ramdomized clinincal trials for the prophylactic treatment of Ta, T1 bladder cancer. *J Urol* **156**: 1934–1941.
46. Popert RJM, Goodall J, Coptcoat PM, Parmar MKB and Masters JRW (1994) Superficial bladder cancer: the response of a marker tumour to a single intravesical instillation of epirubicin. *Br J Urol* **74**: 195–199.
47. Raitanen MP and Lukkarinen O (1995) A controlled study of intravesical epirubicin with or without alpha-2b-Interferon as prophylaxis for recurrent superficial transitional cell carcinoma of the bladder. *Br J Urol* **76**: 697–701.
48. Reading J, Hall RR and Parmar MKB (1995) The application of a prognostic factor analysis for Ta, T1 bladder cancer in routine urological practice. *Br J Urol* **75**: 604–607.
49. Rintala E, Jauhiainen K, Alfthan O *et al.* (1991) Intravesical chemotherapy (mitomycin C) versus immunotherapy (bacille Calmette-Guérin) in superficial bladder cancer. *Eur Urol* **20**: 19–25.
50. Sarosdy MF (1997) A review of clinical studies of Bropirimine immunotherapy on carcinoma in situ of the bladder and upper urinary tract. *Eur Urol* **31** (Suppl. 1): 20–26.
51. Schulman CC (1982) Intravesical chemotherapy for superficial bladder tumours, in *Clinical Bladder Cancer,* (eds L Denis, PH Smith and M Pavone-Maculso), Plenum Press, New York, pp. 101–111.
52. Schulman CC, Robinson M, Denis L *et al.* (1982) Prophylactic chemotherapy of superficial transitional cell bladder carcinoma: an EORTC (30751) randomised trial comparing Thiotepa, an epipodophyllotoxin (VM26) and TUR alone. *Eur Urol* **8**: 207–212.
53. Schwaibold H, Pichlmeier U, Klingenbeager HJ and Huland H (1996) Long-term follow-up of cytostatic intravesical instillation in patients with superficial bladder cancer. Is short-term intensive instillation better than maintenance therapy? *J Urol* **155**: 663A, Abstract 1408.
54. Sharma N and Prescott S (1994) BCG vaccine in superficial bladder cancer. *BMJ* **308:** 801–802.
55. Smith G, Theodorou C, Field G, Hargreave TB and Chisholm GD (1984)

Intravesical methotrexate in the treatment of superficial bladder cancer. *Br J Urol* **56**: 663–667.

56. Smith PH (1981) Chemotherapy of bladder cancer: a review. *Cancer Treat Rep* **65** (Suppl.): 165–173.
57. Smith PH (1983) Treatment of superficial tumours of the bladder, in *A Comprehensive Guide to the Therapeutic Use of Methotrexate in Bladder Cancer*, (ed. RR Hall). *PharmaLibri.*, Sieber & McIntyre, New Jersey, pp. 1–15.
58. Soloway MS (1982) Intravesical chemotherapy and superficial bladder cancer, in *Chemotherapy of Urological Malignancy*, (ed. MSD Spiers), Springer, Berlin, pp. 50–71.
59. Steg A, Leleo C, Debre B, Boccon-Gibod L and Sicard D (1989) Systemic Bacillus Calmette–Guérin infection, 'BCGitis', in patients treated by intravesical Bacillus Calmette–Guérin therapy for bladder cancer. *Eur Urol* **16**: 161–164.
60. Studer UE, Jenzer S, Biedermann C *et al.* (1995) Adjuvant treatment with a vitamin A analogue (etretinate) after transurethral resection of superficial bladder tumours. *Eur Urol* **28**: 284–290.
61. Tolley DA, Hargreave TB, Smith PH *et al.* (1988) Effect of intravesical Mitomycin C on recurrence of newly diagnosed superficial bladder cancer: interim report from the Medical Research Council Subgroup on superficial bladder cancer (urological cancer working party). *BMJ* **296**: 1759–1761.
62. Tolley DA, Parmar MKB, Grigor KM, Lallemand G and the MRC Superficial Bladder Cancer Working Party (1996) The effect of intravesical mitomycin C on recurrence of newly diagnosed superficial bladder cancer: a further report with 7 years follow up. *J Urol* **155**: 1233–1238.
63. Torti FM, Shortliffe LD, Williams RD *et al.* (1988) Alpha-Interferon in superficial bladder cancer: a Northern California Oncology Group Study. *J Clin Oncol* **6**: 476–483.
64. Van der Meijden APM, Debruyne FMJ, Steerenberg PA, De Jong WH and Doesburgh W (1989) BCG-RIVM versus BCG-TICE versus Mitomycin C in superficial bladder cancer. Rationale and design of the trial of the Southeast Co-operative Urological Group, The Netherlands, in *EORTC Genito-urinary Group Monograph No. 6. BCG in Superficial Bladder Cancer*, (eds FMJ Debruyne, L Denis and APM van der Meijden), Alan R Liss, New York, pp. 311–323.
65. Van der Meijden APM, Hall RR, Kurth KH, Bouffioux C and Sylvester R (1996) Phase II trials in Ta, T1 bladder cancer. The marker tumour concept. *Br J Urol* **77**: 634–637.
66. Walzer Y and Soloway MS (1983) Should the follow up of patients with bladder cancer include routine excretory urography? *J Urol* **130**: 672–673.
67. Watkins WE, Kozak JA and Flangan MJ (1967) Severe pancytopenia associated with the use of intravesical Thiotepa. *J Urol* **98**: 470–471.

68. Williams R, Gleason DM, Smith AY *et al.* (1996) Pilot study of intravesical alpha-2b-Interferon for treatment of bladder carcinoma in situ following BCG. *J Urol* **155**: 494A.
69. Witjes JA, van der Meijden APM and Witjes WPJ (1993) A randomised prospective study comparing intravesical instillations of mitomycin C, BCG Tice and BCG RIVM in pTa-pT1 tumours and primary carcinoma in situ of the urinary bladder. *Eur J Cancer* **29A** (12): 1672–1676.
70. Zerbib M, Botto H, Mandel E, Rischmann P and Dahmani F (1997) Intravesical (IV) IFN compared to BCG therapy in 'high risk' superficial bladder cancer – results of a prospective multicentre randomised study. *J Urol* **157**: 213, Abstract 832.

Commentary

Donald L. Lamm and Peter R. Auriemma

This chapter describes the major advances in superficial bladder cancer that have occurred in the past decade. To stimulate critical thought and to avoid the possibility of accepting information as dogma, we wish to comment on the following important issues raised by Drs Smith and Hall.

PROGNOSTIC FACTORS

Numerous authors have clarified the factors of prognostic importance for tumour recurrence and progression. We would emphasize that the risk of recurrence in patients with transitional cell carcinoma increases with the duration of follow-up. While the figure of 70% is commonly quoted, in our review of over 1000 patients in published series with long-term follow-up, 88% of patients followed for 15 years had tumour recurrence[13]. As described in the foregoing chapter, neither T category nor grade significantly influenced tumour recurrence. The factors of greatest prognostic significance for tumour recurrence include presence or absence of tumour at 3-month follow-up cystoscopy, and the presence of previous tumour recurrences or multiple tumours. In Heney's review of the US Collaborative Group A data[5], patients presenting with a single tumour had only 51% recurrence at 5 years compared with 91% in patients with multiple tumours. An excellent review of the US experience by Bostwick[1] confirmed the importance of lamina propria invasion and grade as prognostic factors for disease progression. In 693 patients with Ta transitional cell carcinoma in six series reviewed, the incidence of progression in Grade 1 disease was 5% versus 11% and 38% for grades 2 and 3 respectively. Overall, the average incidence of progression in Ta disease was 9%. In 350 patients with T1 disease, stage progression occurred in 12% of patients with G1, 39% with G2 and 38% with G3 cancer. The overall incidence of progression in T1 disease was 29%. It is noteworthy that 35 of the 141 patients (25%) in this series who developed muscle invasion began with Grade 1 disease. Therefore, while the risk of developing muscle invasion in patients with low grade is low, it is clearly not an innocuous disease in every patient and may well require aggressive therapy to prevent disease progression in some cases.

Are there better prognostic factors for disease progression? In fact, a test that has now been all but abandoned appeared to provide better prognostic information than any other test available. In a review of 506 patients with superficial bladder cancer in six publications, Sheinfeld and Herr[19] found the average incidence of muscle progression in ABH blood-group positive tumours to be only 6% (range 0–19%). By contrast, the disease progression rate of ABH negative tumours was 72% (60–98%). Newer prognostic factors such as abnormal p53 expression also correlate with disease progression, but it is unlikely that any will achieve the specificity of that reported for loss of expression of blood-group antigens.

INTRAVESICAL TREATMENT

Clearly, one of the most important prognostic indicators for recurrence and progression is the treatment given. One of the most important contributions to cytotoxic chemotherapy for superficial bladder cancer has been the recognition of the importance of a single early postoperative instillation. Our preference is to instill cytotoxic chemotherapy immediately following tumour resection. The EORTC study[17] showed that a single instillation of epirubicin given within 6 hours of tumour resection reduced tumour recurrence significantly, even in patients with solitary low-grade tumours. These data support Hall's recommendation that all patients be given a single postoperative instillation. With the single exception of the MRC study[16] that used a dilute thiotepa solution (30 mg in 50 ml), optimal results with intravesical chemotherapy had occurred with a single instillation. These include two studies with thiotepa[2,21], one Adriamycin® study[21] and one mitomycin trial[20].

Multiple controlled comparison studies and an extensive review of the published literature[11,15] have revealed that none of the newer and more expensive cytotoxic agents are superior to thiotepa or any of the other commonly used agents. The intravenous dose of thiotepa is 0.5 mg/kg; a dose of 30 mg in 30 ml of water is therefore safe in most patients even if 100% is absorbed.

INTRAVESICAL BCG VERSUS CHEMOTHERAPY

For patients in need of more intensive intravesical therapy, the evidence now suggests that the benefit of intravesical chemotherapy is very limited as currently used, compared with BCG. Currently, BCG is the most effective intravesical agent known for the prevention of the recurrence and progression of bladder cancer[9]. Multiple studies have demonstrated that BCG prophylaxis following complete TUR or fulguration of superficial bladder tumours significantly reduces recurrence, and prolongs the disease-free interval in comparison to TUR alone[7,10]. Out of five prospective controlled comparisons of intravesical BCG and TUR alone involving 496 patients, four studies demonstrated significant reduction in tumour recurrence. The pooled data showed that overall tumour recurrence was reduced from 72% in patients treated with surgery alone, to 32% in patients treated with intravesical BCG[11]. Cookson and Sarosdy[3] reviewed their series of patients with T1 lesions treated with TUR and BCG and found 91% of patients to be free of tumour recurrence with a mean follow-up of 59 months. In this series, 69% remained free of disease after the initial therapy, and an additional 22% required further TUR and courses of BCG before achieving a disease-free state.

The effect of BCG on tumour progression has been investigated in three randomized studies involving 335 patients. Herr *et al.*[8] evaluated 86 patients and found the time to progression to muscle invasion or metastasis to be prolonged significantly following BCG treatment. In this study stage progression occurred in 35% of controls and 28% of patients treated with BCG. If the definition of progression is expanded to include those patients requiring multiple resections and alternative therapy, then progression is seen in 95% of controls and 53% of those treated with BCG. The mortality rate was reduced from 32% to 14% with BCG immunotherapy, and cystectomy was required in 42% of controls compared with 26% of those treated with BCG. In a subsequent report[6] with 3 years additional follow-up, cancer deaths were reduced from 37% to 12% ($P < 0.01$). The South-West Oncology Group (SWOG) has compared BCG with doxorubicin and found that an increase in stage or extent of disease occurred in 37% of patients receiving doxorubicin compared with 15% of those receiving BCG[12]. Pagano *et al.*[18] studied 133 randomized patients and found progression to stage T2 or higher in 17% of control patients compared with 4% of those treated with BCG immunotherapy. All of these studies show significant reduction in disease progression with the use of intravesical BCG, with the mean rate of progression being 28% for controls and 14% for those receiving BCG.

When comparing the benefit of BCG with surgery alone relative to that of intravesical chemotherapy to surgery, it is clear that intravesical BCG is superior to intravesical chemotherapy for the prevention of both disease recurrence and disease progression. More importantly, direct randomized comparisons have demonstrated that BCG prophylaxis is superior to thiotepa, doxorubicin and mitomycin C. The reasons for the advantage of BCG immunotherapy over chemotherapy in the long-term prevention of tumour recurrence and progression to muscle invasion or metastasis are unknown, but several concepts to support the observed advantage of immunotherapy over chemotherapy can be theorized. Unlike chemotherapy which promptly decreases in tissue concentration and efficacy after instillation, BCG produces a mucosal infection that can persist for many months. Therefore, the continued replication of organisms provides a much longer duration of action than chemotherapy. Furthermore, the continued persistence of BCG may permit continued immune stimulation and anti-tumour activity. Unlike cytotoxic chemotherapy which suppresses immunity and is active only against tumours present at the time of instillation, BCG immunotherapy has the potential of inducing specific immunity, thereby preventing the development of future tumours. Finally, the concentration of the instilled chemotherapeutic agent is critically important since it determines the depth of penetration because these agents are dependent on simple diffusion into the tissue. Penetration to the depths of tumour extension may not be dependable. On the other hand, with BCG immunotherapy, penetration does not occur by passive diffusion but by attachment of the organisms to receptors and subsequent bacterial invasion. BCG can therefore penetrate to the depths of the detrusor muscle and has even been found in pelvic lymph nodes. This difference in penetration allows BCG the benefit of producing a response at greater depths, eradicating tumour cells not reached by chemotherapy. Immunotherapy therefore has the potential for eradicating muscle invasive disease. In fact, although there are no reports of response of muscle invasive tumours to intravesical chemotherapy, histologically documented muscle invasive transitional cell carcinoma has been reported to respond completely to BCG immunotherapy. A study by Garden *et al.*[4] has shown that BCG inhibits the motility and invasive capability of human transitional carcinoma cells in an artificial basement membrane model of invasion.

ORAL PROPHYLAXIS

Oral therapy is clearly more acceptable to most patients than intravesical instillation. Moreover, oral treatment has the potential to treat upper-tract tumours. Our double-blind randomized evaluation of high doses of vitamins A, B_6, C and E plus zinc has been extended to a 6-year follow-up. The reduction in tumour recurrence with high-dose vitamins remains at a highly significant 37% ($P < 0.01$). While the study included only 65 patients, its power was sufficient to yield a highly significant difference. In contrast, for example, the study of smokers in Finland using vitamin E and β-carotene was inadequately powered despite recruiting thousands of patients. Vitamin E appeared to reduce the incidence of bladder cancer from 81 to 74 cases and β-carotene from 79 to 76 occurrences. Studies are under way to confirm these findings by both the Finbladder Group and the Canadian Urologic Oncology Group.

FOLLOW-UP

The suggestion that the upper tracts be followed in a more cost-effective manner using sonography in some patients is well taken. Clearly, sonography does not provide the detail that can be obtained from intravenous urography. However, dilatation can be seen readily. As stated, the definitive recommendation for the frequency of upper-tract evaluation has not yet been made. However, we believe it is most important to consider the prognostic factors, particularly tumour grade, when considering the need for upper-tract evaluation. A very occasional upper-tract sonogram may be all that is needed in a patient with low-grade bladder tumours without recurrence. However, in patients with high-grade bladder cancers, even those who remain tumour-free for an extended period, upper-tract recurrence is both frequent and threatening. For example, in patients with poorly differentiated TCC or carcinoma in situ treated with BCG immunotherapy, upper-tract recurrence is seen in more than 20% when followed long term. In our opinion, these patients deserve upper-tract cytology and retrograde pyelograms annually for an extended period of time.

COMMENTARY REFERENCES

1. Bostwick D (1992) Natural history of early bladder cancer. *J Cell Biochem* **161** (S): 31–38.
2. Burnand KG, Boyd PJR, Mayo ME *et al.* (1976) Single dose intravesical thiotepa as an adjuvant to cystodiathermy in the treatment of transitional cell bladder carcinoma. *Br J Urol* **48**: 55–57.
3. Cookson MS and Sarosdy MF (1992) Management of stage T1 superficial bladder cancer with intravesical BCG therapy. *J Urol* **148**: 797–801.
4. Garden JR, Liu BCS, Redwood SM *et al.* (1992) Bacillus Calmette–Guerin abrogates *in vitro* invasion and motility of human bladder tumour cells via fibronectin interaction. *J Urol* **148**: 900–906.
5. Heney NM (1992) Natural history of superficial bladder cancer: prognostic features and long-term disease course. *Urol Clin N Am* **19**: 429–434.
6. Herr HW (1991) Transurethral resection and intravesical therapy of superficial bladder tumours. *Urol Clin N Am* **18**: 525–528.
7. Herr HW (1992) Use of bacillus Calmette–Guerin vaccine: indications and results, in *Problems in Urology: Transitional Cell Malignancy,* (eds MS Soloway and DF Paulson), Vol. 6, pp. 484–492.
8. Herr HW, Laudone VP, Badalmant RA *et al.* (1988) Bacillus Calmette–Guerin therapy alters the progression of superficial bladder cancer. *J Clin Oncol* **6**: 1450–1456.
9. Herr HW, Laudone VP and Whitmore WF (1987) An overview of intravesical therapy for superficial bladder tumours. *J Urol* **139**: 1363–1369.
10. Lamm DL (1985) Bacillus Calmette–Guerin immunotherapy for bladder cancer. *J Urol* **134**: 40–47.
11. Lamm DL (1992) Long term results of intravesical therapy for superficial bladder cancer. *Urol Clin N Am* **19**: 573–580.
12. Lamm DL, Crissman J, Blumenstein B *et al.* (1989) Adriamycin versus BCG in superficial bladder cancer: a Southwest Oncology Group Study. *Prog Clin Biol Res* **310**: 263–270.
13. Lamm DL and Griffith JG (1992) Intravesical therapy: does it affect the natural history of superficial bladder cancer? *Sem Urol* **10**: 39–44.
14. Lamm DL, Riggs DR, Shriver JS *et al.* (1994) Megadose vitamins in bladder cancer: a double-blind clinical trial. *J Urol* **151**: 21–26.
15. Lamm DL and Torti F (1996) Bladder cancer 1996. *Ca-A Cancer J Clinic* **46**: 103–112.

16. Medical Research Council (MRC) Working Party on Urologic Cancer (1994) The effect of intravesical thiotepa on tumour recurrence after endoscopic treatment of newly diagnosed superficial bladder cancer; further report of the long term follow up of a MRC randomised trial. *Br J Urol* **73**: 632–638.
17. Oosterlink W, Kurth KH, Schroder F *et al.* (1993) A prospective European Organisation for Research and Treatment of Cancer Genito-urinary Group randomised trial comparing transurethral resection followed by a single intravesical instillation of epirubicin or water in single stage Ta, T1 papillary carcinoma of the bladder. *J Urol* **149**: 749–752.
18. Pagano F, Bassi P, Milani C *et al.* (1991) A low dose bacillus Calmette–Guerin regimen in superficial bladder cancer therapy: is it effective? *J Urol* **146**: 32–37.
19. Sheinfeld J and Herr HW (1992) Intervention strategies for chemoprevention of bladder cancer. *J Cell Biochem* **161**: 173–174.
20. Tolley DA, Hargreave TB, Smith PH *et al.* (1989) The effect of intravesical mitomycin C on recurrence of newly diagnosed superficial bladder cancer: interim report from the Medical Research Council Subgroup on Superficial Bladder Cancer. *BMJ* **296**: 1759–1763.
21. Zincke H, Utz DC, Taylor WF *et al.* (1993) Influence of thiotepa and doxorubicin instillation at the time of transurethral surgical treatment of bladder cancer on tumour recurrence: a propsective randomised double blind controlled trial. *J Urol* **129**: 505–508.

5

Bacillus Calmette–Guérin (BCG) for transitional cell carcinoma of the bladder

Adrian P. M. van der Meijden

INTRODUCTION

Immunotherapy is an approach to cancer treatment which attempts to modulate the host's immune system in order to overcome the tumour burden. In most cases this implies immunostimulation. The agents that are capable of inducing or accelerating immunostimulation are called **biological response modifiers** (BRM). Such an agent is the attenuated bovine tuberculous bacillus, Bacillus Calmette–Guérin (BCG).

To develop a vaccine against tuberculosis, Calmette and Guérin cultured a highly virulent strain of bovine tuberculous bacilli and transplanted one culture after another to lose its noxious characteristics. Thirteen years and 231 transplantings later, the bacillus had been tamed[14]. It no longer caused tuberculosis in animals but its antigenic properties were unimpaired. In 1921 the first human was vaccinated with BCG and by 1980 more than 500 million vaccinations had been performed [50].

The possible protective effect of tuberculosis against cancer was first reported in 1929 by the pathologist Pearl[45]. In an autopsy study he observed that a group of patients infected with mycobacterium tuberculosis showed less malignant tumours than a control group. Thirty years previously, Coley had reported that if different tumours were treated with bacterial toxins, provoking an acute infection in or nearby the tumour, regression of that tumour could occur[40]. Holmgren in Sweden was the first to report the use of BCG to treat cancer in humans in 1935[10]. Some successes occurred in 28 patients but this report was largely ignored. Mathé *et al.* reported encouraging results with the use of BCG as adjuvant therapy for acute lymphoblastic leukaemia[36]. In 1938 Rosenthal showed that BCG administered to animals by various routes caused generalized stimulation of the immune apparatus as well as a specific local response[49]. In 1970 Morton and co-workers were the first to use BCG successfully in malignant melanoma by injection directly into the tumour[18,38]. Subsequent studies in dogs confirmed that delayed-type hypersensitivity responses could be induced with BCG injection in the bladder[1]. Lamm and associates initiated the controlled evaluation of BCG immunotherapy using a rat bladder tumour model and demonstrated that repeated BCG treatment reduced tumour growth compared with controls[27].

After this initial success, trials were started in several human tumours but the success rate was disappointing, mainly because the basic principles of immunotherapy with BCG were not understood. These basic principles were developed by Zbar, Rapp and co-workers in a guinea-pig hepatocarcinoma system[58] which they discovered could be cured by intratumoural injection of BCG. Not only the primary neoplasm but also metastases in regional lymph nodes were eradicated. Moreover, the existence of tumour immunity could be demonstrated as BCG-cured animals rejected a second tumour cell challenge. Optimum results were obtained if all principles were fulfilled:

- tumour burden must be small;
- direct contact between tumour cells and BCG is essential;
- adequate dose is necessary; and
- tumours respond better when confined to the parent organ, and when metastases are present, the spread is limited to regional lymph nodes.

Morales *et al.*[37] and Martinez-Pineiro and Muntanola[35] noticed that most of the basic principles of successful BCG immunotherapy were ideally applicable in superficial bladder tumours. Tumour burden is small, direct contact of BCG with tumour cells is possible by intravesical instillation, and tumours are confined to the parent organ. Intravesical BCG is now used throughout the world as a highly effective prophylactic treatment for superficial bladder tumours and is the most effective treatment for carcinoma in situ, other then cystectomy. At the present time bladder cancer is the only solid tumour for which BCG immunotherapy has become established routine treatment.

ROUTE OF ADMINISTRATION

Although it seemed a brilliant idea to bring BCG into direct contact with the bladder wall by intravesical instillation, it was far from clear whether this was the optimal route of administration; besides intravesical administration, Morales *et al.* advocated combination with intradermal BCG scarification[37]. Five different routes of BCG administration have been investigated:

- percutaneous administration alone;
- intralesional injection;

- oral administration;
- intravesical administration combined with percutaneous scarification; and
- intravesical instillation alone.

Percutaneous scarification was used in only a few patients by Martinez-Pineiro and Muntanola[35]. Although an anti-tumour effect was observed, no conclusive data justified the use of this treatment and it was abandoned. The same investigators used **intralesional injections**, as has been tested in other malignant tumours[33]. One patient developed a serious hypersensitivity reaction resulting in allergic shock and this way of administration was discontinued. It may well be that this serious side-effect resembled 'BCG sepsis' which may be observed rarely after intravesical instillation. Also it is unknown whether the complication is exceptional or frequent. From a theoretical standpoint intralesional injection could be the best route of administration and may be worthy of further attention.

The anti-tumour effect of **oral BCG** was studied in patients with muscle invasive bladder cancer who could not be treated with cystectomy[41]. Although tumour regression was observed the results were not impressive. BCG has not been evaluated sufficiently in muscle invasive tumours but such cancers have responded completely to BCG therapy in anecdotal cases. In superficial bladder cancer oral application of BCG was investigated by Lamm *et al.*[27,28]. No measurable anti-tumour effect was found after oral BCG 200 mg three times per week.

The most thoroughly investigated route of administration has been **intravesical instillation**, initially combined with percutaneous administration. Morales developed an empirically based therapeutic regimen which was adopted by most investigators. BCG was instilled in the bladder dissolved in 50 ml of normal saline once a week for 6 consecutive weeks. Simultaneously, BCG was administered percutaneously on the thigh using a multiple puncture apparatus. The additional anti-tumour effect of the percutaneous application remained unclear until excellent results were reported for intravesical instillation only by Brosman in 1982[7]. He observed that all his patients treated by intravesical instillation alone converted from a negative PPD skin reaction to a positive one after a number of instillations, and did not require presensitization nor intradermal innoculations of BCG. Subsequently randomized trials[29,32] suggested that intradermal BCG did not contribute to the response achieved by intravesical instillation and intravesical instillation alone has been adopted as standard practice.

THE BEST BCG STRAIN

All of the known BCG strains are derived from the Pasteur strain developed by Calmette and Guérin in Lille, France almost a century ago. Cultures of the Pasteur strain have been exported to many countries and substrains have been obtained which show resemblance to, but differences from, the original Pasteur strain. These are due to the use of different culture methods and to genetic drifting. Comparative studies of different substrains have been performed for several experimental tumours[13]. However, for superficial bladder cancer very few studies have been carried out[16,57].

The efficacy of BCG in bladder cancer depends on at least two factors: the number of living bacilli, and their viability. Dead BCG bacilli provoke no immunological reaction when instilled in the bladder. The viability is the ability of the bacilli to multiply *in vivo*. The effect of BCG viability on treatment results was reported by Kelley *et al.*[17]. Extreme differences in the number of culturable particles, ranging from 5×10^6 to 1×10^{12} per ampoule were noted. This was not only true for different strains but even between different lots of the same strain. The variable quality of this living product must have led to different outcomes in efficacy as well as in side-effects. At present all manufacturers are able to produce vials with constant quality, viability and number of colony forming units, but vials have to be stored with care and solutions containing BCG have to be prepared fresh in order to maintain their anti-tumour potential.

Statements about the efficacy and inefficacy of certain substrains made a decade ago no longer hold true and comparable results with respect to efficacy and side-effects have been reported. It seems that differences in culture method and even genetic differences play no major role in the application of BCG in superficial bladder cancer. The results of using the following substrains have been published: Pasteur (France), Armand-Frappier (Canada), Tice (USA), Connaught (Canada), Evans (UK), Moreau (Brazil), RIVM (The Netherlands) and Tokyo (Japan). Other countries in Eastern Europe and in Asia have begun to use their own substrains as well. Very few randomized prospective studies comparing different substrains have been performed.

In the UK the Pasteur strain was compared with the Evans strain in patients with a so-called marker lesion[12]. In this small randomized trial the ablative effect of the two substrains, as well as the adverse effects, were comparable. In The Netherlands BCG-Tice was compared with the Dutch RIVM strain in a large prospective randomized study[57]. No differences were found between the two BCG strains with respect to efficacy or with respect to local and systemic side-effects.

DOSE OF BCG

The need for BCG organisms to be present in the instillate has been questioned by van Vooren *et al.*[56]. These authors found that BCG culture filtrate was as effective as the living bacilli in enhancing lymphocyte mediated toxicity against bladder tumour cells in culture. Dead bacilli showed no activity. Until the significance of this finding has been explored in depth, BCG vaccines in their present form will need to be used.

As BCG is a living preparation, storage of BCG would be expected to lead to reduction in the number of viable micro-organisms. However, in a retrospective review Blumenstein and co-workers found no evidence of reduced efficacy with reduced dose, as estimated by the decline of colony forming units (CFU) which occurs with shelf storage[2]. The dose of most preparations is given by manufacturers in milligrams and/or in CFUs. It is advised to use per instillation: 150 mg of Pasteur, 120 mg of Armand-Frappier, 120 mg of Connaught, 50 mg of Tice and 40–80 mg of Japanese BCG. The number of CFUs in one ampoule varies between manufacturers and even between lots of the same substrain. The optimal dose at present ranges between about 3×10^8 CFUs and 1×10^9 CFUs.

The living bacteria are present as solitary bacilli but also as clumps containing five, 50 or even 500 bacilli. These clumps grow in culture as if they were solitary micro-organisms, ultimately forming one colony in the culture disc. In practice, the number of CFUs does not represent the number of bacilli present. It is also unknown whether a clump of bacilli act differently compared with a solitary bacillus, with regard to immune stimulation, when it adheres to the bladder wall. In view of these uncertainties and known variables it is questionable whether a dose reduction of 50% would have any impact at all. Probably a dose reduction of fourfold or more should be investigated to address the question of dose reduction.

The choice of the 'standard' dose for BCG was empirical. Several studies are now ongoing to investigate the efficacy and toxicity of lower doses. Pagano *et al.*[44] have compared 75 mg of the Pasteur strain to the standard dose of 150 mg. The preliminary results indicated similar response rates in 87 patients. The recurrence-free rates were 90.5% (low dose) and 86.4% (standard dose). The incidence of local and systemic side-effects, however, were significantly less in the low-dose group. Cystitis was observed in 31% of low-dose and 54% of standard-dose patients; fever was noted in 14% and 26% respectively. In a study by the Spanish co-operative group CUETO, toxicity was reduced significantly in patients given one-third of standard dose Connaught BCG[34]. There was no apparent difference in efficacy at 16.8 months follow-up but longer follow-up will be needed to confirm this finding.

MAINTENANCE OR BOOSTER THERAPY

The optimal treatment regimen of BCG has not been determined. The best schedule probably depends on the prognostic factors of the tumour to be treated as well as the immunomodulation of the host. Two different concepts of treatment have been proposed. Morales and co-workers started with a single course of six consecutive weekly instillations and stopped thereafter[37]. Others have claimed that six weekly instillations were suboptimal and have introduced maintenance therapy. The theory behind these repeated instillations (even if the patient is free of tumour) is the assumption that 'booster' instillations evoke a renewed immune response against residual malignant cells in the patient[24]. This is supported by the finding of Zlotta *et al.*[59] that a single 6-week course of treatment was unable to induce a sustained systemic cellular immune activation against several BCG antigens. Reactivation of lymphoproliferative response was observed when a second BCG course was administered. The maintenance schedule has not been defined exactly and the number of repeated instillations as well, as the time frame in which they should be administered, is controversial.

The maintenance schedule which has been investigated most thoroughly was developed by Lamm and co-workers[24]. After an induction course of six consecutive weekly instillations, 'booster' instillations (3 consecutive weeks) were given at 3 months, at 6 months and thereafter every 6 months up to 36 months[20]. This means that a patient who remains free of tumour after the initial TUR will receive a total of 27 instillations of BCG over a period of 3 years. The benefit of this intense and prolonged maintenance schedule compared with a single 6-week course of BCG was evaluated by a prospective randomized trial of the South-West Oncology Group (SWOG). Analysis of 385 patients randomized between non-maintenance and maintenance showed significant superiority of the 3-year maintenance regimen. This was observed both for patients with Ta and T1 tumours and for patients with carcinoma in situ[23,25].

The efficacy of repeated 6-week courses, after failure of the first induction course, was investigated by Catalona *et al.*[8]. Of the 100 consecutive patients with superficial bladder cancer, 44 had complete remissions after one 6-week course of BCG. Of 49 patients not rendered tumour-free by the first course and treated with a second 6-week course, 19 were consequently tumour-free. However, there were some patients who had to be treated with three or more courses, and among these selected patients the risk of subsequent cancer (30%) or metastases (50%) was unacceptably high. The risk of developing progression exceeded the prospect of eradicating the superficial tumour in this group. Furthermore, increased toxicity of BCG was noted with repeated courses of treatment. Long-term follow-up of this series of patients[39] has shown that patients treated successfully 10 years ago have no guarantee of remaining free of tumour for the rest of their lives. Patients treated with one or two 6-week courses of BCG, and who were free of tumour at 2 years, experience recurrence rates of 36% and 33%, respectively, during follow-up of 2–11 years. Although 70% of patients were tumour-free after 2 years, this tumour-free status drops to 35% if follow-up is extended to 11 years. Patients treated with BCG need lifelong follow-up.

ADVERSE REACTIONS TO BCG

Intravesical therapy with BCG is considered to be more effective than most chemotherapeutic agents in the treatment and prophylaxis of carcinoma in situ and high-risk T1 bladder tumours. However, compared with intravesical chemotherapy, BCG seems to provoke more local and systemic reactions[5,21,30]. Besides irritative symptoms, BCG may cause systemic side-effects ranging from mild malaise and

Table 5.1 Complications in 2602 patients according to substrains of BCG

	Total no. (%) (2602 pts)	%Armand-Frappier (718 pts)	%Tice (726 pts)	%Connaught (353 pts)	%Pasteur (325 pts)	RIVM (129 pts)
Fever	75 (2.9)	3.8	4.7	4.7	0.6	2.1
Granulomatous prostatitis	23 (0.9)	1.8	1.0	0.2	0.6	0.0
Pneumonitis/hepatitis	18 (0.7)	0.4	0.8	0.6	1.2	0.8
Arthralgia	12 (0.5)	0.7	0.1	0.6	1.8	0.0
Haematuria	24 (1.0)	0.3	0.6	4.4	1.0	0.4
Rash	8 (0.3)	0.4	0.0	0.9	0.0	0.0
Ureteral obstruction	8 (0.3)	0.6	0.4	0.2	0.0	0.0
Epididymitis	10 (0.4)	0.4	0.0	0.2	1.2	0.8
Contracted bladder	6 (0.2)	0.0	0.3	0.2	0.6	0.0
Renal abscess	2 (0.1)	0.0	0.0	0.4	0.0	0.0
Sepsis	10 (0.4)	0.1	0.4	0.9	0.2	0.0
Cytopenia	2 (0.1)	0.0	0.3	0.0	0.0	0.0

Since definitions of side-effects differ from one series to another and all complications were not considered in each series, some assumptions have been made to estimate the overall incidence of complications.

fever to life-threatening or even fatal sepsis in rare instances[51]. These side-effects were reported early in the published literature when little was known about the optimal time of first instillation after transurethral resection and not every BCG manufacturer could guarantee constant quality, dose or viability of this living bacterial product. Although a cause for reasonable concern, clinical observation and basic research have enabled serious toxicity to be minimized. The incidence and type of side-effects have been reviewed extensively[31] and the complications of using different strains of BCG are summarized in Table 5.1.

Local side-effects

Local side-effects of BCG instillation therapy are defined as those confined to the bladder and the organs that are in contact with BCG-contaminated urine, namely the prostate, urethra, epididymis, ureter and the kidney when reflux exists. Reported side-effects include bacterial cystitis (not BCG related), BCG cystitis, haematuria, bladder contracture, granulomatous prostatitis, epididymo-orchitis, ureteral obstruction and renal abscess.

BCG-induced urinary tract infection typically consists on biopsy of acute and chronic inflammation with or without formation of non-caseating granulomas. Patients have the same symptoms as in bacterial (non-BCG related) urinary tract infections, such as urinary frequency, urgency, bladder pain and sometimes haematuria. Granulomatous prostatitis, epididymo-orchitis, ureteral obstruction and renal abscess caused by BCG may present in the same way as if caused by other bacteria. Granulomatous prostatitis is most often asymptomatic and the induration associated with it may be difficult to distinguish by rectal examination from prostate cancer.

Of all the local side-effects BCG cystitis is the most common and occurs in 30–60% of patients[57]. A certain degree of BCG cystitis is considered by most investigators as a normal reaction to BCG therapy. In fact, when the patient does not report any bladder irritability the efficacy of BCG is doubted by some investigators who may then intensify therapy. BCG organisms can persist in the urinary tract for at least 16 months after completion of instillations. Of 125 patients treated with intravesical BCG, five patients were found to have persisting acid-fast mycobacteria in their urine or bladder biopsies, one of whom underwent cystectomy which revealed numerous caseating granulomata containing acid-fast bacilli on histological examination[6]. La Fontaine *et al.*[21] looked for acid-fast bacilli in the operative specimen of patients undergoing cystectomy after intravesical BCG. Nine of 12 patients had granulomatous prostatitis, seven of whom had acid-fast bacilli demonstrable by Ziehl–Neelsen staining. As BCG is instilled by a transurethral catheter, it is also possible that other bacteria are introduced, causing symptomatic bacterial cystitis. Routine urine culture before every instillation demonstrated bacterial cystitis in 18–27% of patients[57]. If unrecognized and untreated, delay of BCG instillations or even cessation of therapy may result.

Systemic side-effects

Fever, influenza-like symptoms, malaise and chills, pneumonitis, hepatitis, arthralgia, myalgia, rash, cytopenia and fatal sepsis have all been described as caused by intravesical BCG. Of all systemic adverse events, low-grade fever (<38.5°C) in combination with flu-like symptoms is encountered most frequently in about half of the patients. This combination of symptoms is probably the result of the immune response to BCG. As reported for BCG-induced cystitis, the presentation of mild systemic side-effects such as slight elevation of temperature and malaise are considered by some as a favourable indication that BCG evokes immunological effects. Conversely, others stop further treatment immediately.

The development of systemic reactions may be explained in two ways. Living bacilli may pass the bladder wall and spread to any organ. Living *Mycobacterium bovis* has been isolated and cultured from granulomatous lesions in the liver and BCG bacilli have been found in the blood[46]. However, in most instances no living BCG is found in the granulomas. This is in contrast to the caseating granulomas caused by *Mycobacterium tuberculosis*, in which the bacterium is usually present. The second possibility is that the systemic BCG symptoms and the formation of granulomata may be the result of a delayed hypersensitivity reaction rather than the effect of direct bacterial spread[31].

Generalized BCG infection manifested by granulomatous pneumonitis or hepatitis is a serious complication, resulting in severe illness and high fever. The diagnosis is made by chest X-ray, showing appearances similar to disseminated pulmonary tuberculosis. Hepatitis is diagnosed by the combination of liver tenderness, elevated liver enzymes and needle biopsy of the liver. Arthralgia, myalgia and rash are considered to be typical allergic reactions. Joints may be swollen and inflamed. Culture of fluids from these joints never shows BCG bacilli. Sepsis following intravesical BCG instillation is the most serious known complication and has been fatal in at least seven patients. Although BCG is the attenuated strain of *Mycobacterium bovis*, the decreased virulence has not rendered it completely harmless. The massive dose administered in the bladder is potentially lethal when given intravenously, especially in immunocompromised patients.

PREVENTION OF ADVERSE EFFECTS

Intravesical BCG therapy should be avoided in those patients with a higher than usual expectation of toxicity. It should not be used in patients suffering from active tuberculosis or in patients with congenital or acquired immunodeficiency syndromes, leukaemia or Hodgkin's disease. Transplant recipients and patients with positive human immunodeficiency virus serology, with or without clinical manifestation of AIDS, should also not be treated. BCG should not be administered during pregnancy, lactation and in cases of intractable urinary infection. In patients with culture-proven bacterial cystitis, BCG should be withheld until the culture becomes negative with antibiotic therapy. *In vitro* culture experiments indicate that BCG micro-organisms can be eradicated by a number of antibiotics, including trimethoprim, sulphamethoxazole and norfloxacin. Therefore it is logical to expect that concomitant administration of antibiotics can influence the efficacy of BCG. There are no reports suggesting that the BCG bacilli can be transmitted by sexual intercourse, but the use of a condom is advised during coitus for 1 week after BCG therapy.

Correct instillation procedure is a prerequisite to the prevention of side-effects. The barrier normally preventing massive systemic spread can be breached when there is fresh bleeding in the urethra, a recent TUR of the bladder or prostate, or when the bladder urothelium is severely inflamed. Almost all cases of reported BCG sepsis were associated with the intravenous absorption of BCG due to traumatic catheterization, increased bladder pressure during instillation or large areas of damaged or inflamed mucosa. Therefore **BCG should not be given until at least 1 week after tumour resection**. At that time fibrin clots will have covered the resection area in the bladder wall. These fibrin clots are essential for a proper adherence of BCG bacilli to the bladder wall which is essential for the start of the immunological cascade[42].

Intravesical BCG will cause some unwanted effects in the majority of patients, which will be severe and even life threatening in a very few. Therefore a risk to benefit analysis has to be made for every patient. Patients with favourable prognostic factors of their superficial bladder cancer (solitary, primary or recurrent pTa, Grade 1 or Grade 2) should not be treated with BCG. The risks, albeit small, are not justified. However, patients with intermediate or high-risk superficial bladder cancer (multiple tumours, T1 tumours, Grade 3 tumours and CIS) are excellent candidates for BCG instillation therapy.

TREATMENT OF SIDE-EFFECTS

Fortunately the great majority of patients treated with BCG experience only mild side-effects[31]. Although irritative bladder symptoms are encountered in between 30 and 60%, in most cases no therapy is needed[57]. Symptoms generally start 2–4 hours after instillation and will subside without any therapy within 6–48 hours. Relief is produced by symptomatic medication such as phenazopyridine or oxybutynin. Also, no therapy is needed when slight elevation of temperature (<38.5°C) occurs or mild malaise or influenza-like symptoms are noted. Repeated BCG administration may increase the incidence and severity of toxicity, so instillations should be delayed until side-effects have subsided.

About 5% of the patients present with moderate to severe local or systemic adverse events[31]. Those patients should be investigated and treated without delay. If the symptoms are recognized in time, very effective treatment is possible and virtually all patients will recover. Although BCG is the attenuated strain of *Mycobacterium bovis*, the sensitivity for anti-tuberculous drugs remains unimpaired and we have found no difference in the sensitivity of different BCG strains for all known anti-tuberculous drugs[54]. All drugs were very effective except pyrazinamide (Table 5.2). In case of severe toxicity the combination of two or three anti-tuberculous drugs is advised for 3–6 months. Treatment recommendations for BCG-related complications are depicted in Table 5.3, varying from withholding BCG in mild cases, to intensive anti-tuberculous therapy in BCG sepsis. Of all anti-tuberculous drugs, isoniazid is used most frequently. Although administration of isoniazid is safe, it is occasionally associated with adverse reactions, the most significant of which is hepatitis. Elevation of serum transaminase activity occurs in 10–20% of patients[31]. Enzyme levels return to normal in most patients despite continuation of medication. Alcohol consumption may enhance the symptoms of hepatitis. If one of the liver function tests exceeds three to five times the upper limit of normal, discontinuation of isoniazid is recommended.

The use of corticosteroids in the treatment of BCG-related severe complications is controversial, probably because little experience has been reported. The absence of BCG organisms in patients with severe septic reactions may indicate that this is the result of a Type IV hypersensitivity reaction. In such cases favourable results

Table 5.2 Sensitivity of BCG strains to tuberculostatic and antibiotic agents. Susceptibility of different BCG strains was tested quantitatively by determining the minimal concentration of drugs needed for inhibition of growth of BCG bacilli. This is the minimal inhibition concentration (MIC value given in the table). The higher the the MIC value the less susceptible BCG bacilli are for a certain tuberculostatic or antibiotic drug

Antimicrobial agent	RIVM	Tice	Pasteur	Connaught	Susceptibility limits
Tuberculostatics					
Capreomycin	≤5	≤5	≤5	≤5	20
Clofazimine	0.5	0.5	0.5	1	2
Cycloserine	10	10	10	10	20
Ethambutol	1	1	1	1	10
Ethionamide	1	1	1	2	5
INH	0.1	≤0.05	≤0.05	0.1	0.2
PAS	≤1	≤1	≤1	≤1	1
Pyrazinamide#	>100	n.g.	>100	n.g.	50
Rifabutin	≤0.2	≤0.2	≤0.2	≤0.2	2
Rifampicin	≤0.1	≤0.1	≤0.1	≤0.1	1
Thiacetazone	≤2	≤2	≤2	≤2	2
Antibiotics					
Beta-Lactams					
Amoxicillin	128	64	128	256	32
Amox/clav. acid	≤16*	≤16	≤16	≤16	32
Cephalothin	128	64	128	128	32
Cefuroxime	128	64	128	128	16
Piperacillin	>256	>256	>256	>256	64
Aminoglycosides					
Amikacin	≤1	≤1	≤1	≤1	16
Gentamycin	4	2	4	4	16
Kanamycin	≤1	≤1	≤1	2	16
Streptomycin	≤1	≤1	≤1	≤1	16
Tobramycin	>4	>4	>4	>4	16
Other drugs					
Doxycycline	≤2	≤2	≤2	≤2	16
Minocycline	≤2	≤2	≤2	≤2	16
Sulphamethoxazole	4	4	4	4	128
Trimethoprim	>64	64	>64	>64	8
Pipemidic acid	32	32	64	32	8
Norfloxacin	1	2	2	2	8
Nitrofurantoin	8	8	8	16	32

n.g., No growth.

BCG strains are by nature resistant to pyrazinamide. MICs have to be determined at low pH (5.5) at which growth is often poor.

* The value pertains to the amoxicillin concentration in the presence of 4 mg/l of clavulanic acid.

Table 5.3 Treatment recommendations for BCG-related complications

Fever < 38.5°C, BCG cystitis, mild malaise	No treatment. Delay BCG until symptoms resolve
Fever > 38.5°C for 12–24 hours	Isoniazid 300 mg daily for 3 months. May resume BCG when asymptomatic
Allergic reactions: arthralgia, myalgia, rash	Isoniazid 300 mg. Further BCG is indicated only if benefit exceeds risk
Acute severe illness, local or systemic pneumonitis, hepatitis, prostatitis, ureteral obstruction, renal abscess, persisting high fever > 39°C	Isoniazid 300 mg, rifampicin 600 mg, ethambutol 1200 mg daily for 6 months, no further BCG
BCG sepsis	Isoniazid 300 mg, rifampin 600 mg, ethambutol 1200 mg. Cycloserine 500 mg orally twice daily. Consider prednisolone 40 mg intravenously immediately

have been reported from corticosteroids administered to treat shock symptoms. Most anti-tuberculous drugs require 2–7 days to inhibit BCG growth *in vitro*. Peak serum concentrations of cycloserine are achieved within 3–4 hours after oral administration[31]. **Cycloserine, therefore, may be life saving in patients with BCG sepsis**. Those patients should be treated with triple tuberculostatic therapy plus 500 mg cycloserine orally twice daily for 5 days (Table 5.3).

FURTHER IMPROVEMENT OF BCG THERAPY

The viability of BCG is considered crucial for its immunostimulation. It is therefore important that nothing associated with its administration should impair this viability. Bohle *et al.*[4] have reported that lubricants (and the bladder urine after the use of lubricants) inhibited the *in vitro* growth of Connaught BCG and recommend that catheter lubricants should be used sparingly or avoided.

Although side-effects, if recognized correctly, can be treated successfully, the question remains whether it will be possible to reduce the adverse effects without reducing the anti-tumour effects. Reducing the dose of BCG could be a possibility and has been discussed already.

To diminish the local irritative bladder symptoms and probably to prevent systemic side-effects, some investigators advise administration of isoniazid (INH) 300 mg for a few days, starting 1 day before instillation. Isoniazid did not influence the tumour growth of urothelial neoplasms in animal models[3]. Minimal inhibition concentration (MIC) values *in vitro* are compared with the concentrations of INH which can be detected in the urine of patients. In the guinea-pig the systemic reaction to BCG can be measured using the PPD skin reaction. This reaction was delayed and reduced when these animals were treated concomitantly with INH and intravesical BCG. Also the number of granulomas in the bladder wall were reduced when INH was administered[3]. Two randomized trials have evaluated the benefit, and possible disadvantages, of prophylactic INH. Colleen *et al.*[9] observed in 159 patients that INH (300 mg orally daily for 3 days, starting the day of BCG instillation) reduced local side-effects but had no impact on systemic toxicity or recurrence rates. An EORTC trial[55] of the same dose of INH started the day before BCG, found no benefit in 579 patients randomized to BCG-Tice alone or with INH. The incidence of bladder symptoms, systemic symptoms and the need to stop BCG because of toxicity were equivalent in both arms of the study. Furthermore, INH itself was associated with an increased incidence of liver function disturbance. Further follow-up is needed to determine any reduction in long-term BCG morbidity.

HOW DOES BCG WORK?

The exact anti-tumour mechanism of BCG therapy in superficial bladder cancer has not been elucidated. This does not mean that no knowledge exists about the complicated cascade of immunological events. Ratliff *et al.*[48] have shown that it is necessary to bring living bacteria in contact with the bladder mucosa. Consequently attachment, retention and internalization of BCG bacilli must take place. Thereafter the induction of immunological events follows and this may lead ultimately to tumour destruction. Ratliff and co-workers divided the mechanisms of action into two fundamental components: initiation events and effector events. The **initiation events**, defined as bacterial attachment, may not be directly associated with the destruction of tumour cells, while **effector events** are immunological features leading to domination of tumour cells. The attachment of BCG to the bladder wall is essential. Without attachment, antigen is not retained and induction of the anti-tumour response will not be established. The mechanisms of attachment have been shown to be direct attachment to epithelial cells, mediated by fibronectin[48] and possibly by glycosaminoglycans[47]. Morales and Nickel demonstrated that an exopolysaccaride capsule produced by the live, metabolically active mycobacteria represents the primary interaction between BCG and the bladder mucosa. It has been suggested that the immunological events are mediated by this bacterial cell surface[42]. The immunological response of the host after contact with BCG is expressed by a delayed-type hypersensitivity reaction (DTH). Without proper attachment this DTH reaction, expressed as the conversion of the PPD skin reaction from negative to positive, is not observed.

Not all effector mechanisms in intravesical BCG therapy are immunological events. As BCG provokes an inflammation of the bladder mucosa, non-immunological reactions may follow as in any bacterial cystitis. Products of this inflammation may kill tumour cells directly without antigenic recognition.

The immunological reactions can be divided into two fundamental types[47]. The first is direct recognition of tumour cell antigens which results in the production of cytotoxic T lymphocytes (TC). Furthermore, helper T cells (Th) will produce cytokines or tumour-specific antibodies. The second is a cytokine-mediated event initiated by the recognition of non-tumour antigens (i.e. BCG antigens). Tumour cells can be eliminated by direct cell-mediated lysis, through TC or Th cells. Killing of cells can be evoked directly by cytokines or cytokines may activate other immunological killer cells (e.g. natural killer [NK] cells) which then eliminate the tumour. Because of the complexity of many possible immunological events, it is difficult to separate the immune response to BCG itself from the relevant anti-tumour immune mechanisms.

The response to BCG expressed as the DTH response is manifest as mononuclear cell infiltrates in the bladder wall. Later on these infiltrates progress to granulomatous inflammation typical of mycobacterial infections[53]. Clinical studies have shown a low correlation between the presence of granulomatous inflammation and anti-tumour effect of BCG. There is also no clear correlation between the absence or presence of a measurable DTH reaction (positive or negative PPD skin reaction) and anti-tumour effect of BCG[47].

Optimal attachment of BCG occurs when it is diluted with preservative and bacteriocidal-free normal saline at a pH of 7.4. Attachment peaks at 2 hours after instillation. With treatments repeated weekly, immunostimulation peaks at 6 weeks. It is remarkable that the original treatment schedule designed empirically by Morales *et al.*[37] appears to be the most effective in clinical practice and has been supported by immunological research.

In animal studies it was demonstrated that intralesional injection of BCG resulted not only in the elimination of tumour and lymph node metastases, but also that these animals were resistant to subsequent challenge with the tumour that was originally injected[58]. This implied that a protective immunity had been developed but this could not be demonstrated conclusively in animal bladder tumour systems. These animal studies are consistent with clinical observations in BCG therapy and may explain why 'booster' instillations result in better clinical outcome than a 6-week induction course alone, assuming that no protective immunity develops after induction therapy. The efficacy of BCG appears to be restricted to tumours that are in direct contact with BCG. Intravesical instillations do not eliminate tumours that are present in the renal pelvis or ureters. However, these tumours are not resistant to topical BCG therapy, as they react favourably when it is applied directly to them through a percutaneous nephrostomy[15].

BCG OR INTRAVESICAL CHEMOTHERAPY?

BCG has become standard first-choice treatment for carcinoma in situ of the bladder (CIS) by common consensus based on experience and the results of a few randomized trials. As discussed in Chapter 6, CIS is a high-risk disease, with local progression, metastasis and death the likely result if first-line treatment fails. In this circumstance tumour control is the first priority and the small risk of serious toxicity from BCG is a secondary consideration.

For Ta and most T1 bladder tumours the situation is different. With the exception of some pT1 G3 tumours the main problem is the frequency of superficial recurrence, not life-threatening disease progression. In the opinion of many urologists even the very small risk of serious systemic toxicity from BCG is not justified in these patients unless intravesical chemotherapy fails. If the nature and likelihood of side-effects from chemotherapy and BCG are explained to patients, many will choose not to receive BCG because they prefer to avoid the systemic effects which, although mild, may interfere with work or pleasure and do not occur with intravesical chemotherapy. The results of randomized trials that have compared BCG with chemotherapy for Ta T1 transitional cell carcinoma have been inconsistent[11,19,26,57].

The strongest argument in favour of BCG is its apparent impact on disease progression in high-risk T1 tumours[22,24]. For low-risk Ta and T1 G1 and G2 tumours, this potential benefit does not apply[20] and single-dose chemotherapy is sufficient prophylaxis, especially in view of its long-lasting efficacy[43,52]. For medium-risk patients the results of ongoing trials should provide clearer evidence to inform the choice of patients between BCG and chemotherapy.

SUMMARY

Intravesical BCG immunotherapy is the treatment of choice for carcinoma in situ of the bladder at the present time, and is the preferred option of some urologists for Ta and T1 bladder cancer. Adverse effects and severe complications are more marked than in intravesical cytotoxic chemotherapy. However, most patients tolerate BCG instillation well without undue adverse effects. It is essential that urologists and general physicians learn to distinguish mild 'normal' therapy-related symptoms from those that pressage systemic 'BCG-osis', for which immediate treatment is essential.

REFERENCES

1. Bloomberg SD, Brosman SA, Husman MS, Cohen A and Battenburg JD (1975) The effect of BCG on the dog bladder. *Invest Urol* **12**: 423–426.
2. Blumenstein BA, Lamm DL, Jewett MA *et al.* (1990) Effect of colony-forming unit dose of Connaught BCG outcome to immunotherapy in superficial bladder cancer: a SWOG study. *J Urol* **143**: 340A, Abstract 608.
3. de Boer EC, Steerenberg PA, Van der Meijden APM *et al.* (1992) Impaired immune response by isoniazid treatment during intravesical BCG administration in the guinea pig. *J Urol* **148**: 1577–1582.

4. Bohle A, Rusch-Gerofs S, Ulmer AJ, Braasch H and Jocham D (1996) The effect of lubricants on viability of Bacille Calmette–Guérin for intravesical immunotherapy against bladder cancer. *J Urol* **155**: 1892–1896.
5. Bouffioux C (1991) Intravesical adjuvant treatment in superficial bladder cancer. A review of the question after fifteen years of experience with the EORTC Genitourinary Group. *Scan J Urol Nephrol Suppl* **138**: 167–177.
6. Bowyer L, Hall RR, Reading J and Marsh MM (1995) The persistence of Bacille Calmette–Guérin in the bladder after intravesical treatment for bladder cancer. *Br J Urol* **75**: 188–192.
7. Brosman SA (1982) Experience with Bacillus Calmette–Guérin in patients with superficial bladder carcinoma. *J Urol* **128**: 27–30.
8. Catalona WJ, Hudson M, Liss A, Gillen DP *et al.* (1987) Risk and benefits of repeated courses of intravesical Bacillus Calmette–Guérin therapy for superficial bladder cancer. *J Urol* **137**: 220–224.
9. Colleen S, Elfving P, al Khalifa M *et al.* (1997) The impact of concomitant isoniazid (INH) administration on side effects and anti-tumour effect of Bacillus Calmette–Guérin (BCG). *J Urol* **157**: 161, Abstract 623.
10. Crispen R (1989) History of BCG and its substrains, in *BCG in Superficial Bladder Cancer: EORTC Genitorurinary Group Monograph 6,* (eds FMS Debruyne, L Denis and APM Van der Meijden), Alan R Liss, New York, pp. 35–50.
11. Debruyne FMJ, Van der Meijden APM and Geboers ADH (1988) BCG-RIVM versus mitomycin C instravesical therapy in patients with superficial bladder cancer. *Urology* **31** (Suppl.): 20–25.
12. Fellows GJ, Parmar MKB, Grigor KM *et al.* (1994) Marker tumour response to Evans and Pasteur bacille Calmette–Guérin in multiple recurrent pTa, pT1 bladder tumours: report from the Medical Research Council Sub-group on superficial bladder cancer (Urological Working Party). *Br J Urol* **73**: 639–644.
13. Gheorghiu M and Lagrange PH (1983) Viability, heat stability and immunogenicity of four BCG vaccines prepared from four different BCG strains. *Ann Immunol (Paris)* **134C**: 125–147.
14. Guerin C (1980) The history of BCG, in *BCG Vaccine: Tuberculosis–Cancer,* (ed. SR Rosenthal), PSG, Littlejohn, p. 37.
15. Herr HW (1985) Durable response of a carcinoma in situ of the renal pelvis to topical bacillus Calmette–Guérin. *J Urol* **134**: 531–532.
16. Kaisary AV (1987) Intravesical BCG therapy in the management of multiple superficial bladder carcinoma. Comparison between Glaxo and Pasteur strains. *Br J Urol* **59**: 554–558.
17. Kelley DR, Ratliff TL, Catalona WJ *et al.* (1985) Intravesical Bacillus Calmette–Guérin therapy for superfical bladder cancer: effect of Bacillus Calmette–Guérin viability on treatment results. *J Urol* **134**: 48–53.
18. de Kernion JJB, Golub SH, Gupta RK, Silverstein M and Morton DL (1975)

Successful transurethral interlesional BCG therapy of bladder melanoma. *Cancer* **36**: 1662–1664.

19. Krege S, Rubben H and members of the Registry of Urinary Tract Tumours (1991) Prospective randomised study of adjuvant therapy after complete resection of superficial bladder cancer: mitomycin C versus BCG Connaught versus TUR alone. *J Urol* **145**: 426A, Abstract 856.
20. Kurth KH, Denis L, ten Kate FJW *et al.* (1992) Prognostic factors in superficial bladder tumours. *Probl Urol* **6** (3): 471–483.
21. La Fontaine PD, Middleman BR, Graham SD and Sanders WH (1997) Incidence of granulomatous prostatitis and acid-fast bacilli after intravesical BCG immunotherapy. *Urology* **49**: 363–366.
22. Lamm DL (1992) Long term results of intravesical therapy for superficial bladder cancer. *Urol Clin N Am* **19** (3): 573–580.
23. Lamm D, Crawford ED, Blumenstein B *et al.* (1992) Maintenance BCG immunotherapy of superficial bladder cancer: a randomised prospective Southwest Oncology Group study. *J Urol* **147**: 274A, Abstract 242.
24. Lamm DL, Blumenstein BA, Crawford ED, Montie JE and Scardino P (1991) A randomised trial of intravesical doxorubicin and immunotherapy with bacille Calmette–Guérin for transitional cell carcinoma of the bladder. *N Engl J Med* **325**: 1205–1209.
25. Lamm DL, Blumenstein BA, Sarosdy M *et al.* (1997) Significant long term patient benefit with BCG maintenance therapy: a South-west Oncology Group study. *J Urol* **157**: 213, Abstract 831.
26. Lamm DL, Crawford ED, Blumenstein BA *et al.* (1993) SWOG 8795: a randomised comparison of bacillus Calmette–Guérin and mitomycin C prophylaxis in stage Ta and T1 transitional cell bladder carcinoma of the bladder. *J Urol,* **149**: 282A, Abstract 275.
27. Lamm DL, Harris SC and Gittes RF (1977) Bacillus Calmette–Guérin and dinitro-chlorobenzene immunotherapy of chemically induced bladder tumors. *Invest Urol* **14**: 369–372.
28. Lamm DL, de Haven JI, Shriver J *et al.* (1990) A randomised prospective comparison of oral versus intravesical and percutaneous BCG for superficial bladder cancer. *J Urol* **144**: 65–67.
29. Lamm DL, Pilot S and Sardosi MF (1986) Oral Bacillus Calmette–Guérin versus intravesical plus percutaneous Bacillus Calmette–Guérin in superficial transitional cell carcinoma. *J Urol* **135**: 186A, Abstract 328.
30. Lamm DL, Stogdill VD, Stogdill BJ and Crispen RG (1986) Complications of bacillus Calmette–Guérin immunotherapy in 1278 patients with bladder cancer. *J Urol* **135**: 272–274.
31. Lamm DL, Van der Meijden APM, Morales A *et al.* (1992) Incidence and treatment of complications of bacillus Calmette–Guérin intravesical therapy in superficial bladder cancer. *J Urol* **147**: 596–600.

32. Luftenegger W, Ackerman DK and Futterlieb A *et al.* (1996) Intravesical versus intravesical plus intradermal BCG: a prospective randomised study in patients with recurrent superficial bladder tumours. *J Urol* **155**: 483–487.
33. Martinez-Pineiro JA (1984) BCG vaccine in superficial bladder tumors: eight years later. *Eur Urol* **10**: 93–100.
34. Martinez-Pineiro JA, Flores N, Isorna S *et al.* (1996) Comparison between a standard BCG dose (81mg.) versus a threefold reduced dose (27mg.) in superficial bladder cancer. *J Urol* **155**: 493A, Abstract 370.
35. Martinez-Pineiro JA and Muntanola P (1977) Non-specific immunotherapy with BCG vaccine in bladder tumours. A preliminary report. *Eur Urol* **3**: 11–22.
36. Mathe G, Amiel J, Schwartzenberg L *et al.* (1969) Immunotherapy for acute lymphoblastic leukemia. *Lancet* **1**: 697–699.
37. Morales A, Eidinger D and Bruce AW (1976) Intracavity Bacillus Calmette–Guérin in the treatment of superficial bladder tumors. *J Urol* **116**: 180–183.
38. Morton DL, Eilber FR and Malmgren RA (1970) Immunologic factors which influence response to immunotherapy in malignant melanoma. *Surgery* **68**: 158–163.
39. Nadler RB, Catalona WJ, Hudon M'Liss A and Ratliff TL (1994) Durability of the tumor-free response for intravesical Bacillus Calmette–Guérin therapy. *J Urol* **152**: 367.
40. Nauts HC, Fowler GA and Bogatka F (1953) A review of the influence of bacterial infection and of bacterial products (Coley's toxins) on malignant tumours in man. *Acta Med Scand* **145** (Suppl. 276): 1–103.
41. Netto NR, Jr and Lemos CG (1984) Bacillus Calmette–Guérin immunotherapy of infiltrating bladder cancer. *J Urol* **132**: 675–677.
42. Nickel JC, Morales A, Heaton JPW *et al.* (1985) Ultrasound study of the interaction of BCG with bladder mucosa after intravesical treatment of bladder cancer. *J Urol* **133**: 268A, Abstract 620.
43. Oosterlinck W, Kurth KH, Schröder F *et al.* (1993) A prospective European Organisation for Research and Treatment of Cancer Genitourinary Group randomised trial comparing transurethral resection followed by a single intravesical instillation of epirubicin or water in single stage Ta, T1 papillary carcinoma of the bladder. *J Urol* **149**: 749–752.
44. Pagano F, Bassi P, Milani C *et al.* (1991) A low dose bacillus Calmette–Guérin regimen in superficial bladder cancer therapy: is it effective? *J Urol* **146**: 32–35.
45. Pearl R (1929) Cancer and tuberculosis. *Am J Hyg* **9**: 97.
46. Pinsky CM, Hirshault Y and Oettgen H (1973) Treatment of malignant melanoma by intra-tumoral injection of BCG. *Natl Cancer Inst* Monograph 39: 225–228.
47. Ratliff TL (1992) Role of the immune response in BCG for bladder cancer. *Eur Urol* **21** (Suppl. 2): 17–21.

48. Ratliff TL, Palmer JO, McGarr JA and Brown EJ (1987) Intravesical bacillus Calmette–Guérin therapy for murine bladder tumors: initiation of the response by fibronectin-mediated attachment of bacillus Calmette–Guérin. *Cancer Res* **47**: 1762–1766.
49. Rosenthal SR (1938) The general tissue and humeral response to an avirulent tubercle bacillus, in *Illinois Medical and Dental Monographs,* University of Illinois Press, Urbana.
50. Rosenthal SR (1980) *BCG Vaccine: Tuberculosis–Cancer,* PSG, Littleton, Foreword.
51. Steg A, Leleu C, Debré B, Boccon-Gibod L and Sicard D (1989) Systemic bacillus Calmette–Guérin infection. BCG-itis in patients treated by intravesical bacillus Calmette–Guérin therapy for bladder cancer. *Eur Urol* **16**: 161–164.
52. Tolley DA, Parmar MKB, Grigor KM, Lallemand G and the MRC Superficial Bladder Cancer Working Party (1996) The effect of intravesical mitomycin C on the recurrence of newly diagnosed superficial bladder cancer: a further report with seven years follow up. *J Urol* **155**: 1233–1238.
53. Van der Meijden APM, Steerenberg PA, de Jong WH *et al.* (1986) The effects of intravesical and intradermal application of a new BCG on the dog bladder. *Urol Res* **14**: 207–210.
54. Van der Meijden APM, van Klingeren B, Steerenberg PA *et al.* (1991) The possible influence of antibiotics on results of bacillus Calmette–Guérin intravesical therapy for superficial bladder cancer. *J Urol* **146**: 444–446.
55. Vegt PDJ, Van der Meijden APM, Sylvester R *et al.* (1997) Does isoniazid reduce side effects of intravesical Bacillus Calmette–Guérin therapy in superficial bladder cancer? Interim results of EORTC protocol 30911. *J Urol* **157**: 1246–1249.
56. van Vooren JP, Zlotta AR, Drowart A *et al.* (1997) Culture filtrate from bacillus Calmette–Guérin (BCG) is as effective as live BCG to enhance lymphocyte-mediated bladder tumour cell killing in man and in mice. *J Urol* **157**: 383A, Abstract 1501.
57. Witjes JA, Van der Meijden APM, Witjes WPJ and the members of the Dutch South-East Co-operative Urological Group (1993) A randomised propsective study comparing intravesical instillations of mitomycin C, BCG-Tice and BCG-RIVM in pTa pT1 tumours and primary carcinoma in situ of the urinary bladder. *Eur J Cancer* **29A**: 1672–1676.
58. Zbar B, Bernstein ID, Bartlett GL, Hanna MG, Jr and Rapp HJ (1972) Immunotherapy of cancer: regression of intradermal tumors and prevention of growth of lymph node metastases after intralesional injection of living mycobacterium bovis. *J Natl Cancer Inst* **49**: 119–130.
59. Zlotta R, Drowart A, van Vooren JP *et al.* (1997) The need of a maintenance BCG therapy is sugggested by the inability of intravesical BCG to promote a long term systemic immune T-cell activation. *J Urol* **157**: 383A, Abstract 1499.

Commentary

Donald L. Lamm

Adrian van der Meijden has presented an erudite and authoritative review of BCG immunotherapy for bladder cancer. This review can serve as a guide to the use of BCG and, as such, will significantly improve the management of bladder cancer. Rather than reiterate the important principles presented, let me raise some points of controversy to stimulate thought, further research, and hopefully further progress in BCG immunotherapy.

The principles of BCG immunotherapy had been largely defined prior to Dr Morales' successful initiation of BCG in the treatment of bladder cancer. Dr van der Meijden has listed and discussed these important principles: small tumour burden, direct juxtaposition of BCG and tumour cells, adequate BCG dose, as well as localized tumour and tumour antigenicity. Remarkably, while 100 000 is the maximum number of cells that can be consistently eradicated using BCG immunotherapy in mouse models, sizeable tumours (as large as 1–2 cm) have been observed to regress following intravesical BCG administration. It is clear that the immune stimulation induced by BCG is primarily a local phenomenon, but effective systemic responses have also been reported. For example, on rare occasions intralesional injection of BCG into cutaneous melanoma can result in the regression of uninjected tumours. Adequate BCG dose is essential but the therapeutic window of 3×10^8 or less to more than 1×10^9 CFU is impressively wide. As pointed out in the chapter, intravesical BCG administration most effectively fulfils the requirements for optimal BCG instillation. The high success of intravesical BCG administration makes it very difficult to demonstrate statistically significant improvements in efficacy with alternative strategies.

PERCUTANEOUS BCG

It is correct that the addition of percutaneous BCG has not significantly improved the results of intravesical treatment, in both our randomized trial[5] and that of Luftnegger *et al.*[6]. Patients who had received both intravesical and percutaneous BCG had a lower, albeit not statistically significant, incidence of tumour recurrence. Therefore, our failure to demonstrate the improved efficacy of percutaneous BCG administration may simply relate to the low power of these studies. Similarly, while we found no evidence of anti-tumour effect using oral BCG[4], Netto and Lemos found that high doses of oral BCG significantly reduced bladder tumour recurrence[7]. Immunization of children and healthy young adults using percutaneous BCG results in a very high percentage of PPD skin test conversion. However, in patients with bladder cancer, only about two-thirds of patients will convert from PPD skin test negative to positive. The previous studies have suggested that PPD skin test conversion does not reliably predict response to BCG therapy, but my unpublished experience in much larger multicentre clinical trials leads me to believe that PPD skin test conversion does significantly increase a patient's chance of remaining free of tumour recurrence. Therefore, our next-generation protocols will use percutaneous BCG administration initially, followed 1–2 weeks later by intravesical instillation.

DOSE AND SCHEDULE OF BCG

Controversy remains regarding the optimal dose of BCG administration and I agree that a quarter dose reduction would be worthy of investigation. However, I do believe that we have conclusively demonstrated that my 3-week maintenance BCG schedule at 3, 16, 12, 18, 24, 30 and 36 months is the current optimal regimen. To summarize, in patients followed for more than 5 years, intravesical chemotherapy reduces tumour recurrence by about 7%. Even suboptimal BCG schedules reduce tumour recurrence by about 20%. To date, treatment with successive 6-week courses of BCG, monthly maintenance BCG, or single instillations at 3-monthly intervals have failed to improve results significantly. In contrast, the 3-week maintenance schedule reduces long-term recurrence by an additional 27% when compared with patients receiving a standard 6-week induction alone. In my opinion the most common mistake made today with BCG immunotherapy is the widespread adoption of the repeated 6-week course regimen for BCG immunotherapy[1]. As noted by Dr van der Meijden, a second 6-week course of BCG increased complete response rate from 56% to 61% (not statistically significant). The second 6-week induction is not based on considerations of tumour biology or host immune response. The risk for tumour recurrence is lifelong in most patients. It is known that the stimulation induced by BCG wanes significantly with time. However, in patients who have been previously immunized with BCG, a second stimulation in these patients typically occurs following the third intravesical instillation and repeated instillations commonly result in immune suppression with the fourth, fifth and sixth instillations. Therefore it is preferable to give only three instillations separated by periods of 3–6 months, depending on the risk of tumour recurrence and the side-effects observed with treatment. Importantly, treatments should be individualized and given to stimulate effective immune response rather than heightened or prolonged toxicity. In patients with increased side-effects, discontinuation of the second or third instillation resulted in no apparent reduction in anti-tumour response. Indeed, only 16% of patients in the South-West Oncology Group Study received the full complement of eight intravesical courses of BCG.

SIDE-EFFECTS

I find it useful to consider BCG immunotherapy as exercise for the immune system. Vigorous exercise can result in pain, but precautions should be taken to avoid any permanent injury. It is very helpful to explain to the patients that BCG induces an immune response that may result in mild urinary frequency, dysuria, low-grade fever and mild malaise. Such symptoms are associated with an improved response to BCG immunotherapy. Indeed, in my unpublished experience, the use of isoniazid to treat patients who have increased side-effects resulted in a statistically significant reduction in tumour recurrence when compared with patients who did not receive isoniazid treatment. These results are to be expected. Administration of excessively high doses of BCG actually suppresses the immune response[8]. In patients who have increased side-effects from BCG, continued growth or excessive attachment of BCG may have occurred. These patients are therefore at risk of receiving excess BCG and treatment with anti-tubercular antibiotics would be expected to improve the results. Fortunately life-threatening systemic reactions to BCG are rare. We originally proposed the use of cycloserine in the management of these patients since most other anti-tubercular antibiotics take several days to inhibit the growth of the organism. In our review of the published cases of BCG sepsis, we have seen no deaths among patients who received cycloserine as part of their initial management for BCG sepsis. However, BCG is in fact relatively resistant to cycloserine. The BCG septic reaction has been considered by some to be largely hypersensitive in nature. Steg *et al.* reported five cases of systemic BCG reaction that were treated successfully with isoniazid, rifampin and prednisone 40 mg daily[9]. In our controlled animal studies we have confirmed that the addition of prednisone to isoniazid and rifampin is superior to cycloserine[2,3]. Since that time we have consistently recommended the use of prednisone, isoniazid and rifampin with good results.

BCG is consistently the optimal treatment of carcinoma in situ and is the treatment of choice for aggressive transitional cell carcinoma that is Grade 3, stage Ta or T1, rapidly recurrent or chemotherapy refractory disease. With appropriate knowledge and experience BCG immunotherapy can be used safely to improve markedly the treatment of superficial bladder cancer.

COMMENTARY REFERENCES

1. Catalona WJ, Hudson MA, Gillen DP *et al.* (1987) Risks and benefits of repeated courses of intravesical Bacillus Calmette–Guérin therapy for superficial bladder cancer. *J Urol* **137**: 220–224.
2. DeHaven JI, Traynelis CT, Riggs DR *et al.* (1992) Antibiotic and steroid therapy of massive systemic Bacillus Calmette–Guérin toxicity. *J Urol* **147**: 738–742.
3. Koukol SC, DeHaven JI, Riggs DR and Lamm DL (1995) Drug therapy of bacillus Calmette–Guérin sepsis. *Urol Res* **22**: 273–275.
4. Lamm DL, DeHaven JI, Shriver J *et al.* (1990) A randomized prospective comparison of oral versus intravesical and percutaneous BCG for superficial bladder cancer. *J Urol* **144**: 64–67.

5. Lamm DL, DeHaven JI, Shriver J *et al.* (1991) Prospective randomized comparison of intravesical with percutaneous Bacillus Calmette–Guérin versus intravesical Bacillus Calmette–Guérin in superficial bladder cancer. *J Urol* **145**: 738–740.
6. Luftnegger W, Ackermann DK, Futterlieb A *et al.* (1996) Intravesical versus intravesical plus intradermal bacillus Calmette–Guérin: a prospective randomized study in patients with recurrent superficial bladder tumors. *J Urol* **155**: 483–487.
7. Netto RR, Jr and Lemos CG (1983) A comparison of treatment methods of prophylaxis of recurrence superficial bladder tumors. *J Urol* **129**: 33–34.
8. Richert DF and Lamm DL (1984) Long term protection in bladder cancer following interlesional immunotherapy. *J Urol* **132**: 570–573.
9. Steg A, Leleu C, Debra B *et al.* (1989) Systemic Bacillus Calmette–Guérin infection in patients treated by intravesical BCG therapy for superficial bladder cancer. *Prog Clin Biol Res* **310**: 325–334.

6

Carcinoma in situ

Gerhard Jakse

INTRODUCTION

Carcinoma in situ (CIS) is a histological entity which was described initially in 1912 by Bowen[4] indicating a flat, reddish lesion of the skin. The characteristic appearance of CIS of the cervix and vagina was defined by Broders[7] in 1932. It was Melicow[68] who described CIS of the urinary bladder in areas adjacent to visible tumours, in 1952. He had already suggested that CIS might be a pre-neoplastic lesion which developed into recurrent bladder cancers. Melicow and Hollowell[69] demonstrated that CIS may be multifocal within the bladder, in the renal pelvis and in the urethra as well. Melick *et al.*[70], Melamed *et al.*[66] and Koss *et al.*[54,55] gave clear evidence that CIS was the precursor lesion of overt solid or papillary transitional cell carcinomas induced by the carcinogen *para*-aminodiphenyl. Moreover, they were able to demonstrate that malignant cells could be detected in the urine of these industrial workers before the tumours could be seen on endoscopy[55]. Utz and co-workers[106,107,108] drew attention to the unusual and frequently aggressive behaviour of CIS. In the following years careful mapping studies comparing cystoscopy with cystectomy showed that CIS occurs alone, with dysplasia or in association with overt tumours[22,39,56].

CLINICAL DEFINITION AND BEHAVIOUR OF CIS

The histology of CIS has been described in Chapter 2. It is now generally accepted that a considerable proportion of bladder cancers developed from a long-lasting continuous spectrum of pre-neoplastic and pre-invasive change in the urothelium, of which CIS is a critical stage[22,29,35,67]. About 20–30% of patients with bladder cancer already have an invasive tumour at first presentation[49]. Some of these patients have symptoms for a considerable period of time suggesting a long period of growth. Others are without symptoms until diagnosis, indicating that the CIS phase must have evolved very rapidly into an invasive neoplasm. Evidence of a pre-existing CIS phase, both temporally and spatially, is extensive[19,22,23,36,48,80,89,97]. The 1993 Antwerp International Bladder Cancer Consensus Conference proposed a definition that separated 'primary' from 'secondary' CIS[60]:

1. **primary CIS:** no previous or concurrent bladder tumour;
2. **secondary CIS:**
   - *concurrent,* CIS found in association with visible tumour;
   - *subsequent,* CIS detected during follow-up of previous TCC.

Most CIS cases are secondary and are asymptomatic, being diagnosed at the same time as an overt tumour or during routine follow-up of previous TCC. Primary CIS occurs less frequently but most are symptomatic[22,89].

The position of CIS in an evolving dysplastic process has been challenged by Orozco *et al.*[81], who summarized two schools of thought from their review of published literature:

1. CIS is an aggressive cancer whose appearance identifies sites of future invasion and adverse outcome.
2. CIS is an indolent lesion that appears *in situ* because it cannot invade, evidence for the latter being its long clinical course and adverse outcome only when a widespread disease.

Supporting the second of these two possibilities Orozco *et al.*[81] cite their experience of 29 patients with primary CIS and 73 with secondary CIS, most of the latter also having poorly differentiated TCC, many of them invasive. Of the 29 primary cases 21 (72%) developed no progression in 5 years; the overall risk of progression was 30% and only 7% died of bladder cancer, 'In most instances a patient presenting with carcinoma in situ will die of another disease.'

In a similar review of 71 patients with CIS (20 primary and 51 secondary) Jenkins *et al.*[46] noted a better prognosis for the primary disease, with 95% 3-year survival when treated conservatively. Riddle *et al.*[89] had observed that patients with localized CIS (a single visible lesion or only one biopsy positive for CIS out of several) had a better prognosis than multifocal CIS. This has also been noted by Ovesen *et al.*[82]: progression occurred in only 1/13 of their patients with primary CIS compared with 27/47 with secondary disease, all treated with BCG.

For many patients CIS is a multifocal disease in the bladder. More than 25% of the bladder mucosa was replaced by CIS in 70 cystectomy specimens evaluated by Farrow and Utz[22]. Farrow *et al.*[20] found isolated clusters of two to three malignant cells with normal looking urothelium up to 4 mm away from areas of CIS. Furthermore, coherent sheets of malignant cells may expand laterally into adjacent normal urothelium, thus partially explaining the extension of CIS into the ureter, posterior urethra and even the seminal vesicles[43].

In a remarkable Danish study[115] of the natural history of CIS, 31 patients with Grade 3 CIS were followed by intensive surveillance but received no treatment other than TUR unless severe bladder symptoms or muscle invasion occurred. Twenty-six of the 31 patients had previous or concurrent T1 TCC (secondary CIS) and with 14 years follow-up it became evident that CIS was highly predictive of eventual muscle invasion. However, 7/31 died of unrelated disease, and of those who progressed, half did not do so for 8 years. The authors drew two conclusions. Although the long-term outlook of CIS was very serious, 'many patients may live a comfortable life with their bladder intact'. Second, in assessing the efficacy of treatment for CIS, long-term follow-up is essential.

Orozco *et al.*[81] were of the opinion that CIS of itself is not a particularly hazardous disease; it is the associated T1 and G3 cancers that so often arise in association with CIS that are the cause of trouble later. Current practice reveals, however, that many urologists disagree, to the extent that immediate cystectomy is recommended even for primary CIS. These differing views will reflect personal experience to a certain extent and suggest that CIS is indeed 'a neoplasm of uncertain biologic behaviour'[81]. Weinstein *et al.*[112] termed the disease the 'pathobiology of a paradox'. The explanation probably lies in factors such as those demonstrated by Hofstaeder *et al.*[32,33] namely, that areas of CIS are heterogeneous and consist of a different amount of normal (diploid) and aneuploid cells. CIS must be seen as a conglomerate of cell populations of various differentiation, clonally related, but phenotypically distinct[80]. This fact may be reason why the clinical course of CIS is so variable. Immunopathological studies demonstrated changes of antigenicity of the urothelial cell membrane analogous to those seen in overt papillary or solid bladder cancers[13,17,34,111]. Moreover, it was shown that the lymphoid infiltration of the lamina propria shows a reduction of T-helper cells and increase of T-suppressor cells compared to normal[39]. Furthermore growth factors such as epidermal growth factor (EGF) have an important role in progression of high-grade urothelial malignancies[71,78]. The amount of EGF receptor may differentiate highly aggressive CIS from those cases with a prolonged course of the disease.

SYMPTOMS AND DIAGNOSIS

Symptoms

Riddle *et al.*[89] described CIS as causing symptoms of 'bladder outflow obstruction' but usually associated with **penile, perineal or suprapubic pain**. Utz *et al.*[106] drew attention to the fact that **bladder irritability** may be the first sign of CIS or invasive TCC. Utz and Zincke[107] reviewed 486 patients who were treated for so-called interstitial cystitis; in 23% of the men and 1.3% of the women CIS was the ultimate and correct diagnosis. As a basic rule, any man with unexplained bladder instability should be suspected of having CIS. The intensity of symptoms usually parallels the extent of the disease[20]. In most patients symptoms such as pain, frequency and urgency are relieved immediately if visible areas of CIS are destroyed endoscopically, or resolve more slowly but completely following successful intravesical treatment. With the increasing awareness of clinicians of secondary CIS and the use of urine and bladder washing cytology and mucosal biopsies, cystoscopically invisible CIS is being diagnosed more frequently.

In about 70% of CIS patients **microhaematuria** is present[20,35]. This, however, is only of importance in patients without a previous history of bladder cancer, whose investigation has been discussed in Chapter 1.

Cytology

Urinary cytology is the most useful tool in the diagnosis and follow-up of patients with CIS. The cohesiveness of urothelial cells is reduced significantly due to the loss of intracellular junctions. Malignant cells are shed easily and will be found abundantly in the urine[76]. The accuracy of urine cytology for CIS (as distinct from Ta and T1, G1 and G2 tumours) varies between 80 and 100%[18,19,20,21,35,36,55,66,75,106]. Farrow and Utz[22] identified CIS in 190/203 patients who had positive cytology but negative cystoscopy. Additional information may be gained by flow cytometry and automated image analysis[24,28,50,51,85,104]. Badalament *et al.*[1] demonstrated that by means of flow cytometry future tumour recurrence could be predicted much earlier than by conventional cytology and cystoscopy. Furthermore it can be used as a prognostic factor for the efficacy of BCG[1,51]. Whether voided urine or bladder washings provide better results is unclear at present[18,57,76]. In patients with symptoms suggestive of CIS, urine cytology should be examined in case it is positive when bladder biopsies are negative[60]. Routine cytology of the upper urinary tract before starting organ-preserving therapy is debatable and only usually employed to clarify a suspicious lesion seen on intravenous urography (IVU).

Cystoscopy

The classical description of CIS on cystoscopy is of an area of mucosa that is reddish, velvety and granular, due to increased lymphoid infiltration of the lamina propria and significant neoangiogenesis[108]. Most of the time CIS is patchy and ill-defined and sometimes bleeds easily, thus it is important to inspect the bladder when only half distended. Sophisticated methods such as tetracycline ultraviolet fluorescence[113], selective surface staining by methylene blue[25] or laser detection[60] may be helpful but have gained no clinical acceptance. Recent developments in fluorescence cystoscopy using α-amino-levulinic-acid (ALA) are promising, since it seems that there is a selective uptake not only in the visible tumour but also in pre-neoplastic lesions[58]. Although a xenon light source is necessary, the instruments are less costly than lasers. However, as noted in Chapter 2, this technique may lead to overdiagnosis of mild and moderate dysplasia and at the present time ALA is not approved by the authorities for medical use. Further studies are currently supported by the Urological Group of the German Cancer Society (AUO-DGK) to establish the true benefit of this diagnostic method.

Biopsy

The problem posed by pre-symptomatic primary CIS, as may occur in workers exposed to industrial carcinogens, and asymptomatic secondary CIS, is that it is usually invisible to cystoscopy; cytology is positive but cystoscopy normal. The dilemma was 'discovered' by Melik *et al.*[70], Koss *et al.*[54,55] and Melamed *et al.*[66] who investigated more than 500 workers exposed to benzidine. It became evident that almost all patients with positive cytology and normal cystoscopic findings eventually developed visible tumours. The time to this event varied significantly. These findings were recently updated by Crosby *et al.*[11]. For these rare patients, multiple random biopsies of the normal-looking epithelium, repeated if necessary at regular intervals, are the only alternative to awaiting the development of a visible tumour.

The role of random biopsies at the time of diagnosis or during follow-up of Ta and T1 bladder tumours has been discussed in Chapter 1. Their importance in patients with pT1 G3 tumours[42] has been emphasized by Vicente *et al.*[109] who demonstrated that, following treatment with intravesical chemo- or immunotherapy, T1 G3 tumours associated with CIS behaved quite differently from those without. Patients with CIS progressed in 65% to muscle invasive cancer; those without progressed only in 10%. Similar results were reported by Solsona *et al.*[98] and Nogueira-March[79].

The need for routine multiple biopsies to document the extent of disease in patients known to have CIS[36], and to evaluate response to conservative treatment of CIS, is also well established. Similarly, the need to confirm an abnormal finding in only one biopsy of normal-looking urothelium has been considered in Chapter 1.

Imaging

Intravenous urography (IVU) should be done on a routine basis at the time of diagnosing CIS and at intervals during follow-up to exclude upper urinary tract tumours. Solsona *et al.*[99] reviewed the association of upper-tract TCC with CIS. Of 138 patients with CIS, 4.3% had upper-tract tumours at diagnosis. During follow-up 34 (25%) developed upper-tract tumours after a mean interval of 38 months. This upper-tract recurrence rate was significantly higher in patients with CIS than patients with Ta T1 tumours, and was also significantly higher in patients treated with cystectomy for CIS compared with those undergoing cystectomy for invasive bladder cancer. Furthermore, Herr and Whitmore[31] clearly demonstrated that the incidence of CIS in the distal ureter is similar in patients undergoing cystectomy for CIS compared with patients after BCG therapy for CIS, namely about 30%. Moreover, Zincke *et al.*[116] reported that 9% of patients undergoing cystectomy for CIS will develop upper urinary tract tumours.

Ultrasound, CT scan and MRI are of no importance in CIS patients, since the incidence of lymph node or distant metastases is insignificant, and the sensitivity as well as specificity of these investigations are low.

TREATMENT

The treatment of CIS confined to the bladder is influenced by several factors, such as previous or concurrent papillary or solid tumours, or simple factors concerning the general health of the patient. The following aspects have to be considered especially:

1. CIS is in most cases a panurothelial disease; the natural course of the disease cannot be predicted. Although the lesion by itself contains anaplastic cells, CIS can be quiescent for years, which means we do not know the actual duration of the disease and at what time the lesion may become invasive.
2. The extent of the disease is of importance and can be estimated roughly by endoscopy and multiple biopsies, bearing in mind that endoscopically visible and invisible lesions are associated very closely. Also, that CIS as a solitary lesion may be highly aggressive: we know that about 80% of solitary invasive tumours present themselves already as invasive and these tumours must have had an asymptomatic or minimally symptomatic *in situ* phase[49].
3. CIS associated with Ta and T1 TCC is a bad prognostic sign if TUR is the only treatment[35,86,87]. This natural course is significantly altered by adequate intravesical treatment.
4. Although it is quite clear that factors such as the patient's performance status are important if we are considering cystectomy, the risk of tumour progression may be more crucially important than organ-preserving treatment. Lifelong close follow-up is mandatory and cystectomy may become necessary at a later time.
5. Although several authors have recommended fulguration or TUR for asymptomatic primary CIS as the only treatment[24,89], because of the known unpredictability of its biological behaviour and the availability of effective intravesical therapy, this recommendation is no longer valid.

TUR alone

Koss *et al.*[55] reported that 7/13 asymptomatic patients with primary CIS developed an invasive tumour within 1–13 years. Utz and Farrow[108] reported on 62 patients with CIS treated by fulguration or TUR alone, 37 (60%) of whom developed invasive cancer within 2 years and 24 (38%) of whom died due to progressive disease. Although Riddle *et al.*[89] observed progression in only one of 13 patients with unifocal, minimally symptomatic CIS followed for 7 years after fulguration, 18/23 patients with diffuse, symptomatic CIS developed invasive cancer. Jakse *et al.*[35] treated 20 patients with TUR alone, of whom 11 died due to progressive cancer within 6 months to 4.2 years. Prout *et al.*[87] followed 12 patients with CIS who did not want or could not tolerate intravesical chemotherapy and were treated by TUR alone. In all but one patient radical cystectomy was performed and/or tumour progression occurred within 5 years.

It is certainly true that CIS may be detected while still a small unifocal lesion and that these lesions can be controlled by fulguration or TUR alone. Also, persistently positive urine cytology after TUR, with no visible cystoscopic abnormality, may be followed closely without additional treatment for many years. However, as the risk of progression is high, and unpredictable, I would argue strongly for adjuvant intravesical treatment as long as there are no prognostic indicators to separate the good from the bad.

Radiotherapy

There is very little published evidence about the effectiveness of radiotherapy for CIS. External beam irradiation has been reported to delay tumour progression[20,35] but Riddle *et al.*[89] found that 7/11 patients treated with radiotherapy died of progressive TCC within 1 year.

Intravesical chemotherapy

Intravesical **thiotepa** was found to be effective against CIS by Farrow *et al.* in 1977[20]; 14/20 patients remained tumour-free for up to 67 months. Similar results were reported by the National Bladder Cancer Collaborative Group A[77], Koontz *et al.*[52] and Prout *et al.*[87], who achieved complete remission of CIS in 18/40 patients. Of the responders only 1/18 died of TCC with average follow-up of 66 months, compared with 8/22 non-responders, despite cystectomy in five. **Epodyl** is no longer available but Robinson *et al.*[91] reported only one complete response in six patients with CIS.

Mitomycin C and **doxorubicin (Adriamycin®)** have been used more extensively for CIS. Jauhiainen *et al.*[45] reported complete response in 14/17 patients treated with mitomycin C. Soloway[96] used mitomycin C 40 mg weekly for 8 weeks in 12 patients, five of whom had been treated previously with thiotepa. Another National Bladder Cancer Collaborative Group[53] trial tested mitomycin in patients who had failed thiotepa, and achieved complete response in 9/20 patients. In a more recent study,

Stricker *et al.*[101] have reported a 79% complete response rate following mitomycin, the mean duration of remission being 21 months. However, a lower complete response rate for mitomycin C was reported by Lamm *et al.*[62]. Only 16 of 35 patients achieved complete remission. Dose–response studies have not been reported and a higher dose intensity may improve the overall response rate. Four early phase II trials of doxorubicin reported complete response rates between 9% and 85%[14,15,37,38,59]. In one of these, Jakse *et al.*[37] treated 46 patients with primary and secondary CIS with doxorubicin, 40 mg in 20 ml of saline every 2 weeks or 80 mg diluted in 40 ml of saline monthly. This higher concentration of doxorubicin produced significantly higher response rates[38] and complete remission was obtained in 73%, 76% and 85%, respectively, in concomitant, secondary and primary CIS. After mean follow-up of 45 months, 47% of complete responders had developed a recurrent tumour, with or without progression[41]. Six of the original 46 patients died due to progression, the high mortality being attributed to incorrect estimation of the tumour extent, irregular follow-up, old age or refusal for cystectomy. In the South-West Oncology Group randomized trial[61] comparing BCG with doxorubicin, complete response was observed in 23 (34%) of 67 patients with CIS treated with doxorubicin[62]. An estimated 53% of patients with complete response remained free of disease for 5 years, but it has to be noted that the dose as well as the intensity of the protocol used was significantly less than in other studies[37,38,59].

In a small, randomized study, Solsona *et al.*[98] treated 40 patients with secondary CIS with mitomycin C or doxorubicin. Overall a 70% complete response rate was achieved, 52.5% after the first cycle and an additional 17.5% following a second cycle. Forty-three per cent of responders recurred with an average follow-up of 50 months, but only four patients developed muscle invasion during this time. An interesting study by Kurth *et al.*[59] has suggested that response of CIS to intravesical therapy may be dose dependent. Using intravescial epirubicin with increasing doses of 30 mg, 50 mg and 80 mg, complete response was observed in 3/7, 4/5 and 9/10 patients, respectively.

Intravesical immunotherapy

Carcinoma in situ is the ideal disease to treat with local immunotherapy because:

1. the tumour load is low;
2. the immunotherapy agent is in intimate contact with tumour cells;
3. there is usually a large number of immunocompetent cells in the lamina propria; and
4. the bladder is as immunoreactive as the skin.

Bacillus Calmette–Guérin (BCG) has become the most popular form of immunotherapy for CIS. It induces a local inflammatory reaction and a significant *in situ* immune response, as indicated by the investigations of El-Demiry *et al.*[16], Böhle *et al.*[3] and Jakse *et al.*[40]. BCG appears to act on CIS in a manner similar to its action on papillary bladder carcinoma, by inducing a specific immune reaction as well as tumour destruction by a local inflammatory reaction[12]. As discussed in Chapter 5 it is the long-term effect of the specific immune reaction that can be exploited to fullest advantage in patients with CIS.

Following the earlier reports of success with BCG against papillary tumours[73], Morales reported BCG activity in 5/7 patients with CIS[74] treated with weekly intravesical and percutaneous BCG. Although a dose–response study for BCG was never performed, a cycle of six, weekly instillations of BCG has become established practice for CIS, yielding a complete response rate of about 70%. An immediate second cycle of six instillations increases the total response rate to approximately 80%[8,44,83]. Similar results were reported by Lamm *et al.*[63,64] who gave a boost of three additional instillations at 3 months. Numerous non-randomized studies have demonstrated response rates of this order and have established BCG as the treatment of choice for patients with carcinoma in situ. Steg *et al.*[100] demonstrated that primary and secondary CIS responded equally well to BCG. Similar results were obtained in EORTC protocol 30861[40] in which primary, secondary and concomitant CIS were distinguished. With a total of 103 patients the response rates of the three different types of CIS appeared to be equal[44]. Moreover, it is important to note that a planned second cycle after 4 weeks salvaged approximately 50% of the patients who had not achieved complete response initially. Merz *et al.*[72] also demonstrated that the response of primary and secondary CIS was similar (83% and 81%) although recurrence with a median follow-up of 40 months was 19% for secondary CIS compared with 10% for the primary disease. Overall 19% of patients relapsed within 9 months, of which more than half had invasive cancer. These latter emphasize the need for frequent endoscopic follow-up following initially successful treatment. Another important aspect of BCG treatment is that it may eliminate CIS when previous intravesical chemotherapy has failed. A retrospective multicentre audit by Reitsma *et al.*[88] demonstrated 66% complete response in 36 patients treated previously with thiotepa and/or mitomycin C. A similar response has been documented for patients treated previously with doxorubicin[61]. Conversely, it is of interest that patients who did not respond to BCG could be managed effectively by intravesical doxorubicin, with a reported response rate of 60%[61].

Many studies have demonstrated that a high proportion of patients with CIS in the bladder also have CIS either in the prostatic urethra or lower ureters[6,9,19,41,48,84,94]. Due to the high efficacy of intravesical therapy for CIS and high-grade Ta T1 bladder cancers, TCC of the prostatic urethra is being encountered more frequently during follow-up. Although this information has led to considerable debate about the need to investigate these sites prior to cystectomy and to confirm ‘negative margins’ where the ureters are transected, this factor has been largely ignored when patients have been treated with BCG. Intravesical BCG with and without TUR of the prostate has been used in several small studies which suggest that intravesical BCG can eliminate CIS in the prostatic urethra and prostatic ducts, even when the bladder neck is intact[84]. Bretton *et al.*[6] reported that 44% of patients treated with TUR and BCG suffered local bladder recurrence or had metastatic progression. The cystectomy rate was 30%. Similarly, Schellhammer *et al.*[94] reported a local failure rate of 29% and a cystectomy rate of 41% of 17 patients with CIS. In these conservatively managed patients the prognosis depended mainly on the concurrent bladder cancer rather than on the prostatic mucosal or ductal involvement. Whether BCG has a similar effect on CIS in the lower ureters can only be surmised and the anatomic extent of BCG activity must be assumed to be the bladder itself and tissues immediately adjacent. This assumption was endorsed by Solsona *et al.*[99] and Brosman[9], in whose series of 48 patients the majority of treatment failures occurred outside the bladder, including the upper ureters and renal pelvis. General experience suggests that this reported risk of upper-tract recurrence may be unusually high, but the fact that CIS treatment failures do occur in the prostatic urethra and the upper tracts is noted by virtually all publications that comment on long-term outcome, and emphasizes the need for constant vigilance of the whole urinary tract in patients treated with either intravesical therapy or cystectomy. The possibility of treating upper-tract CIS by percutaneous perfusion with BCG[103] should not be overlooked.

The other major cause of concern is the likelihood of relapse of CIS within the bladder and its progression to muscle invasion, despite the apparent complete elimination of disease by BCG or other intravesical treatment in the short term. Reported experience has been rather variable. No studies have yet reported the very long-term follow-up recommended by Wolf *et al.*[115] before the true benefit of BCG can be evaluated. In a review by Bowyer *et al.*[5] of 54 patients treated with BCG, 26% with primary CIS (9/35) and 53% with secondary CIS (10/19) had progressed by 3 years. In Brosman's study[8] of 48 complete responses after BCG, only 42% of patients (20/48) remained tumour-free beyond 2 years. More recently, in the South-West Oncology Group[61] trial that compared BCG with doxorubicin, 44.7% of patients treated with BCG were predicted to be disease-free at 5 years. In a non-randomized study of 60 patients with primary or secondary CIS treated with two 6-week cycles of BCG (the second only as necessary), Ovesen[82] reported a 64% complete response rate with relapse occurring in only 12%, so that 52% of patients remained in complete remission after 4 years. Benefit was most striking in primary CIS, where muscle invasion occurred in only 8% compared with 57% for secondary CIS. These authors also noted that the likelihood of progression was related to the initial success of BCG therapy; progression occurred in 40% of patients who responded to one cycle of BCG, 62% of those treated successfully with two cycles and 89% in those patients in whom two cycles of BCG failed to eliminate CIS. In EORTC protocol 30861[44], 25 (32%) of 77 patients have recurred within the bladder after average follow-up of 4 years. In addition, seven patients have developed local regional extravesical disease and two metastatic disease. The Dutch randomized trial[26] that compared mitomycin C with BCG-RIVM and BCG-Tice included 52 patients with primary or secondary CIS. Numbers were too small to permit any comparisons between the three treatments, but the overall complete response rate was 65% for patients with CIS. Eighteen per cent progressed and radical cystectomy was performed in 21% after a median follow-up of 29 months. Fifty-six per cent remained disease-free with median follow-up of 45 months. As noted by Brosman[9] BCG appears to be a very effective treatment for CIS that avoids cystectomy, but for about 50% of patients cystectomy is delayed rather than prevented. Those who develop cancer in their renal pelvis or ureters would have developed the recurrence even if early cystectomy had been performed.

Only much longer follow-up of the few multicentre studies that have been conducted will demonstrate how many patients with CIS are actually cured by BCG. Some indication of the long-term outcome following BCG therapy has been provided by the most recent follow-up analysis of the Memorial Sloan-Kettering Cancer Center study[10,30] of 86 patients with high-risk bladder cancer treated between 1978 and 1981. These patients were randomized to treatment by TUR alone or TUR plus BCG. The small number of patients preclude comparisons between the two treatments, and in any case half of the patients randomized to TUR alone received BCG subsequently at the time of relapse. Also, the patients were a mixture of CIS, T1 and a few with a previous history of muscle invasion. None the less, as 81% had CIS it is reasonable to draw some broad conclusions. With median follow-up of surviving patients of 15 years, 53% (46/86) progressed, 36% underwent cystectomy, 21% developed upper-tract tumours at a median of 7.3 years, and 19% developed TCC in the prostatic urethra or ducts. The overall outcome at 15 years revealed 34% of patients dead of bladder cancer, 27% dead of other causes and only 27% alive with an intact functioning bladder. As these authors remark, 'while significant decreases in tumour recurrences in available randomised trials are indisputable, the evidence of durable progression-free survival is less apparent. Most studies that demonstrate a significant impact of BCG on stage progression are remarkable for lack of long term follow up.'[10]

The possibility of improving the long-term success of BCG may be achieved by the administration of **maintenance or 'booster' instillations** following successful initial treatment. The benefits of the 'six + three' regimen recommended by Lamm *et al.*[64] (Chapter 5) appear to apply equally to CIS patients as well as those with papillary tumours. The 5-year disease-free status in patients treated by maintenance BCG was over 75%. For most urologists this schedule has now become accepted as the standard treatment regimen. Whether it is the optimal regimen remains to be seen, but given the serious nature of CIS the toxicity of the standard dose and booster instillations is outweighed by the advantage of bladder preservation in the majority of patients. On the other hand, some urologists and patients alike are concerned by the toxicity of BCG and would welcome ways to reduce this. As discussed in Chapter 5, a number of studies have explored the possibility of reducing toxicity by using reduced doses of BCG. Preliminary results suggest that toxicity may be reduced while efficacy remains unimpaired[2,83]. Further studies are needed with prolonged follow-up in larger numbers of patients, particularly with CIS.

BCG or intravesical chemotherapy?

A Finnish multicentre study[90] compared mitomycin C alone with mitomycin alternating with BCG. Of 68 patients randomized, complete response rates were 45% and 71% at 3 months and 47% and 74% after 2 years for mitomycin alone versus the combination with BCG. However, the statistical power is low and the follow-up is short, therefore the interpretation of these results has to be confirmed by a larger study with longer follow-up. The choice between BCG and chemotherapy has been discussed in considerable detail in Chapter 4. The arguments apply equally to carcinoma in situ. The South-West Oncology Group study[61] suggested a clear advantage for BCG compared with doxorubicin for patients with CIS. EORTC protocol 30906 is currently comparing the benefits of intravesical BCG with intravesical epirubicin, booster instillations of both agents being given at 6-monthly intervals up to 3 years. Although the risk of life-threatening toxicity from BCG is very low, the systemic side-effects affect a high proportion of patients and if the long-term complete remission rate with epirubicin is not significantly less than that for BCG, it would be a practical proposition to use epirubicin as initial treatment (and avoid systemic symptoms thereby), reserving BCG for those who failed to respond to an initial 6-week course of chemotherapy. The result of this study is awaited with interest.

Bropirimine

Bropirimine is an oral immunomodulator whose initial success was reported against carcinoma in situ. The review of Sarosdy *et al.*[93] summarized the findings of clinical trials conducted so far. These used bropirimine 3 g/day for three consecutive days each week for up to 1 year. A requirement for all the studies was that voided urinary cytology should be positive after initial assessment and endoscopic surgical treatment of visible lesions in the bladder. In the phase II trial of bladder CIS, 61% (20/33) of patients had a complete response, including patients who had failed previous BCG therapy. Responses were also seen in 10/21 evaluable patients with upper-tract CIS. The main drawback of bropirimine, which is an interferon inducer, is the significant systemic side-effects, which are similar to those of other interferon therapies. For this reason it was unlikely to become the first-line treatment in CIS, but might be used in BCG failures or CIS of the upper urinary tract. Two further international trials were commenced that tested bropirimine specifically in BCG-resistant and BCG-intolerant CIS, and compared oral bropirimine to intravesical BCG in patients with newly diagnosed bladder CIS. As noted previously, both these trials were closed in 1997 and further clinical trials of bropirimine are not expected in the immediate future.

α-Interferon (α-IFN)

Several phase II studies have investigated the activity of α-IFN in superficial bladder cancer[27,95,105] but experience with CIS has been very limited. In 1990 Glashan[27] reported from a small multicentre study that high-dose intravesical IFN (10 × 8 IU) gave a 43% response rate, compared with only 5% following a lower dose (10 × 7 IU). Local and systemic side-effects were minimal. More recently, a pilot study[114] in 29 patients treated previously by BCG but having failed by intolerance, no response or relapse after initial response, were treated with α-2b interferon for 12 weeks and then monthly for 9 months. Of 22 patients evaluable for response, 10 were disease-free with follow-up less than 1 year. A pilot study by Stricker *et al.*[102] tested a combination of α-2b IFN with BCG in seven patients with CIS using 60 mg BCG with increasing doses of IFN. The combination was well tolerated, six patients responded completely and no tumour progression had occurred by 12 months. The rationale and expected additional benefit of this combination requires further consideration.

Keyhole limpet haemocyanin (KLH)

KLH is a natural product of a water snail. It is a potent antigen and immune stimulating agent. Jurincic-Winkler *et al.*[47] reported a 52% complete response rate in 21 patients with CIS. In contrast to other immunotherapy, the toxicity is negligible. These promising results need to be confirmed by others.

CYSTECTOMY

Intravesical BCG has displaced cystectomy as the initial treatment of choice for most patients with CIS. For patients with primary CIS who have been assessed by resection of visible lesions and multiple biopsies of normal epithelium, and in whom meticulous histological examination shows no evidence of micro-invasion, there can be little argument. However, for patients whose secondary CIS is widespread or associated with multiple pT1 G3 tumours, cystectomy may be preferred. In some of these patients careful examination of the cystectomy specimen[19] reveals unsuspected invasion, sometimes into superficial muscle, which would account in part for the much higher risk of disease progression if such patients are treated conservatively. None the less, in most cases a realistic assessment of the likely outcome of treatment options can be discussed with the patient for whom the trial of one cycle of intravesical BCG is usually an attractive and reasonable choice.

Except in those patients who have had multiple treatments for CIS and previous resections for Ta or T1 papillary tumours, the likelihood of lymph node metastases is low[19,41]. Thus, when performing cystectomy specifically for CIS, a limited pelvic lymphadenectomy for staging purposes only may be all that is required, and a full pelvic dissection as described in Chapter 7 for the treatment of invasive bladder cancer may be omitted. However, the problem of extension of CIS into the lower ureters and the prostatic urethra require modifications to cystectomy. Although *en bloc* simultaneous urethrectomy has been standard practice for all patients with CIS in the past, preservation of the urethra for orthotopic bladder substitution may be possible for some patients, as discussed in Chapter 7. Although suspected micro-invasion may be present, nerve-sparing cystectomy[110] would not usually be contra-indicated. If the urethra is preserved, the results should be similar to radical prostatectomy as resection of the membranous urethra is not usually performed[65]. Several other technical points are particularly relevant to cystectomy for carcinoma in situ[20,41,60]:

- The ureters should be transected above the crossing of the iliac vessels to minimize the risk of residual CIS in the ureter. Many urologists recommend confirmation of a negative ureteric margin by taking frozen sections. However, the difficulties of detecting CIS on frozen section are considerable and the possibility of an incorrect intra-operative report must not be overlooked. In general, if the ureters are divided high enough, frozen section is not necessary. In either circumstance the ureters will be short and unsuitable for ureterosigmoidoscopy.
- The presence of CIS in the prostatic urethra or prostate ducts should be excluded in all patients with CIS planned for cystectomy, by taking TUR biopsies of the prostatic urethra and underlying prostate stroma at 5 and 7 o'clock, extending from the bladder neck to the distal end of the verumontanum[92].
- If the extent or location of CIS indicates the need for simultaneous urethrectomy, this should be performed with great care, preferably through a separate perineal incision, to ensure that the bulbar and membranous urethra is not opened.

ASSESSMENT OF RESPONSE AND FOLLOW-UP

The safety of bladder-conserving treatment for patients with CIS lies in the accuracy of assessing response to intravesical treatment. Although intravesical therapy is used principally for prophylaxis in patients with papillary Ta and T1 TCC, it is used initially with curative intent for CIS. Because of the multicentric nature of CIS, it has to be assumed that endoscopic resection and fulguration is almost certainly incomplete in the majority of patients. Hence the need for intravescial therapy. For the same reason, it is impossible to be 100% certain that intravesical treatment has been successful without examining the whole of the urothelium. However, clinical experience confirms that a normal, visible appearance of the bladder mucosa on cystoscopy, negative biopsies of any suspicious area and at least four mucosal biopsies from normal-looking urothelium, a normal bimanual palpation combined with negative bladder washing cytology (or postoperative voided urine cytology) together provide a sufficiently reliable indication of complete response to treatment. In addition, just as the prostatic urethra should have been biopsied before commencing treatment, it should be re-biopsied at least at the time of the first post-treatment evaluation. Thereafter cystoscopy with bladder washing or voided urine cytology should be performed at 3-monthly intervals for 2 years. Provided the bladder remains normal, less frequent cystoscopy may be considered but because of the long-term risk of recurrence, patients with CIS should probably not be cystoscoped at annual intervals until they have been tumour-free for at least 4 or 5 years.

If bladder washing or voided urine cytology should become positive in the presence of a normal-looking bladder, further urine samples should be examined and if the presence of malignant cells is confirmed, IVU and possibly retrograde ureterograms with 'washings' of the upper urinary tract for cytology should be performed. The presence of positive samples from the upper urinary tract without evidence of mucosal abnormality on either ureteroscopy or contrast imaging may be treated by BCG administered either by the percutaneous or retrograde approach. Initial experience suggests that this treatment can be successful, but careful and thorough follow-up by ureteroscopy will be essential. In some patients, urine cytology remains persistently positive with no abnormality detectable in the upper urinary tract, bladder or urethra. In this circumstance further treatment with BCG may be considered as the time scale for the appearance of the inevitable visible lesions within the bladder cannot be predicted.

REFERENCES

1. Badalament RA, Gay H, Cibas ES *et al.* (1987) Monitoring intravesical Bacillus Calmette–Guerin treatment of superficial bladder carcinoma by postoperative urinary cytology. *J Urol* **138**: 763–765.
2. Bassi P, Milani C, Meneghini A *et al.* (1992) Dose response of bacillus Calmette–Guérin (BCG) in superficial bladder cancer: a phase III randomized trial low-dose vs standard dose regimen. *J Urol* **147**: 273A.
3. Böhle A, Gerdes J, Ulmer AJ *et al.* (1990) Effects of local bacillus Calmette–Guérin therapy in patients with bladder carcinoma on immunocompetent cells of the bladder wall. *J Urol* **144:** 53–58.
4. Bowen JT (1912) Precancerous dermatoses: a study of two cases of chronic atypical epithelial proliferation. *J Cutan Cis* **30**: 241–255.
5. Bowyer L, Hall RR, Reading J and Marsh MM (1995) The persistence of Bacille Calmette–Guérin in the bladder after intravesical treatment for bladder cancer. *Br J Urol* **75**: 188–192.
6. Bretton PR, Herr HW, Whitmore WF *et al.* (1989) Intravesical Bacillus Calmette–Guerin therapy for *in situ* transitional cell carcinoma involving the prostatic urethra. *J Urol* **141**: 853–856.

7. Broders A (1932) Carcinoma in situ contrasted with benign penetrating epithelium. *J Am Med Ass* **99**: 1670–1674.
8. Brosman SS (1985) The use of Bacille Calmette–Guerin in the therapy of bladder carcinoma in situ. *J Urol* **134**: 39–42.
9. Brosman S (1989) The influence of Tice strain BCG treatment in patients with transitional cell carcinoma in situ, in *EORTC Genito-urinary Group Monograph 6: BCG in Superficial Bladder Cancer*, (eds FMJ Debruyne, L Denis and APM van der Meijden), Alan R Liss, New York, pp. 193–205.
10. Cookson MS, Herr HW, Zhang ZF *et al.* (1997) The treated natural history of high risk superficial bladder cancer: fifteen year outcome. *J Urol* **158**: 62–67.
11. Crosby JH, Allsbrook WC Jr, Koss LG *et al.* (1990) Cytologic detection of urothelial cancer and other abnormalities in a cohort of workers exposed to aromatic amines. *Acta Cytol* **35**: 263–268.
12. De Boer EC, De Jong WH, van der Meijden APM *et al.* (1991) Presence of activated lymphocytes in the urine of patients with superficial bladder cancer after intravesical immunotherapy with bacillus Calmette–Guérin. *Cancer Immunol Immunother* **33**: 411–416.
13. Decenzo JM, Howard P and Irish CE (1975) Antigenic deletion and prognosis of patients with a stage A transitional cell bladder carcinoma. *J Urol* **114**: 874–878.
14. Edsmyr F, Berlin IM, Boman J *et al.* (1980) Intravesical therapy with Adriamycin in patients with superficial bladder tumours. *Eur Urol* **6**: 132–136.
15. Edsmyr F, Andersson L and Esposti P (1984) Intravesical chemotherapy of carcinoma in situ in bladder cancer. *Urology* (Suppl.) **34**: 37–39.
16. El-Demiry MIM, Smith G, Ritchie AWS *et al.* (1987) Local immune responses after intravesical BCG treatment of carcinoma in situ. *Br J Urol* **60**: 543–548.
17. Emmott RC, Droller MJ and Javadpour N (1981) Studies of A, B or O (H) surface antigen specificity: carcinoma in situ and non-malignant lesions of the bladder. *J Urol* **125**: 32–35.
18. Esposti PL and Zajicek J (1972) Grading of transitional cell neoplasms of the urinary bladder from smears of bladder washings: a critical review of 326 tumours. *Acta Cytol* **16**: 529–537.
19. Farrow GM, Utz DC and Rife CC (1976) Morphological and clinical observations of patients with early bladder cancer treated with total cystectomy. *Cancer Res* **36**: 2495–2501.
20. Farrow GM, Utz DC and Rife CC (1977) Clinical observations on 69 cases of in situ carcinoma of the urinary bladder. *Cancer Res* **37**: 2794–2802.
21. Farrow GM (1979) Pathologists role in bladder cancer. *Sem Oncol* **6**: 198–206.
22. Farrow GM and Utz DC (1982) Observations on micro-invasive transitional cell carcinoma of the urinary bladder. *Clin Oncol* **1** (2): 609–615.

23. Farrow GM, Barlebo H and Enjoji M (1986) Transitional cell carcinoma in situ. Progress in clinical and biological research, in *Developments in Bladder Cancer*, (eds L Denis, T Niijima, G Prout and F Schroder), Alan R Liss, New York, pp. 85–96.
24. Farsund T, Laerum OD and Hostmark J (1983) Ploidy disturbance of normal-appearing bladder mucosa in patients with urothelial cancer: relationship to morphology. *J Urol* **130**: 1076–1082.
25. Gill WB, Huffman JL, Lyon ES *et al.* (1984) Selective surface staining of bladder tumors by intravesical methylene blue with enhanced endoscopic indentification. *Cancer* **53**: 2724–2427.
26. Gils-Geilen van RJM, Debruyne FMJ, Witjes WPJ *et al.* (1994) Risk factors in carcinoma in situ of the urinary bladder. *Urology* **45**: 581–586.
27. Glashan RW (1990) A randomizied controlled study of intravesical a-2b-Interferon in carcinoma in situ of the bladder. *J Urol* **144**: 658–661.
28. Gustafson H, Tribukait B and Esposti PL (1982) The prognostic value of DNA analysis in primary carcinoma in situ of the urinary bladder. *Scand J Urol Nephrol* **16**: 141–146.
29. Herr HW (1983) Carcinoma in situ of the bladder. *Sem Urol* **1**: 15–22.
30. Herr HW, Pinskiy CM, Whitmore WF *et al.* (1986) Long term effect of intravesical Bacillus Calmette–Guerin on flat carcinoma in situ of the bladder. *J Urol* **135**: 265–267.
31. Herr HW and Whitmore WF (1987) Ureteral carcinoma in situ after successful intravesical therapy for superficial bladder tumors; incidence, possible pathogenesis and management. *J Urol* **138**: 292–294.
32. Hofstaeder F, Jakse G, Lederer B *et al.* (1980) Cytophotometric investigations of DNA content in transitional cell tumors of the bladder: comparison of results with clinical follow-up. *Pathol Res Pract* **167**: 254–264.
33. Hofstaeder F, Delgado R, Jakse G *et al.* (1986) Urothelial dysplasia and carcinoma in situ of the bladder. *Cancer* **57**: 356–361.
34. Jakse G and Hofstaeder F (1978) Further experience with the specific red cell adherence test (SRCA) in bladder cancer. *Eur Urol* **4**: 356–360.
35. Jakse G, Hofstaeder F, Leitner G *et al.* (1980) Carcinoma in situ der Harnblase. Eine diagnostische und therapeutische Herausforderung. *Urologe A* **19**: 93–99.
36. Jakse G, Hofstaeder F and Marberger H (1980) Wert der Harnblasenzytologie und Quadrantenbiopsie bei oberflächlichen Blasenkarzinomen. *Aktuelle Urologie* **11**: 309–313.
37. Jakse G, Hofstaeder F and Marberger H (1981) Intracavity doxorubicin hydrochloride therapy for carcinoma in situ of the bladder. *J Urol* **125**: 185–190.
38. Jakse G (1984) Intravesical chemotherapy for carcinoma in situ of the urinary bladder: 5 years later. *Eur Urol* **10**: 289–293.

39. Jakse G, Rammal E and Hofstaeder F. (1985) Local immune response in preneoplasia and in situ carcinoma of the urinary bladder, in *Testicular Cancer and Other Tumours of the Genitourinary Tract*, (eds M Pavone-Macaluso, PH Smith and MA Bagshaw), Plenum Press, New York, pp. 351–354.
40. Jakse G and members of the EORTC-GU Group (1989) Intravesical instillation of BCG in carcinoma in situ of the urinary bladder. EORTC Protocol 30861, in *BCG in Superficial Bladder Cancer*, (eds DMS Debruyne, L Denis and APM van der Meijden), EORTC Gentioutinary Group, Monograph 6, Alan R Liss, New York, pp. 187–192.
41. Jakse G, Putz A and Feichtinger J (1989) Cystectomy: the treatment of choice in patients with carcinoma in situ of the urinary bladder? *Eur J Surg Oncol* **15**: 211–216.
42. Jakse G, Loidl W, Seeber G *et al.* (1987) Stage T1, grade 3 transitional cell carcinoma of the bladder: an unfavorable tumor? *J Urol* **137**; 39–43.
43. Jakse G, Putz A and Hofstaeder F (1987) Carcinoma in situ of the bladder extending into the seminal vesicles. *J Urol* **137**: 44–45.
44. Jakse G, Hall RR, Bono A *et al.* (1989) Intravesical BCG treatment of carcinoma in situ of the urinary bladder. EORTC phase II protocol 30861. *J Urol* **141** (Suppl.): 229A, Abstract 240.
45. Jauhiainen K, Sotarauta M, Permi J *et al.* (1986) Effect of Mitomycin C and Doxorubicin instillation on carcinoma in situ of the urinary bladder. A Finnish Multicenter Study. *Eur Urol* **12**: 32–37.
46. Jenkins BJ, England HR, Fowler CG *et al.* (1988) Chemotherapy for carcinoma in situ of the bladder. *Br J Urol* **61**: 326–329.
47. Jurnicic-Winkler C, Metz KA, Beuth J *et al.* (1995) Effect of keyhole limpet hemocyanin (KLH) and bacillus Calmette–Guérin (BCG) instillation on carcinoma in situ of the urinary bladder. *Anticancer Res* **15**: 2771–2776.
48. Kakizoe T, Matsumoto K, Nishio Y *et al.* (1984) Analysis of 90 tep-sectioned cytstectomized specimens in bladder cancer. *J Urol* **131**: 467–472.
49. Kaye KW and Lange PH (1982) Mode of presentation of invasive bladder cancer: reassessment of the problem. *J Urol* **128**: 31–33.
50. Klein FA, Herr HW, Whitmore WF *et al.* (1982) An evaluation of automated flow cytometry (FCM) in detection. *Cancer* **50**: 1003–1008.
51. Klein FA, Herr HW and Sogani PC (1982) Detection and follow up of carcinoma of the urinary bladder by flow cytometry. *Cancer* **50**: 389–391.
52. Koontz W, Prout G, Jr and Smith W (1981) The use of intravesical thio-tepa in the management of noninvasive bladder carcinoma in the bladder. *J Urol* **125**: 307–312.
53. Koontz W, Heney NM, Soloway MS *et al.* (1985) Mitomycin C for patients who have failed on thio-tepa. *Urology* **14**: 221–225.
54. Koss LG, Melamed MR, Ricci A *et al.* (1965) Carcinogenesis in the human urinary bladder. Observations after exposure to *para*-aminodiphenyl. *N Engl J Med* **272**: 767–770.

55. Koss LG, Melamed MR and Kelly RE (1969) Further cytologic and histologic studies of bladder lesions in workers exposed to *para*-aminodiphenyl: progress report. *J Natl Cancer Inst* **43**: 233–249.
56. Koss LG, Tiamson EM and Robins MA (1974) Mapping cancerous and precancerous bladder changes. A study of the urothelium in 10 surgically removed bladders. *J Am Med Ass* **227**: 281–285.
57. Koss LG, Deitch G, Ramanathan R *et al.* (1985) Diagnostic value of cytology of voided urine. *Acta Cytol* **29**: 810–814.
58. Kriegmeier M. Stepp H, Baumgarten R *et al.* (1996) Fluorescence transurethral resection of bladder cancer following intravesical application of 5-aminolevulinic acid. *J Urol* **155** (Suppl.): 665A, Abstract 1418.
59. Kurth KH, van der Vigh WJF, ten Kate F *et al.* (1991) Phase 1/2 study of intravesical epirubicin in patients with carcinoma in situ of the bladder. *J Urol* **146**: 1508–1513.
60. Kurth KH, Schelhammer PF, Okajima E *et al.* (1995) Current methods of assessing and treating carcinoma in situ of the bladder with or without involvement of the prostatic urethra. *Int J Urol* **2** (Suppl. 2): 8–22.
61. Lamm DL, Blumenstein BA, Crawford ED *et al.* (1991) A randomised trial of intravesical doxorubicin and immunotherapy with Bacille Calmette–Guérin for transitional cell carcinoma of the bladder. *N Engl J Med* **325**: 1205–1209.
62. Lamm DL, Blumenstein BA, Crawford ED *et al.* (1995) Randomized intergroup comparison of bacillus Calmette–Guérin immunotherapy and Mitomycin C chemotherapy prophylaxis in superficial transitional cell carcinoma of the bladder. A Southwest Oncology Group study. *Urol Oncol* **1**: 119–126.
63. Lamm DL, Crawford ED, Blumenstein BA *et al.* (1992) Maintenance BCG immunotherapy of superficial bladder cancer: a randomized prospective Southwest Oncology Group Study. *J Urol* **147**: 242A.
64. Lamm DL, Blumenstein BA, Sarosdy MF *et al.* (1997) Significant long-term patient benefit with BCG maintenance therapy. *J Urol* **157** (Suppl. 4): 831A.
65. Lepor H, Gregerman M, Crosby R *et al.* (1985) Precise localisation of the autonomic nerves from the pelvic plexus to the corpora cavernosa: a detailed anatomical study of the adult male pelvis. *J Urol* **133**: 207–212.
66. Melamed MR, Koss LG, Ricci A *et al.* (1960) Cystohistological observations on developing carcinoma of the urinary bladder in man. *Cancer* **13**: 67–74.
67. Melamed MR, Voutsa NG and Grabstald H (1964) Natural history and clinical behavior of in situ carcinoma of the human urinary bladder. *Cancer* **17**: 1533–1545.
68. Melicow MM (1952) Histological study of vesical urothelium intervening between gross neoplams in total cystectomy. *J Urol* **68**: 261–279.
69. Melicow MM and Hollowell JW (1952) Intra-urothelial cancer: carcinoma-in-situ, Bowen's disease of the urinary system; discussion of thirty cases. *J Urol* **68**: 763–772.

70. Melick WF, Escue HM, Naryka JJ *et al.* (1955) The first reported cases of human bladder tumors due to a new carcinogen – xenylamine. *J Urol* **74**: 760–766.
71. Mellon K, Wright C, Kelly P *et al.* (1995) Long-term outcome related to epidermal growth factor receptor status in bladder cancer. *J Urol* **153**: 919–925.
72. Merz VW, Marth F, Zingg EJ *et al.* (1993) Analysis of early failures after intravesical instillation therapy with Bacille Calmette–Guérin for carcinoma in situ of the bladder. *J Urol* **149**: 283A.
73. Morales A, Eidinger D and Bruce AW (1976) Intracavity Bacillus Calmette–Guérin in the treatment of superficial bladder tumors. *J Urol* **116**: 180–183.
74. Morales A (1980) Treatment of carcinoma in situ of the bladder with BCG. *Cancer Immunol Immunother* **9**: 69–74.
75. Müller F, Kraft R and Zingg EJ (1985) Exfoliative cytology after transurethral resection of superficial bladder tumors. *Br J Urol* **57**: 530–534.
76. Murphy WM, Soloway MS, Jukkola AF *et al.* (1984) Urinary cytology and bladder cancer. The cellular features of transitional cell neoplasms. *Cancer* **53**: 1555–1565.
77. National Bladder Collaborative Group A (1977) The role of intravesical thiotepa in the management of superficial bladder cancer. *Cancer Res* **37**: 2916–2917.
78. Neal DE, Sharples L, Smith K *et al.* (1990) The epidermal growth factor receptor and the prognosis of bladder cancer. *Cancer* **65**: 1619–1625.
79. Nogueira-March JL (1991) Superficial bladder carcinoma. The pT1 G3 tumours: progression, recurrence and survival. Treatment, in *Evaluation of Chemotherapy in Bladder Cancer*, (eds H Villavicencio and WR Fair), SIU Reports 6, Churchill Livingstone, pp. 45–59.
80. Norming U, Tribukait B, Gustafson H *et al.* (1982) Deoxyribnucleic acid profile and tumor progression in primary carcinoma in situ of the bladder: a study of 63 patients with Grade 3 lesions. *J Urol* **147**: 11–15.
81. Orozco RE, Martin AA and Murphy WM (1994) Carcinoma in situ of the urinary bladder. *Cancer* **74**: 115–122.
82. Ovesen H, Horn T and Steven K (1997) Long term efficacy of intravesical bacillus Calmette–Guérin for carcinoma in situ: relationship of progression to histological response and p53 nuclear accumulation. *J Urol* **157**: 1655–1659.
83. Pagano F, Bassi P, Milani C *et al.* (1991) A low dose Bacillus Calmette–Guerin regimen in superficial bladder cancer therapy: is it effective? *J Urol* **146**: 32–36.
84. Palou J, Xavier B, Laguna P *et al.* (1996) *In situ* transitional cell carcinoma involvement of prostatic urethra: Bacillus Calmette–Guérin therapy without previous transurethral resection of the prostate. *Urology* **47**: 482–484.

85. van der Poel HG, Oosterhof GON, Debruyne FMJ *et al.* (1993) Image analysis in superficial transitional cell carcinoma of the bladder. *Sem Urol* **XI** (3): 164–170.
86. Prout GR, Griffin PP, Daly JJ *et al.* (1986) Carcinoma in situ of the urinary bladder with and without associated vesical neoplasms. *Cancer* **52**: 524–532.
87. Prout GR, Griffin PP and Daly JJ (1987) The outcome of conservative treatment of carcinoma in situ of the bladder. *J Urol* **138**: 766–770.
88. Reitsma DJ, Guinan P, Lamm DL *et al.* (1989) Long-term effect of intravesical Bacillus Calmette–Guerin (BCG). Tice strain on flat carcinoma in situ of the bladder, in *BCG in Superficial Bladder Cancer*, (eds DL Debruyne and APM van der Meijden), Progression in clinical and biological research, Alan R Liss, New York, Vol. 310, pp. 171–185.
89. Riddle PR, Chisholm GD, Trott PA and Pugh RCB (1976) Flat carcinoma in situ of bladder. *Br J Urol* **47**: 829–833.
90. Rintala E, Jauhiainen K, Rajala P *et al.* (1995) Alternating mitomycin C and Bacillus Calmette–Guérin instillation therapy for carcinoma in situ of the bladder. *J Urol* **154**: 2050–2053.
91. Robinson MRG, Shetty MB, Richards B *et al.* (1977) Intravesical epodyl in the management of bladder tumours: combined experience of the Yorkshire Urological Cancer Research Group. *J Urol* **188**: 972–973.
92. Sakomoto N, Tsuneyoshi M, Naito S *et al.* (1993) An adequate sampling of the prostate to identify prostatic involvement by urothelial carcinoma in bladder cancer patients. *J Urol* **149**: 318–321.
93. Sarosdy MF, Lowe BA, Schellhammer PF *et al.* (1996) Oral bropirimine immunotherapy of carcinoma in situ of the bladder: results of a phase II trial. *Urology* **48**: 21–27.
94. Schellhammer PF, Bean MA and Whitmore WF (1977) Prostatic involvement by transitional cell carcinoma: pathogenesis patterns and prognosis. *J Urol* **118**: 399–403.
95. Schmitz-Dräger BJ, Ebert T and Ackerman R (1996) Intravesical treatment of superficial bladder carcinoma with inferons. *World J Urol* **3**: 218–223.
96. Soloway MA (1985) Treatment of superficial bladder cancer with intravesical Mitomycin C: an analysis of immediate and long-term response in 70 patients. *J Urol* **134**: 1107–1109.
97. Solsona E, Iborra I, Vecentericos J *et al.* (1990) Compartimento biologico del carcinoma in situ vesical no tratado. *Arch Esp Urol* **43**: 643–645.
98. Solsona E, Iborra I, Ries JW *et al.* (1996) Extravesical involvement in patients with bladder carcinoma in situ: biological therapy implications. *J Urol* **155**: 895–900.
99. Solsona E, Iborra I, Ricos JV *et al.* (1997) Upper urinary tract involvement in patients with bladder carcinoma in situ [TIS]: its impact on management. *Urology* **49**: 347–352.

100. Steg A, Belas M, Lenen CH *et al.* (1989) Intravesical BCG therapy in patients with superficial bladder tumours. *EORTC GU Group Monograph 6*: pp. 153–160.
101. Stricker PD, Grant ABF, Hosken BM *et al.* (1990) Topical Mitomycin C therapy for carcinoma in situ of the bladder: a follow-up. *J Urol* **143**: 34–35.
102. Stricker PD, Pryor K, Nicholson T *et al.* (1996) Bacillus Calmette–Guérin plus intravesical Interferon alpha-2b in patients with superficial bladder cancer. *Urology* **48**: 957–962.
103. Studer UE, Casanova G, Kraft R *et al.* (1989) Percutaneous BCG perfusion of the upper urinary tract for carcinoma in situ. *J Urol* **142**: 975–977.
104. Tanke BB, Brusee JAM, Schelvis-Knepfle CFHM *et al.* (1984) Automatisierte Urin-Zytoligie mit Hilfe des LEYTAS Bildanalysesystems, in *Das Harnblasenkarzinom* (eds KH Bickler and R Harzman), Springer, Berlin.
105. Torti FM, Shortliffe LD, Williams RD *et al.* (1988) Alpha-interferon in superficial bladder cancer: a Northern California Oncology Group Study. *J Clin Oncol* **6**: 476–481.
106. Utz DC, Hanah KA and Farrow G (1970) The plight of the patient with carcinoma in situ of the bladder. *J Urol* **103**: 160–164.
107. Utz DC and Zincke H (1974) The masquerade of bladder cancer *in situ* as interstitial cystitis. *J Urol* **111**: 160–161.
108. Utz DC and Farrow GM (1980) Management of carcinoma in situ of the bladder. The case for surgical management. *Urol Clin N Am* **7**: 533–542.
109. Vicente J, Laguna MP, Duarte D *et al.* (1991) Carcinoma in situ as a prognostic factor for G3 pT1 bladder tumours. *Br J Urol* **68**: 380–382.
110. Walsh PC, Lepor H and Eggleston JC (1983) Radical prostatectomy with preservation of sexual function: anatomical and pathological considerations. *Prostate* **4**: 473–475.
111. Weinstein RS, Alroy J, Farrow GM *et al.* (1979) Blood group isoantigen detection in carcinoma in situ of the urinary bladder. *Am Cancer Soc* **43**: 661–668.
112. Weinstein R, Miller AW and Pauli BU (1980) Carcinoma in situ: comments on the pathobiology of a paradox. *Urol Clin N Am* **7**: 523–531.
113. Whitmore WF, Bush IM and Esquivel E (1964) Tetracycline ultraviolet fluorescence in bladder carcinoma. *Cancer* **12**: 1528–1532.
114. Williams RD (1996) Pilot study of intravesical alpha-2b Interferon for treatment of bladder carcinoma in situ following BCG failure. *J Urol* **155**: 494A , Abstract 735.
115. Wolf H, Melsen F, Pedersen SE and Nielsen KT Natural history of carcinoma in situ of the urinary bladder. *Scand J Urol Nephrol* (Suppl. 157): 147–151.
116. Zincke H, Garbeff PJ and Beahrs RJ (1984) Upper urinary tract transitional cell cancer after radical cystectomy for bladder cancer. *J Urol* **131**: 50–52.

7

Cystectomy, bladder reconstruction, urinary diversion and stomatherapy

William H. Turner and Urs E. Studer

In the USA in 1994 51 200 new cases and 10 600 deaths from bladder cancer were expected and there were 5094 deaths from bladder cancer in England and Wales in 1992[14,135]. Despite a huge recent increase in the knowledge of the genetic and molecular mechanisms of bladder cancer, this has not yet been translated into clinical benefit, either in improved staging or treatment[80,100,151,161,162,164,186,205]. In the absence of new treatment options, there is no dispute that cystectomy is the most effective local treatment for bladder cancer, although it is the most aggressive.

Debate and controversy relate to the **indications** for cystectomy, i.e. a low or high threshold. This is because of uncertainty, lack of predictability or difference of opinion about the natural history of bladder cancer, the progression rate after other treatment modalities, the risks of cystectomy and the quality of life after cystectomy, which depends on the diversion or bladder replacement used.

There is also considerable difference of opinion and practice regarding urinary diversion and bladder substitution. Numerous forms of both have been described and no long-term follow-up data exist yet for continent diversion or orthotopic bladder substitutes, so it is not possible to show advantages of one particular method over the others. Urologists are therefore justified at present in using a method that is successful in their hands and when used on their patients. There are, however, some important common principles that underlie the use of these techniques and common potential short- and long-term complications.

We will attempt to give an outline of the issues involved in cystectomy, urinary diversion and bladder substitution, indicating the opposing views. We will also give our own views and indicate our own management policy.

CYSTECTOMY

Indications and preparation

Cystectomy is generally indicated for patients with muscle-invasive tumours[165,198], although this must be judged by the urologist in view of the characteristics of the tumour, the general condition of the patient and the available facilities. The two other possible primary surgical approaches are radical transurethral resection[58,64] or partial cystectomy[196]. Radical transurethral resection is appropriate for a patient with a small T2 tumour which can be completely resected. A second resection of the tumour site, quadrantic bladder biopsies and prostatic urethral biopsies at 2–3 weeks are mandatory and must all be negative, otherwise we proceed to cystectomy. Partial cystectomy is indicated for solitary tumours situated in the mobile posterior part of the bladder or in the dome and at least 3 cm from the bladder neck, with normal surrounding urothelium allowing resection of a 2–3 cm cuff of bladder. The bladder capacity must also be sufficient to avoid producing a small capacity postoperatively. Two particular indications for partial cystectomy are a urachal tumour arising in the dome[62,67] and a tumour arising in a diverticulum: otherwise the indications are rarely met. A potential problem with partial cystectomy is circumferential spread around the bladder wall. This occurs in relation to the depth of the tumour, and with deep muscle invasion there may be at least 60% circumferential involvement of the bladder microscopically[8].

Multifocal carcinoma in situ (CIS) or single T1 G3 lesions are not indications for primary cystectomy in our view. Intravesical BCG can preserve the bladder in around 80% of these patients[120]. However, patients with persistent or early recurrent T1 G3 disease after a single treatment of BCG, given appropriately, have a high tendency to progression, either within or outwith the bladder[66,120]. Such patients should be investigated aggressively to determine the appropriate surgical therapy[73], because the yield from a second course of BCG is low[120].

The upper tracts should be assessed by intravenous urography and any lesion should be managed on its own merits: this may or may not influence the form of urinary diversion deployed after cystectomy. The prostatic urethra is biopsied preoperatively, bilaterally at the level of the verumontanum. Invasive cancer or CIS here is an indication for urethrectomy and consequently contra-indicates orthotopic bladder substitution.

Candidates for cystectomy must be operable and without evidence of lymphatic or distant metastases (N0, M0). We perform biplanar chest radiography, liver ultrasound, bone scan in patients with raised alkaline phosphatase and computed tomography of the pelvis. Although the accuracy of loco-regional staging with computed tomography is less than ideal, magnetic resonance imaging has not yet been shown to be preferable[21,74,154,206]. CT-guided aspiration cytology of suspicious pelvic lymph nodes is performed. We judge suitability on biological rather than chronological grounds.

Patients with demonstrable pelvic lymph node metastases (N+) diagnosed preoperatively are offered MVAC (methotrexate, vinblastine, Adriamycin®, cisplatin) or CMV (cisplatin, methotrexate and vinblastine) chemotherapy[207] and complete responders are advised to have a radical cystectomy. This is because both the presence of residual carcinoma in the bladder, despite negative biopsies and cytology, is common and the risk of relapse without cystectomy is high[156,174,175]. Furthermore, patients found to have histologically proven lymph node metastases (pN+) after radical cystectomy and whose renal function allows, are offered adjuvant MVAC to try to improve their chances of survival.

Preoperative preparation

All patients, regardless of the planned urinary diversion, are seen by a stoma nurse and wear a stoma bag for 24 hours to allow optimal assessment of a stoma site, which is then marked indelibly. An enema on the evening before surgery, 48 hours of perioperative ampicillin, gentamicin and metronidazole and subcutaneous heparin, from after the insertion of an epidural catheter with anaesthesia until fully mobile postoperatively, are given. The patient is positioned with the hips slightly hyperextended and slightly abducted and a 22Ch catheter is placed (to aid palpation of the urethra during surgery).

Surgical technique

The steps of radical cystectomy are:

1. laparotomy and exclusion of disseminated intra-abdominal disease;
2. pelvic lymphadenectomy;
3. ligation and division of the anterolateral and dorsolateral vesical pedicles and ligation of Santorini's plexus;
4. mobilization of the ureters and transection at the level of the iliac vessels;
5. incision of the peritoneum in the rectovesical pouch and blunt mobilization of the bladder posteriorly between Denonvillier's fascia and the rectum;
6. ligation and transection of the dorsomedial pedicles;
7. transection of Santorini's plexus and the membranous urethra; and
8. retrograde mobilization of the prostate and excision of the specimen.

Lymphadenectomy

Pelvic lymphadenectomy involves meticulous removal of all lymphatic and fibroareolar tissue within the following boundaries: laterally, the genitofemoral nerve; distally, the femoral canal; proximally, the bifurcation of the common iliac artery; inferiorly, the side wall and floor of the obturator fossa down to the level of the superior and inferior vesical arteries; and medially, the side wall of the bladder. The obturator vessels are resected after identification and preservation of the obturator nerve. We excise nodal tissue from the external iliac, internal iliac and obturator sites separately to facilitate precise assessment of the location of any unsuspected nodal involvement. We do not dissect proximally to the iliac bifurcation for two reasons: first, the prognosis with proximal nodal involvement is dismal and this probably does not occur without distal involvement[172] and, second, dissection medial to the ureters could endanger the sympathetic supply of the membranous urethra and impair continence following orthotopic bladder substitution.

Pelvic lymphadenectomy requires around 30–45 minutes and we do it for three reasons. First, some patients with lymph node involvement might be cured if these nodes are removed[92,94,195]. Second, accurate staging is only possible after lymphadenectomy and, third, a far more anatomical cystectomy is possible after a thorough lymphadenectomy and this allows safer dissection, particularly of the vascular pedicles.

The presence of gross lymph node metastases has been viewed as evidence of systemic disease and survival amongst such patients, despite a meticulous lymphadenectomy, is rare[192,204]. However, 5-year survivals of 30% or more following radical cystectomy and pelvic lymphadenectomy for N+ disease have been reported[34,92,213]. Most of these patients had either microscopic disease and/or disease only in one or two lymph nodes and these patients with low-volume nodal disease are targetted when performing lymphadenectomy. However, since no details of the site of nodal metastasis within the pelvis were given, it is possibile that the long-term survivors were those whose positive nodes were close to the bladder. Only in this case would a less radical lymphadenectomy have been sufficient. But because there is no clear evidence to support this idea, a meticulous lymphadenectomy, as described above, remains necessary. The possibility that some of Skinner's survivors with microscopic nodal metastases had benefited from chemotherapy cannot be excluded.

Therefore for these reasons, despite the absence of randomized, controlled data showing an advantage, we continue to perform a meticulous lymphadenectomy, as described above, while recognizing that for our cystectomy population as a whole, it probably has only limited impact on survival[65].

If there is gross, unsuspected intraoperative nodal disease, then we do not proceed with surgery and treat the patient with chemotherapy. If there is limited nodal involvement, the patient has significant bladder symptoms and cystectomy with resection of a margin of uninvolved tissue is possible, then we proceed with surgery and give postoperative chemotherapy.

Nerve-sparing cystectomy

This has been popularized by Walsh and colleagues[158]. In order to preserve erection, the neurovascular bundle must not only be preserved dorsolateral to the prostate as described by Walsh, but also more cranially, in the angle between the prostate, bladder base and seminal vesicle. This requires ligation and division of the dorsomedial pedicle close to the seminal vesicle, more ventrally than usually done. This reduces the excision margin and potentially the radicality of resection. Furthermore, cadaver studies suggested that this may compromise, in particular, the completeness of lymph node excision[143]. However, in an admittedly low-risk group of 76 patients (47 of whose tumours were stage T2 or less), the actuarial 5-year local recurrence rate was only 7.6%[18]. No patient had positive margins at the site of nerve-sparing and no patient with other positive margins had local recurrence. Our view is that it seems unwise to spare the bundle on the affected side of a unilateral tumour and so it must be completely resected on that side, but we attempt to spare the bundle on the contralateral side. Nerve-sparing may have more impact on continence after orthotopic bladder substitution, than it does on potency.

In an attempt to preserve maximal urethral length, having first fully mobilized the prostatic apex on both sides of the urethra, we transect the urethra proximally within the ring that the prostatic apex makes around the urethra, rather than flush at the distal limit of the apex.

Ureteric carcinoma in situ and carcinoma

Ureteric CIS occurs in around 9% of patients having cystectomy, but the likelihood of subsequent invasive ureteric cancer in these patients seems to be only 3–4%[98]. None the less, it seems intuitively wise to resect the ureters at a safe distance from the bladder and to exclude ureteric CIS by frozen section, so that ureteric anastomoses are done using healthy ureter. Ureteric or renal pelvic cancer develops in about 3% of patients after cystectomy, but seems to be more common in those with carcinoma in situ of the bladder, those with involved intravesical ureters and those with urethral involvement[104,220]. The prognosis in patients with upper-tract cancer after cystectomy is generally poor and so careful follow-up of patients with the above risk factors is advocated.

The ureteroileal anastomoses either for an ileal conduit or for an orthotopic bladder substitute are done end-to-side with Nesbit's technique[128] using continuous Vicryl®. We use ureteric stents, brought out either through the conduit or through the wall of the bladder substitute and silicon tube drains both to the ureteroileal anastomoses and to the pouch–urethral anastomosis (in the case of a bladder substitute).

Urethrectomy

Synchronous urethrectomy is indicated in patients with either carcinoma or carcinoma in situ in prostatic urethral biopsies[147,218], but in our view neither multifocal bladder tumours[185] nor CIS of the bladder are indications. Overall, the risk of urethral recurrence seems to be between 4% and 10%[9,59,93,153,155]. In 100 of our patients with orthotopic ileal bladder substitutes, 30 had CIS in the cystectomy specimen. Four of these patients had urethral recurrences after a median of 30 months follow-up; two of the four had small foci of invasive urethral cancer, but died of synchronous metastases, and the other two had urethral CIS which was treated conservatively[194]. Urethral cytology[214] allows early detection of urethral recurrence and urethrectomy if appropriate. Secondary urethrectomy is indicated in patients who have either overt urethral cancer at the apex of the prostate in their cysto-prostatectomy specimen (which should have been diagnosed preoperatively) or those who subsequently develop invasive urethral cancer, particularly when conservative treatment has failed.

Female patients

In women undergoing cystectomy, the procedure usually involves *en bloc* resection of the uterus, both tubes and ovaries together with most of the vagina and the entire urethra[115]. In carefully selected cases, it may be possible to preserve the urethra in order to allow reconstructive surgery (see below). We feel it is also reasonable to leave the contralateral ovary in younger women with lateralized, unifocal tumours.

Morbidity and mortality

The safety of cystectomy has undoubtedly increased over the past 25 years. This is probably due to a small extent to improved surgical technique, in particular, better training in adequate control of the vascular supply of the bladder and more appropriate drainage, but more so to improved perioperative care and aggressive management of complications. A review of 675 cystectomies over a 20-year period reported a 2.5% 30-day mortality rate[44] and this is similar to current figures of around 2%[60,138,169,194].

The figures for early and late morbidity in the Duke series are representative of good results from the literature[44]. The major early complications were: prolonged ileus (6%), wound infection (6%), sepsis (5%) and pelvic abscess (5%). The incidence of other significant complications were: small bowel obstruction or fistula, 5%; rectal injury, 2%; pulmonary embolus, 2%; deep vein thrombosis, 2%; myocardial infarction, 2%; and ureterointestinal anastomotic disruption, 1%. Thirty-one per cent of patients had one or more early complications. The major late complications in this series were small bowel obstruction (7%) and ureterointestinal stricture (7%).

Postoperatively we use epidural analgesia, chest physiotherapy and incentive spirometry, mobilization on the first postoperative day and parenteral nutrition. We use daily body weight as a check on fluid status and we supplement with albumin until total serum protein is 50 mg/100 ml. We maintain serum osmolality at around 285 mosmol, usually by supplementing with intravenous 5% glucose, to counteract postoperative ADH secretion caused by intraoperative crystalloid administration and pulmonary and abdominal loss of free water. We have seen less postoperative ileus since replacing systemic opiates with postoperative epidural analgesia.

The occasional cystectomist

Excellent surgical results can be obtained by inexperienced surgeons when assisted by trained urologists[148] and this underlines the safety of training under expert supervision. Whether or not such results can be obtained by experienced urologists who perform few cystectomies is not documented, but intuitively it seems unlikely, and our view is that probably at least 8–10 cystectomies per year are required to get good results. This throughput is also important to maintain the experience of the entire team involved with peri- and postoperative care.

Results

The results of cystectomy depend overwhelmingly on the tumour stage of the patient, and so the results of different series are generally a function of patient selection, which in turn dictates the staging of the series overall. Results from the 1960s and 1970s were generally around 20–40% overall 5-year survival. There are few large series which are exclusively recent, but overall results have probably improved to 50% and better[4,17,49,104,123,124,170,189,192]. As well as the decrease in perioperative morbidity mentioned above, results have probably improved due to more careful preoperative staging, allowing exclusion of M+ patients, which used not to be possible, and probably also improved patient and physician awareness of the need to investigate haematuria.

The dominant influence in cystectomy series of the local tumour extent (pT stage) and the presence or absence of lymph node metastases (pN stage) has been documented[5,45,192]. Other factors that influence survival include positive resection margins and age[45]. In a randomized trial of adjuvant cisplatin, patients with advanced tumours, pT3b and pT4a, had 5-year survival of 40%, unaffected by cisplatin[189]. This figure is surprisingly good and suggests the additional influence of the excellent performance status of the patients, because only patients thought fit enough to tolerate chemotherapy were included.

Local recurrence

This occurs in about 10% of all patients after cystectomy and seems to increase with increasing pT stage; radiotherapy appears not to influence local recurrence[45].

Urothelial cancer follow-up

We follow patients after cystectomy with clinical assessment, annual urography and chest radiography for at least the first 3 years, and urethral lavage cytology in patients who retain their urethra.

Advantages and disadvantages of cystectomy

Cystectomy remains the treatment associated with the highest local cure rate. The disadvantages are the perioperative mortality (albeit now low), loss of bladder function and possibly also continence and potency. There is also the possibility that the operation was overtreatment, in that some patients may also have been cured by a primarily more conservative approach such as partial cystectomy and radical transurethral resection (TURB) and/or BCG. However, of these aggressive local treatments, only BCG can address the multifocal tendency of bladder cancer and even the effect of BCG is limited to superficial disease.

In contrast, the cystectomy patient benefits from improved local control and removal of the possibility of disabling haematuria. If the patient has metastases, cystectomy would only be considered for symptomatic relief of very troublesome local symptoms refractory to other measures.

The uncertainty over selection is caused by the inadequacy of our current pre-cystectomy prognostic factors and their inability to distinguish reliably between patients who really benefit from cystectomy and those who benefit less. It is to be hoped that the new wave of prognostic factors, such as p53, E-cadherin and proliferation indices can improve this situation[20,39,99,152].

URINARY DIVERSION, BLADDER RECONSTRUCTION AND STOMATHERAPY

Following cystectomy, urine can either be diverted into an incontinent stoma, into a continent urinary reservoir which is either catheterized regularly by the patient or is controlled by the anal sphincter or into an orthotopic bladder substitute and the patient can then void urethrally. The early evolution of urinary diversion has been reviewed[190]. It is often felt that the quickest and simplest procedures for the patient are incontinent urinary diversions and that both continent diversion and orthotopic bladder substitutes add to the technical difficulty and length of the operation, as well as potentially increasing the risk of short- and long-term complications. However, it is now clear that there are advantages and disadvantages to all procedures and a simplistic view of this is not possible. Thus careful assessment of each patient is vital to allow an appropriate choice of diversion. The standard against which other techniques are currently compared is the ileal conduit, but there is no doubt that continent diversion and orthotopic bladder substitutes are rapidly increasing in popularity.

The perceived advantages of conduits are that they are simple and safe in the long-term. The patient can have a relatively passive role and stoma care must simply be performed either by the patient or by a carer. The obvious disadvantages of a conduit are that a visible stoma and an external appliance may be psychologically difficult for patients and that the commitment to stoma care must indeed be lifelong. If stoma care fails, the patient simply becomes wet, but the upper tracts are not at immediate risk: this is a fail-safe method. The long-term results, however, are far from ideal[142] (see below).

Conversely, continent diversions and orthotopic bladder substitutes preserve body image better and orthotopic bladder substitutes allow the patient to appear outwardly normal with apparently normal voiding habits. Stoma care is not necessary, although patients with a continent diversion must catheterize their pouch several times per day. However, if catheterization fails or if a patient with an orthotopic bladder substitute fails to void regularly, chronic or even acute retention may develop, with the possibility of pouch damage or rupture and of metabolic disturbances. Although continent diversions and orthotopic bladder substitutes which are not emptied may simply leak, these are not **invariably** fail-safe methods, in contrast to conduits. A major advantage of orthotopic bladder substitutes compared to normally functioning continent diversions is that the urine is almost always sterile.

Furthermore, conduits and continent diversions require either that the patient has the mental and physical capacity to perform stoma care or that a carer can do this: by contrast, impaired manual dexterity, for example, should not contra-indicate an orthotopic bladder substitute. This means that any physical or mental disability may affect the choice of diversion. Social considerations are also important and in some countries the presence of a stoma and external appliance is culturally unacceptable, dictating the need for either a continent rectal reservoir or for an orthotopic bladder substitute.

With all urinary diversions, common principles apply regarding the function of gastrointestinal segments as urinary reservoirs and the most effective means of deploying them. Similarly, there are common principles concerning the effects on the gastrointestinal tract of resection and of incorporation of gastrointestinal segments into the urinary tract, the metabolic effects of the contact of urine and bowel, and the mechanical problems related to the reservoir and the upper tracts. The degree to which these represent a clinical problem varies considerably between different diversions and this can be important in the choice of diversion for a given patient.

Finally, consideration must be given to underlying gastrointestinal or renal disease, which may be relative contra-indications to the use of particular bowel segments, either because of prior abdominal radiotherapy or inflammatory bowel disease, for example, or because of insufficient renal reserve to cope with the metabolic effects of a given diversion.

The use of gastrointestinal segments in urinary diversion

The physical and physiological considerations concerning the selection of bowel segments for use in urinary diversion and the effect of the dimensions of the chosen segment on the initial reservoir volume have been discussed[69,89,191]. Attempts have also been made to quantify the mechanics of reservoir function from a mathematical basis[30,31]. It is generally accepted that gastrointestinal segments should be detubularized, both to prevent synchronous contractions and to maximize reservoir volume[52]. The presence of synchronous contractions in tubular segments is thought to explain the high incidence of, particularly, nocturnal incontinence seen in such diversions[53,144]. In addition, the application of Laplace's law means that spherical reservoirs will have greater compliance than tubular ones. The maximum reservoir volume from any bowel segment is obtained by detubularization and cross-folding as a sphere[55]. This allows a given reservoir volume with the minimum bowel resection.

INCONTINENT URINARY DIVERSION

Nephrostomy

The simplest form of urinary diversion is nephrostomy drainage. This is used as palliation, for locally advanced or metastatic tumours when primary cystectomy is not appropriate. The indications may be intractable haematuria (when ipsilateral ureteric ligation is also required) or ureteric obstruction when chemotherapy is planned. Ureteric obstruction due to advanced bladder cancer is not generally an appropriate indication for nephrostomy if chemotherapy is not planned. Open nephrostomy drainage is now replaced by percutaneous drainage, which has a much lower morbidity[136].

Cutaneous ureterostomy

Cutaneous ureterostomy is seldom employed because the morbidity of an open operation is avoided by percutaneous nephrostomy drainage and the risks of stenosis at the level of the skin are much less with a conduit, if an open operation is planned anyhow[7].

Ureterosigmoidostomy

This is the oldest form of urinary diversion[163]. It was the standard diversion before the ileal conduit was described[209,211,219]. The colon forms a high-pressure reservoir and urine mixes with stool and is inevitably contaminated with faecal flora. If the anal sphincter is competent preoperatively, continence is usually excellent. Ureterosigmoidostomy is a quick and simple diversion and there are undoubtedly patients who have an excellent result from this procedure. However, the operation has become unpopular because of the high incidence of late complications, which include pyelonephritis and upper-tract deterioration, presumed to be due to reflux of inevitably infected urine, hyperchloraemic acidosis and the development of colonic cancer.

Conduits

The ileal conduit was described in 1950 by Bricker and remains the standard urinary diversion against which others are judged. It superseded ureterosigmoidostomy because of the lower incidence of metabolic complications and the reduced incidence of long-term renal deterioration which it was initially believed to have. An advantage is that it requires simple small bowel surgery and the open end-to-side ureteroileal anastomosis is also the simplest such anastomosis[128]. Although Bricker described a flush stoma, now either a protruding end stoma or a Turnbull loop stoma is used[13,28]. Longer-term follow-up has now shown that ileal conduits do indeed have significant physical and psychological morbidity (see below) and this has stimulated the increasing use of continent diversion and orthotopic bladder substitutes. Both colonic[37,76,125] and jejunal conduits[54] have also been used, although we believe that in almost all circumstances, the use of jejunal conduits is contra-indicated (see below).

CONTINENT URINARY DIVERSION

A continent urinary diversion is an intra-abdominal urinary reservoir, catheterizable or with an outlet controlled by the anal sphincter. Continent urinary diversions can be classified according to the afferent segment, the reservoir and the efferent segment. The majority of afferent segments use some form of anti-reflux mechanism. Reservoirs have been constructed from the stomach, the ileum, the ileocaecal segment, the colon and the rectum, and various efferent segments have been used.

Ileal, ileocaecal and colonic reservoirs

Probably the best-studied ileal continent diversion is the Kock pouch[88,168,171]. When used as a catheterizable reservoir, it requires just over 80 cm of ileum and employs a nipple valve of intussuscepted ileum for both the afferent and efferent limbs. This produces an anti-reflux mechanism and a continent, catheterizable stoma. It has been described as tedious, but not technically difficult surgery[171]. The catheterizable Kock pouch is, however, undoubtedly not a simple diversion and the complication and reoperation rate can be high, although technical modifications have been made to reduce this[171]. The desire for improved quality of life has led Kock's and Skinner's groups to adapt the Kock pouch for use as an orthotopic bladder substitute[16,87].

A number of ileocaecal and colonic continent diversions have been described[10,25,101,109,150,199,210]. A perceived advantage of these techniques over the ileal reservoirs is the possibility that simple, tunnelled, anti-reflux ureteric anastomoses can be made in the colonic wall, but not readily in the ileal wall. Continent catheterizable diversions typically have a maximum capacity of around 700 ml and an end filling pressure of less than 25 cmH_2O[25,61,150,167,208].

Rectal and gastric reservoirs

Rectal reservoirs include the MAINZ II pouch and the modified rectal bladder of Ghoneim[41,47]. In both procedures, the rectosigmoid is opened and cross-folded. In Ghoneim's pouch, the ureters are implanted in an anti-reflux fashion using an extramucosal serous-lined tunnel[38]. Detubularized and cross-folded stomach has been used in combination with the appendix according to the Mitrofanoff principle for continent diversion[130]. These continent reservoirs appear to have similar urodynamic characteristics to the ileal and colonic reservoirs[1,41].

Efferent limb techniques

A variety of different continence mechanisms for the catheterizable efferent limb have been described and reviewed by Hinman[70]. There is an important practical difference between flap valves, such as the Mitrofanoff principle[122] or the intussuscepted nipple of the Kock pouch and those with mechanical compression of the outlet, such as the Indiana pouch. Flap valves remain continent with increasing reservoir pressure and guarantee absolute continence: this risks upper-tract damage, overdistension or even reservoir rupture if the reservoir is exposed to high pressure. Conversely, the mechanically compressed valves provide continence to a fixed closure pressure well above the baseline reservoir pressure, but at pressures beyond this fixed level, the efferent limb leaks. This may be inconvenient and embarrassing for the patient, but offers a degree of fail-safe mechanism for the reservoir and upper tracts.

Mitrofanoff used the appendix tunnelled through the bladder wall and brought out as a catheterizable stoma at the skin, but the principle may be used with a variety of other tissues and has also been used to create a catheterizable neourethra[216,217]. Benchekroun and co-workers have described an efferent limb of invaginated ileum, which can be used with any type of reservoir, where the lumen is closed along half of the circumference at the junction between the reservoir and the efferent limb[11]. A suggested advantage of this technique over the Mitrofanoff technique, is that endoscopic access to the reservoir is possible much more easily.

ORTHOTOPIC BLADDER SUBSTITUTION

During the 1950s and 1960s when ileal conduits and ureterosigmoidostomies were totally dominant as the preferred urinary diversion, orthotopic bladder substitutes began to emerge gradually, with the use of ileocaecal and colonic segments anastomosed to the urethra[48,203]. This was an era when the mortality of cystectomy was high and there was a desire for a diversion that was a quick and safe end to a cystectomy. This was an argument for the use of an ileal conduit or a ureterosigmoidostomy. As it has become clearer that the risks of cystectomy in experienced hands are acceptable and that the long-term results of ileal conduit and ureterosigmoidostomy are less good than first thought, so the search has widened for a more satisfactory urinary diversion[166]. The initial enthusiasm for continent diversion as an alternative to an ileal conduit or a ureterosigmoidostomy is gradually being replaced by a desire to provide optimal quality of life with an orthotopic bladder substitute. It cannot be stressed too strongly that no follow-up data exist for orthotopic bladder substitutes and so enthusiasm for their use should be tempered by the need for them to stand the test of time.

Anastomosis of a bladder substitute to the urethra after radical cystectomy is generally only considered in men. This is because a radical cystectomy for urothelial cancer in women almost invariably includes a total urethrectomy, because of the panurothelial nature of the disease and the perceived risk of urethral recurrence in women. There is preliminary evidence that this may be unnecessarily conservative[177,183] and urethra-sparing radical cystectomy in combination with orthotopic bladder substitution is now being attempted in women with tumours in the body of the bladder with negative biopsies of the bladder neck. This is supported by years of experience with subtotal cystectomy and bladder augmentation for certain urothelial and non-urothelial tumours in women. Although there are encouraging preliminary reports of the results of urethra-sparing radical cystectomy in combination with orthotopic bladder substitution for urothelial cancers in women, concerns about urethral recurrence and the adequacy of the resection permitted by nerve-sparing cystectomy in women mean that the technique can only be regarded as investigational at present[29,179,182,202]. Considerable patient years of follow-up are required before this approach can be considered standard treatment. Although referral patterns to centres with super-specialist interests in female bladder substitution may lead to the impression that it is used in many women, outside the setting of a prospective study, a conduit, continent diversion or ureterosigmoidostomy is still used in the majority of women after radical cystectomy.

Orthotopic bladder substitutes have been constructed from stomach[129,130], ileum[3,16,22,60,77,87,91,119,137,188], the ileocaecal segment[3,10,24,97,114,160] and the colon[51,144,197]. Our preference is for a low-pressure, cross-folded bladder substitute made from a 40 cm segment of ileum[194]. An afferent limb of a further 15 cm of intact isoperistaltic ileum from the same isolated segment is used and the ureters are anastomosed to this with an open end-to-side Nesbit anastomosis[128]. The reservoir is hand sewn with Vicryl® and is anastomosed with six interrupted Vicryl® sutures to the urethra. Construction and anastomosis of the reservoir takes about 30 minutes longer than would be needed for formation of an ileal conduit. The ureteroileal anastomoses are stented and the bladder substitute is drained with a 12Ch suprapubic catheter and a 18Ch urethral catheter which is used to irrigate the reservoir to remove mucus. The ureteric stents are removed at 8 and 9 days. A cystogram, including a film after emptying contrast, is done on day 12 and if there is no leak, the suprapubic catheter is then removed. The urethral catheter is removed on day 14 and voiding commences. Residual urine is checked daily, urine is cultured after all tubes are removed and the postoperative antibiotic is routinely changed pending an antibiogram. Venous blood gases are checked daily after voiding begins and bicarbonate is given orally if the negative base excess exceeds 4 mmol/l (see below).

Ileal orthotopic bladder substitutes typically have a long-term functional capacity of 300–500 ml and an end filling pressure of 15–20 cmH_2O[6,26,77,119,137,169] and should be emptied with minimal residual urine[26]. If the reservoir is filled beyond maximum functional capacity during urodynamics, phasic contractions ('pouch instability') may be seen[26], but this is a urodynamic finding with no clinical significance. Available data suggest that detubularized ileocaecal and colonic orthotopic bladder substitutes probably have similar urodynamic characteristics to ileal bladder substitutes[24,144].

Detubularized pouches are emptied by relaxation of the pelvic floor and, if necessary, abdominal straining. This requires careful training, and the aid of a specialist nurse, who has the time to take the patients carefully through the initial phase of learning to empty the pouch, is invaluable. There is no normal sensation of bladder filling, although patients are often aware of the pouch when it is full. Correspondingly there can be no normal feedback to the central nervous system during filling and so urethral pressure cannot rise as it does normally during filling. This means that, particularly at night, the possibility of overfilling and consequent leakage is present. An initial capacity of around 150 ml is desirable and means that overflow incontinence often occurs in the early postoperative weeks. This is not a problem to a patient who has been adequately counselled preoperatively and thus knows that the pouch will expand with time. Because of this possibility of early nocturnal incontinence, we instruct our patients to wake at night twice to void in the first few postoperative months. This is already the established preoperative voiding habit of most of them and is not seen by the patients as any problem. Emptying the reservoir during the night ensures continence in over 80% by the first year. The remainder require a no more than a pad. An ultimate reservoir capacity of around

500 ml seems to be ideal and the capacity increases in the first postoperative year, increasing continence coming with it. Pelvic floor contractions are also taught to the patient and their importance is in the early postoperative phase to preserve continence by day when movement, coughing, sneezing, etc. cause sudden increases in intra-abdominal pressure. When the reservoir is well filled, unless these sudden pressure increases are matched by a voluntary rise in urethral pressure, incontinence may occur in some patients.

Definitions of incontinence after orthotopic bladder substitution vary and this makes it hard to compare different series. It is, however, clear that daytime continence shoud be achieved in almost all patients, whereas night-time continence is slightly less good. The aetiology of incontinence following orthotopic bladder substitution may be multifactorial[96,157]. Clearly it can be due to either an abormality of the reservoir and/or the sphincter mechanism. Urine entering the pouch at night tends to be concentrated and will attract water osmotically, increasing nocturnal urine volume compared to preoperatively. This, coupled with the initial low reservoir capacity may lead to incontinence. The importance of detubularization of the reservoir and the loss of the normal reflex increase in urethral pressure with filling have been emphasized. Careful anatomical dissection of the prostatic apex to preserve the neurovascular supply to the external sphincter mechanism is probably helpful in improving results[46].

EARLY MORBIDITY OF CONDUITS, CONTINENT DIVERSIONS AND BLADDER SUBSTITUTES

The complications of conduits, Kock pouches and other urinary diversions have been reviewed extensively[63,81,95]. Significant early complications following urinary diversion include myocardial infarction, thromboembolism, pneumonia, small bowel obstruction or fistula, intra-abdominal sepsis and ureteroileal leakage. About 20% of patients can be expected to have a significant early postoperative complication following urinary diversion [22,81,121,139,169].

LATE MORBIDITY

True long-term morbidity data only exist for ureterosigmoidostomy and ileal conduit, because there are no long-term data yet for continent diversion or orthotopic bladder substitutes.

Renal

Perhaps the most fundamental goal of urinary diversion after cystectomy is to preserve renal function: this may have a secondary metabolic benefit by preserving renal ability to compensate for metabolic disturbances. Renal function may potentially be at risk from several factors after urinary diversion. These include urinary infection, urinary tract obstruction, ureteric reflux, stone formation and recurrent tumour formation. Following cystectomy and urinary diversion, renal function may be assessed using serum creatinine, intravenous urography, scintigraphically or with a formal clearance measurement. It has been suggested that the most accurate method in this setting is an inulin clearance during diuresis and that creatinine clearance under diuresis approximates to this[117]. Finally, in assessing the effects of urinary diversion on renal function it should be remembered that many series of ureterosigmoidostomy and ileal conduit include mainly children and the results may not be applicable to those in adults with bladder cancer.

Calculi form in about 5% of patients with a ureterosigmoidostomy, a conduit or a continent diversion[95,108,126,141], although calculi in continent diversions can largely be eliminated by avoiding staples. They are usually struvite stones, and risk factors studied in ileal conduit patients include upper-tract dilatation, pyelonephritis, reduced renal function, the presence of urea-splitting organisms, residual urine in the conduit and systemic acidosis. Residual urine in the conduit is believed to promote bicarbonate secretion and hence systemic acidosis, which leads to bone buffering. This generates hypercalciuria and, in the presence of urea-splitting organisms, struvite stone formation occurs[33]. In addition, stone formation may be predisposed to by hyperoxaluria in patients with a urinary diversion involving the terminal ileum. Impaired fat absorption due to bile salt loss after extensive terminal ileal resection, may lead to reduced binding in the intestinal lumen of calcium with oxalate. This allows increased oxalate absorption and consequent hyperoxaluria[36].

Infection can be manifest as bacteruria or as acute or chronic pyelonephritis. The vast majority of conduits have chronic bacteruria[12,76,139,159], the mixture of urine and stool in ureterosigmoidostomy or in any continent rectal diversion is inevitably contaminated and the majority of continent diversions also have chronic bacteruria[113,168]. Conduits are believed to become contaminated by ascending infection from around the stoma flange, whereas it is thought that continent diversions become contaminated by catheterization[107]. The majority of orthotopic bladder substitutes remain sterile, particularly if they are free of residual urine[191]. Acute pyelonephritis occurs in between 5% and 20% of conduit patients[108,125,139,159] and up to 70% of patients with a ureterosigmoidostomy[209].

Urinary tract obstruction in a patient with a urinary diversion has various causes and may also be related to surgical technique: it may be due to ureteric kinking, ureteroileal anastomotic stricture, conduit stenosis (usually subfascial) or at the efferent limb/reservoir–urethral anastomosis, although the most common of these is certainly ureteroileal anastomotic stricture. The roles of radiotherapy and of periureteric fibrosis from early ureteroileal leakage in the aetiology of these strictures are disputed and a frequent cause is believed to be ischaemia after overmobilization of the ureter[105]. The incidence of ureteroileal anastomotic stricture is usually from 3% to 6%[126,139,142], although rates of 17% for ileal conduit[176] and 22% in children with colonic conduits using an anti-reflux implantation have been reported[37]. The combination of infection and obstruction is potentially very damaging to the kidney and this is a much clearer relationship than that of reflux and infected urine. We therefore feel that any ureteric anastomosis should be performed using a technique that yields as low a rate of obstruction as possible, particularly in conduits or continent diversions.

The role of reflux in the genesis of renal damage after urinary diversion is controversial[187]. In children with primary reflux, it remains unclear whether sterile reflux produces renal damage and it seems likely that either infection or high pressure is also required to produce reflux nephropathy[146]. In the absence of data to the contrary, by analogy, it seems likely that in urinary diversion with chronic bacteruria and the possibility of intermittent high-pressure peaks, either some form of anti-reflux measure or surgery to reduce pressure is appropriate. This would apply to patients with a ureterosigmoidostomy, possibly to conduit patients (with the risk of conduit stenosis[127]) and to patients with a continent diversion where body movement in the presence of an absolutely continent outlet flap-valve (nipple or Mitrofanoff) may allow intermittent high-pressure peaks.

Orthotopic bladder substitutes, in contrast, empty without an isolated reservoir pressure rise[193]. We can demonstrate reflux by overfilling the reservoir of our patients with an orthotopic bladder substitute, but when they void by relaxing the pelvic floor and then, if necessary, straining, they get an equal rise in the abdomen and in the retroperitoneum and do not reflux. Coupled with the fact that the reservoir contains sterile urine and maintains a low pressure during filling, we therefore feel justified in using a open end-to-side ureteroileal anastomotic technique (Nesbit) with the lowest stricture rate[128], to try to minimize the risk of long-term renal damage due to ureteroileal obstruction[142].

When upper-tract dilatation is seen during follow-up of a patient with a urinary diversion, the possibility of an upper-tract tumour or one at the site of ureteric implantation must be borne in mind. One study of patients having cystectomy and ileal conduit for bladder cancer found that 5% of patients had developed upper-tract tumours within 56 months of the cystectomy[176]. All of these patients had had distal ureteric carcinoma in situ at the time of cystectomy.

The long-term effects of urinary diversion on renal function are not yet known for continent diversion or orthotopic bladder substitutes. There are data concerning conduits and ureterosigmoidostomy, but these data are unsatisfactory because generally only serum creatinine and/or urographic appearances are examined and these are not optimal methods of assessment[106]. Urography may demonstrate dilatation in the absence of obstruction or renographic deterioration and a normal creatinine does not exclude deterioration in renal function[106]. One study that used serum creatinine, urography and renography, showed a 5–10% decrease in renographic renal function at about 3 years after ileal conduit, colonic conduit and continent caecal diversion[106].

Late mechanical complications and reoperation

These late complications can be considered as those physically related to the diversion and those not.

Late complications related to the diversion requiring reoperation in conduit patients include conduit retraction or stenosis and ureteroileal stenosis: reoperation rates range from 5% to 68%[37,78,141,145,159]. Reoperation for complications unrelated to the stoma is primarily for intestinal obstruction and occurs in up to 11% of patients[78,145,159].

In patients with continent diversion, open reoperation for diversion-related complications is required in 7–16% of patients[149,208,212]. Indications include efferent limb problems such as leakage, difficulty in catheterizing and parastomal hernia, and afferent limb problems including stenosis and reflux. A further 2–4% require open reoperation for complications not related to the reservoir, notably incisional hernia and intestinal obstruction.

So far, up to 9% of patients with an orthotopic bladder substitute have required open reoperation for diversion-related late complications[60,119,160,169,194] and around 2% require reoperation for incisional hernia and intestinal obstruction. The reoperation rate for diversion-related complications appears to be less than that for continent diversion, because the efferent limb of the continent diversions requires relatively high rates of reoperation. In both the continent diversion and the orthotopic bladder substitute groups, there are, in addition, patients with late diversion-related complications who have successful endourological treatment[60,90,178].

Metabolic

Metabolic consequences of urinary diversion may result both from the effects of resection of a gastrointestinal segment and from the effect of the contact of urine with the gastrointestinal segment in its new situation, which generally allows exchange of solutes across the bowel mucosa.

The effects of resection depend on the normal function of the gastrointestinal segments which are used for urinary diversion[180,191]. The stomach secretes hydrochloric acid and intrinsic factor which is required for vitamin B_{12} absorption. The jejunum absorbs nutrients and can secrete sodium and chloride ions to produce an isotonic small bowel content. The ileum has a complex ability to vary its function according to the nature of its contents[201], but it generally absorbs sodium and chloride ions and can secrete hydrogen, potassium and bicarbonate ions. The terminal ileum also absorbs vitamin B_{12} and bile acids for recirculation. The ileocaecal valve, when competent, prolongs intestinal transit time and may thus allow the colon to perform one of its main roles, the passive absorption of water, by means of active reabsorption of sodium and chloride ions. In all gastrointestinal segments, water has a tendency to flow across the mucosa osmotically, mainly between cells through intercellular junctions. However, this tendency and hence the ability of the segment to maintain an osmotic gradient, depends on the 'tightness' of the intercellular junctions in that segment[116]. This ability is greatest in the stomach and colon and least in the jejunum and ileum, where considerable water movement can follow ion secretion into the lumen.

Gastric resection can reduce or prevent secretion of intrinsic factor into the gut. Jejunal resection alone generally has little effect. Resection of less than 100 cm of ileum leads to loss of bile acids, which may cause an osmotic diarrhoea. Resection of more than 100 cm produces a marked reduction in bile acids, therefore fat malabsorption and steatorrhoea[72]. As mentioned above, this may lead to hyperoxaluria and urinary tract stone formation. There is also an increased predisposition to gallstones after ileal resection[68]. Resection of the ileocaecal valve and/or large right colonic segments may reduce colonic reabsorption of water and thus cause diarrhoea. The effects of pre-existing gastrointestinal disease, radiotherapy or prior resections should obviously be considered carefully when the choice of urinary diversion is being made, because the patient's functional reserve may already be reduced.

The effects of contact between urine and bowel depend on three factors: the area of bowel mucosa, the contact time and the solute concentrations in the urine. The first two inevitably increase with increasing reservoir size and so the metabolic complications of urinary diversion due to solute movement across the reservoir's mucosa should be more likely with larger reservoirs. The adverse effects of resection also are likely to depend on the length of gastrointestinal segment resected. These are good a priori grounds for being as conservative as possible in resecting gastrointestinal segments for use in urinary diversion. The metabolic effects are further influenced by the renal compensation. There is a renal tubular exchange mechanism that resorbs sodium for either potassium or hydrogen ions and this can be used to regulate acid–base balance.

Urinary diversion with stomach may cause a hypochloraemic alkalosis due to loss of hydrogen and chloride ions into the urine. Compensatory reduced renal hydrogen ion secretion necessitates increased potassium ion secretion, and hypokalaemia follows. It has been suggested that this makes the stomach particularly suitable for urinary diversion in patients with renal failure and acidosis.

Jejunum has been used for conduits, but with notably bad results because of a hyponatraemic, hypochloraemic, hyperkalaemic acidosis[112]. This results from secretion of sodium and chloride ions into the urine and osmotic loss of water with them. Dehydration follows and consequent aldosterone production increases renal potassium secretion. This reduces the ability of the renal sodium–potassium/hydrogen ion exchanger to secrete hydrogen ions and acidosis develops. The high urinary potassium concentration enhances potassium reabsorption from the conduit and hence hyperkalaemia.

Ileal and colonic urinary diversions are susceptible to hyperchloraemic, hypokalaemic acidosis. This occurs in up to 50% of ureterosigmoidostomies[209], acidosis has been found in 33% of patients with ileal, ileocaecal or colonic urinary diversion[134] and mild acidosis has been found in all of a small group of patients with continent colonic or orthotopic colonic diversions[86]. Other authors have, however, reported minimal or no acidosis[2,95]. The mechanism of acidosis is controversial: although it was originally believed, because of low serum bicarbonate and high serum chloride, to be a consequence of excretion of bicarbonate and absorption of chloride, there is evidence that ammonium absorption provides the source of excess hydrogen ions[84] and that ammonium is absorbed along with chloride, instead of potassium[82,83]. Interestingly, higher serum chloride levels have been found in patients with colonic reservoirs than with ileal reservoirs, many years after surgery[32], suggesting that absorptive adaptation has not occurred, and there is experimental evidence that the morphological changes, which occur with time in gastrointestinal segments used in urinary diversion, are not paralleled by a decrease in absorption in ileal segments[57]. There is also concern about the possible effects of chronic acidosis on bone mineralization[85,118,134], although so far no clinical evidence of this has been found.

About 45% of our orthotopic bladder substitute patients with a reservoir made from 40 cm of ileum have early postoperative acidosis. If the negative base excess is more than 4 mmol/l, we give oral bicarbonate, although this can almost invariably be stopped within 3 months. Subsequent metabolic decompensation can occur if the patient fails to void the reservoir regularly and develops a chronic retention. This is treated primarily by rehydration and a urethral catheter.

Although vitamin B_{12} deficiency has been reported[2,181], neither it nor gallstones have proved so far to be a significant problem in patients with urinary diversion, although bowel dysfunction due to bile acid malabsorption has been reported[35]. The potential problems of loss of the ileocaecal valve have been addressed by the authors of an elegant surgical reconstruction of a valve mechanism[42]. While this may prevent mechanical regurgitation of colonic contents into the ileum and thus prevent bacterial overgrowth, it could not compensate for the possible functional loss of the reabsorptive capacities of the terminal ileum. So the use of this approach seems a curious logic to us, when excellent orthotopic ileal bladder substitutes are available.

The multitude of potential metabolic problems following urinary diversion after cystectomy emphasize the need for long-term follow-up of these patients and for careful prospective study to exclude the possible clinical manifestation of these complications.

Histological changes and tumour formation

There has been interest in the histological changes in urinary diversions both because of the possible metabolic consequences and the possible neoplastic consequences. Ileum used in both conduits and in continent diversions undergoes villous and microvillous atrophy[50,71,140], although caecal mucosa does not seem to atrophy[113]. In patients with continent diversion, an increase in the number of mucus-storing goblet cells in ileal mucosa and a shift from sulphomucin to sialomucin production in caecal mucosa has been seen[71,113]. Conversely, electron microscopic study of the mucosa of caecal reservoirs showed microvillous atrophy of the caecal mucosa, but not the ileal nipple valve mucosa[23].

Adenocarcinoma has been described after ureterosigmoidostomy, ileal and colonic conduits and ileal augmentation cystoplasty[40,75]. No case has yet been reported following continent diversion or orthotopic bladder substitution. The incidence seems to be highest following ureterosigmoidostomy (up to 29%)[184] and lowest following ileal conduit. The median latency of tumour occurrence following ureterosigmoidostomy was 26 years, but benign tumours were seen as soon as 3 years after operation[75]. Tumours usually occur where urothelium and intestinal mucosa meet. Despite radical surgery, the prognosis of these tumours is poor[173].

Although both the extent of the ureterointestinal anastomosis and the length of time that urine is in contact with intestinal mucosa have been suggested as aetiological factors, research has centred on the role of nitrosamines, ornithine decarboxylase and the inflammatory reaction around the suture line[56,200].

Our incomplete understanding of tumour formation after urinary diversion adds a further indication for lifelong follow-up for patients with gastrointestinal segments in the urinary tract. This should ideally include endoscopy and biopsy, possibly with cytology and imaging studies and should begin no later than 10 years after surgery, and perhaps sooner. Ileal conduit patients can probably be exempt.

QUALITY OF LIFE AND STOMATHERAPY

Although patients with an ileal conduit may report a high quality of life[27,43], obvious problems are recognized. Stomatherapists are essential for acceptable care of patients with incontinent or continent urinary diversions, because patients invariably bring their problems to them, rather than to the urologist or community nurse[27]. This relationship should begin preoperatively with the test siting of the stoma bag (see above). This should always be within the patient's view and usually medial to the lateral border of the rectus sheath, below the umbilicus and away from scars, bony prominences and skin creases. This is also an important time for the stomatherapist to try to inform and reassure the patient about what lies ahead. Postoperatively the patient is taught, if at all possible, to care for his or her own stoma and this should preferably be achieved before discharge.

The major long-term problem and fear of conduit patients is leakage[79,133]: local skin rashes and excoriation are also common and often related to poor stoma care routines[131]. The stomatherapist plays a vital role in patients whose stoma is not optimal, to try to customize the pouch system to overcome local problems, the most common of which is a flush or even retracted stoma. In addition, optimizing stoma care routines may be sufficient to deal with some of the more common local skin reactions[215].

The psychological effects of conduits and continent diversions have been compared[15,110,111] and continent diversion scores consistently higher. This is principally because anxiety about leakage is much less and because body image is enhanced by the lack of a bag. Self-catheterization has to be taught to patients with a continent diversion before discharge and the stomatherapist is as vital to them subsequently as to the conduit patient.

Sexual function

Cystectomy has traditionally been viewed as the end of sexual activity in both men and women. A study of men and women undergoing preoperative radiotherapy and radical cysto-prostato-urethrectomy and radical cystectomy respectively, showed that only three of 29 men and one of six women who were sexually active preoperatively, had unchanged sexual activity postoperatively[132]. The authors stressed that the reasons for the reduction in sexual activity are multifactorial, but interestingly they emphasized that 14 of 29 men and five of six women considered their overall sexual life satisfactory postoperatively, despite the decrease in sexual activity.

The realization that damage to the cavernous nerves during cystectomy may cause impotence led to the concept of nerve-sparing cystectomy and this has produced post-cystectomy potency rates of up to 83% in patients not undergoing urethrectomy[158]. A modified urethrectomy technique, intended to preserve the internal pudendal arteries, has been developed and preservation of potency has been recorded in case reports[19]. However, nerve-sparing surgery is not always possible or successful and the management options in these situations have been reviewed recently[102].

CONCLUSIONS

The most effective local treatment for invasive bladder cancer remains cystectomy. To achieve the low complication rates which are now required of urologists, considerable care in patient selection and preparation, operative technique and postoperative care is necessary. Furthermore, meticulous follow-up regarding urothelial cancer is mandatory. This means that patients who are candidates for cystectomy should be managed by urologists with the appropriate experience, commitment and facilities.

Urinary diversion after cystectomy has become a broad field in which there is no clear overall leader at present. Long-term data for continent diversion and orthotopic bladder substitutes, on a wide range of possible complications, remain to be gathered and this means that patients should be followed for life. It must be remembered that current enthusiasm for these techniques is based on their outstanding medium-term functional results. It is to be hoped that they stand the test of time, but at present it should be emphasized that the patient with bladder cancer should be managed first and foremost according to the primary disease and not according to the feasibility or otherwise of bladder reconstruction.

REFERENCES

1. Adams MC, Mitchell ME and Rink RC (1988) Gastrocystoplasty: an alternative solution to the problem of urological reconstruction in the severely compromised patient. *J Urol* **140**: 1152–1156.
2. Åkerlund S, Delin K, Kock NG *et al.* (1989) Renal function and upper urinary tract configuration following urinary diversion to a continent ileal reservoir (Kock pouch): a prospective 5 to 11-year followup after reservoir construction. *J Urol* **142**: 964–968.
3. Alcini E, D'Addessi A, Giustacchini M *et al.* (1988) Bladder reconstruction after cystectomy: use of ileocecal segment and three-loop ileal reservoir. *Urology* **31**: 10–13.
4. Amling CL, Thrasher JB, Frazier HA *et al.* (1994) Radical cystectomy for stages Ta, Tis and T1 transitional cell carcinoma of the bladder. *J Urol* **151**: 31–35.
5. Babiker A, Shearer RJ and Chilvers CE (1989) Prognostic factors in a T3 bladder cancer trial. Co-operative Urological Cancer Group. *Br J Cancer* **59**: 441–444.
6. Bachor R, Frohneberg D, Miller K *et al.* (1990) Continence after total bladder replacement: urodynamic analysis of the ileal neobladder. *Br J Urol* **65**: 462–466.
7. Bachor R and Hautmann R (1993) Options in urinary diversion: a review and critical assessment. *Sem Urol* **11**: 235–250.
8. Baker R (1955) Correlation of circumferential lymphatic spread of vesical cancer with depth of infiltration: relation to present methods of treatment. *J Urol* **73**: 681–690.
9. Beahrs JR, Fleming TR and Zincke H (1984) Risk of local urethral recurrence after radical cystectomy for bladder cancer. *J Urol* **131**: 264–266.
10. Bejany DE and Politano VA (1993) Modified ileocolonic bladder: 5 years of experience. *J Urol* **149**: 1441–1444.
11. Benchekroun A, Essakalli N, Faik M *et al.* (1989) Continent urostomy with hydraulic ileal valve in 136 patients: 13 years of experience. *J Urol* **142**: 46–51.
12. Bernstein IT, Bennicke K, Rordam P *et al.* (1991) Bricker's ileal conduit urinary diversion with a simple non-refluxing uretero ileal anastomosis. *Scand J Urol Nephrol* **25**: 29–33.
13. Bloom DA, Lieskovsky G, Rainwater G *et al.* (1983) The Turnbull loop stoma. *J Urol* **129**: 715–718.

14. Boring CC, Squires TS, Tong T *et al.* (1994) Cancer statistics, 1994. *Can Cancer J Clin* **44**: 7–26.
15. Boyd SD, Feinberg SM, Skinner DG *et al.* (1987) Quality of life survey of urinary diversion patients: comparison of ileal conduits versus continent Kock ileal reservoirs. *J Urol* **138**: 1386–1389.
16. Boyd SD, Lieskovsky G and Skinner DG (1991) Kock pouch bladder replacement. *Urol Clin N Am* **18**: 641–648.
17. Bredael JJ, Croker BP and Glenn JF (1980) The curability of invasive bladder cancer treated by radical cystectomy. *Eur Urol* **6**: 206–210.
18. Brendler CB, Steinberg GD, Marshall FF *et al.* (1990) Local recurrence and survival following nerve-sparing radical cystoprostatectomy. *J Urol* **144**: 1137–1140.
19. Brendler CB, Schlegel PN and Walsh PC (1990) Urethrectomy with preservation of potency. *J Urol* **144**: 270–273.
20. Bringuier PP, Umbas R, Schaafsma HE *et al.* (1993) Decreased E-cadherin immunoreactivity correlates with poor survival in patients with bladder tumors. *Cancer Res* **53**: 3241–3245.
21. Bryan PJ, Butler HE, LiPuma JP *et al.* (1987) CT and MR imaging in staging bladder neoplasms. *J Comput Assist Tomogr* **11**: 96–101.
22. Camey M (1987) Bladder replacement by ileocystoplasty following radical cystectomy. *Sem Urol* **5**: 8–14.
23. Carlen B, Willen R and Månsson W (1990) Mucosal ultrastructure of continent cecal reservoir for urine and its ileal nipple valve 2–9 years after construction. *J Urol* **143**: 372–376.
24. Carroll PR and McAninch JW (1991) Use of the ileocecal segment for bladder substitution or continent urinary diversion. *Urol Int* **46**: 283–289.
25. Carroll PR, Presti JJ, McAninch JW *et al.* (1989) Functional characteristics of the continent ileocecal urinary reservoir: mechanisms of urinary continence. *J Urol* **142**: 1032–1036.
26. Casanova GA, Springer JP, Gerber E *et al.* (1993) Urodynamic and clinical aspects of ileal low pressure bladder substitutes. *Br J Urol* **72**:728–735.
27. Chadwick DJ and Stower MJ (1990) Life with urostomy. *Br J Urol* **65**: 9–191.
28. Chechile G, Klein EA, Bauer L *et al.* (1992) Functional equivalence of end and loop ileal conduit stomas. *J Urol* **147**: 582–586.
29. Chen ME, Pisters LL, Malpica A *et al.* (1997) Risk of urethral, vaginal and cervical involvement in patients undergoing radical cystectomy for bladder cancer: results from a contemporary cystectomy series from M D Anderson Cancer Center. *J Urol* **157**: 2120–2123.
30. Colding-Jørgensen M, Poulsen AL and Steven K (1993) Mechanical characteristics of tubular and detubularised bowel for bladder substitution: theory, urodynamics and clinical results. *Br J Urol* **72**: 586–593.
31. Colding-Jørgensen M and Steven K (1993) A model of the mechanics of

smooth muscle reservoirs applied to the intestinal bladder. *Neurourol Urodyn* **12**: 59–79.

32. Davidsson T, Åkerlund S, Forssell Aronsson E *et al.* (1994) Absorption of sodium and chloride in continent reservoirs for urine: comparison of ileal and colonic reservoirs. *J Urol* **151**: 335–337.
33. Dretler SP (1973) The pathogenesis of urinary tract calculi occurring after ileal conduit diversion. I. Clinical study. II. Conduit study. 3. Prevention. *J Urol* **109**: 204–209.
34. Dretler SP, Ragsdale BD and Leadbetter WF (1973) The value of pelvic lymphadenectomy in the surgical treatment of bladder cancer. *J Urol* **109**: 414–416.
35. Durrans D, Wujanto R, Carroll RN *et al.* (1989) Bile acid malabsorption: a complication of conduit surgery. *Br J Urol* **64**: 485–488.
36. Earnest DL, Johnson G, Williams HE *et al.* (1974) Hyperoxaluria in patients with ileal resection: an abnormality in dietary oxalate absorption. *Gastroenterology* **66**: 1114–1122.
37. Elder DD, Moisey CU and Rees RW (1979) A long-term follow-up of the colonic conduit operation in children. *Br J Urol* **51**: 462–465.
38. El Mekresh MM, Hafez AT, Abol-Enein H and Ghoneim MA (1997) Double folded rectosigmoid bladder with a new ureterocolic antireflux technique. *J Urol* **157**: 2085–2089.
39. Esrig D, Freeman JA, Elmajian D *et al.* (1994) p53 nuclear accumulation: an independent marker of prognosis in transitional cell carcinoma of the bladder. *J Urol* **151**: 442A.
40. Filmer RB and Spencer JR (1990) Malignancies in bladder augmentations and intestinal conduits. *J Urol* **143**: 671–678.
41. Fisch M, Wammack R, Muller SC *et al.* (1993) The Mainz pouch II (sigma rectum pouch). *J Urol* **149**: 258–263.
42. Fisch M, Wammack R, Spies F *et al.* (1994) Ileocecal valve reconstruction during continent urinary diversion. *J Urol* **151**: 861–865.
43. Fosså SD, Reitan JB, Ous S *et al.* (1987) Life with an ileal conduit in cystectomized bladder cancer patients: expectations and experience. *Scan J Urol Nephrol* **21**: 97–101.
44. Frazier HA, Robertson JE and Paulson DF (1992) Complications of radical cystectomy and urinary diversion: a retrospective review of 675 cases in 2 decades. *J Urol* **148**: 1401–1405.
45. Frazier HA, Robertson JE, Dodge RK *et al.* (1993) The value of pathologic factors in predicting cancer-specific survival among patients treated with radical cystectomy for transitional cell carcinoma of the bladder and prostate. *Cancer* **71**: 3993–4001.
46. Gasparini ME, Hinman F, Jr, Presti JC, Jr *et al.* (1992) Continence after radical cystoprostatectomy and total bladder replacement: a urodynamic analysis. *J Urol* **148**: 1861–1864.

47. Ghoneim MA, Ashamallah AK, Mahran MR *et al.* (1992) Further experience with the modified rectal bladder (the augmented and valved rectum) for urine diversion. *J Urol* **147**: 1252–1255.
48. Gil-Vernet JM, Jr (1950) Technique for construction of a functioning artificial bladder. *J Urol* **83**: 39–50.
49. Giuliani L, Giberti C, Martorana G *et al.* (1985) Results of radical cystectomy for primary bladder cancer. Retrospective study of more than 200 cases. *Urology* **26**: 243–248.
50. Goldstein MJ, Melamed MR, Grabstald H *et al.* (1967) Progressive villous atrophy of the ileum used as a urinary conduit. *Gastroenterology* **52**: 859–864.
51. Goldwasser B, Barrett DM and Benson RJ (1986) Bladder replacement with use of a detubularized right colonic segment: preliminary report of a new technique. *Mayo Clin Proc* **61**: 615–621.
52. Goldwasser B, Barrett DM, Webster GD *et al.* (1987) Cystometric properties of ileum and right colon after bladder augmentation, substitution or replacement. *J Urol* **138**: 1007–1008.
53. Goldwasser B, Rife CC, Benson RCJ *et al.* (1987) Urodynamic evaluation of patients after the Camey operation. *J Urol* **138**: 832–835.
54. Golimbu M and Morales P (1975) Jejunal conduits: technique and complications. *J Urol* **113**: 787–795.
55. Goodwin WE, Winter CC and Barker WF (1959) 'Cup-patch' technique of ileocystoplasty for bladder enlargement or partial substitution. *Surg Gynecol Obstet* **108**: 240–244.
56. Groschel J, Riedasch G, Kalble T *et al.* (1992) Nitrosamine excretion in patients with continent ileal reservoirs for urinary diversion. *J Urol* **147**: 1013–1016.
57. Hall MC, Koch MO, Halter SA *et al.* (1993) Morphologic and functional alterations of intestinal segments following urinary diversion. *J Urol* **149**: 664–666.
58. Hall RR, Newling DW, Ramsden PD *et al.* (1984) Treatment of invasive bladder cancer by local resection and high dose methotrexate. *Br J Urol* **56**: 668–672.
59. Hardeman SW and Soloway MS (1990) Urethral recurrence following radical cystectomy. *J Urol* **144**: 666–669.
60. Hautmann RE, Miller K, Steiner U *et al.* (1993) The ileal neobladder: 6 years of experience with more than 200 patients. *J Urol* **150**: 40–45.
61. Hedlund H, Lindstrom K and Månsson W (1984) Dynamics of a continent caecal reservoir for urinary diversion. *Br J Urol* **56**: 366–372.
62. Henly DR, Farrow GM and Zincke H (1993) Urachal cancer: role of conservative surgery. *Urology* **42**: 635–639.
63. Hensle TW and Dean GE (1991) Complications of urinary tract reconstruction. *Urol Clin N Am* **18**: 755–764.

64. Herr HW (1987) Conservative management of muscle-infiltrating bladder cancer: prospective experience. *J Urol* **138**: 1162–1163.
65. Herr HW (1988) Bladder cancer: pelvic lymphadenectomy revisited. *J Surg Oncol* **37**: 242–245.
66. Herr HW (1991) Progression of stage T1 bladder tumors after intravesical bacillus Calmette–Guerin. *J Urol* **145**: 40–43.
67. Herr HW (1994) Urachal carcinoma: the case for extended partial cystectomy. *J Urol* **151**: 365–366.
68. Hill GL, Mair WS and Goligher JC (1975) Gallstones after ileostomy and ileal resection. *Gut* **16**: 932–936.
69. Hinman F, Jr (1988) Selection of intestinal segments for bladder substitution: physical and physiological characteristics. *J Urol* **139**: 519–523.
70. Hinman F, Jr (1990) Functional classification of conduits for continent diversion. *J Urol* **144**: 27–30.
71. Höckenström T, Kock NG, Norlén LJ *et al.* (1986) Morphologic changes in ileal reservoir mucosa after long-term exposure to urine. A study in patients with continent urostomy (Kock pouch). *Scand J Gastroenterol* **21**: 1224–1234.
72. Hofmann AF and Poley JR (1972) Role of bile acid malabsorption in pathogenesis of diarrhea and steatorrhea in patients with ileal resection. I. Response to cholestyramine or replacement of dietary long chain triglyceride by medium chain triglyceride. *Gastroenterology* **62**: 918–934.
73. Hudson MA (1992) When intravesical measures fail. Indications for cystectomy in superficial disease. *Urol Clin N Am* **19**: 601–609.
74. Husband JE, Olliff JF, Williams MP *et al.* (1989) Bladder cancer: staging with CT and MR imaging. *Radiology* **173**: 435–440.
75. Husmann DA and Spence HM (1990) Current status of tumor of the bowel following ureterosigmoidostomy: a review. *J Urol* **144**: 607–610.
76. Husmann DA, McLorie GA and Churchill BM (1989) Nonrefluxing colonic conduits: a long-term life-table analysis. *J Urol* **142**: 1201–1203.
77. Iwakiri J, Gill H, Anderson R *et al.* (1993) Functional and urodynamic characteristics of an ileal neobladder. *J Urol* **149**: 1072–1076.
78. Jaffe BM, Bricker EM and Butcher HJ (1968) Surgical complications of ileal segment urinary diversion. *Ann Surg* **167**: 367–376.
79. Jones MA, Breckman B and Hendry WF (1980) Life with an ileal conduit: results of questionnaire surveys of patients and urological surgeons. *Br J Urol* **52**: 21–25.
80. Jones PA and Droller MJ (1993) Pathways of development and progression in bladder cancer: new correlations between clinical observations and molecular mechanisms. *Sem Urol* **11**: 177–192.
81. Killeen KP and Libertino JA (1988) Management of bowel and urinary tract complications after urinary diversion. *Urol Clin N Am* **15**: 183–194.

82. Koch MO, Gurevitch E, Hill DE *et al.* (1990) Urinary solute transport by intestinal segments: a comparative study of ileum and colon in rats. *J Urol* **14**: 1275–1279.
83. Koch MO and Hall MC (1992) Mechanism of ammonium transport: inhibition by potassium and barium. *J Urol* **148**: 1285–1287.
84. Koch MO and McDougal WS (1985) The pathophysiology of hyperchloremic metabolic acidosis after urinary diversion through intestinal segments. *Surgery* **98**: 561–570.
85. Koch MO, McDougal WS, Hall MC *et al.* (1992) Long-term metabolic effects of urinary diversion: a comparison of myelomeningocele patients managed by clean intermittent catheterization and urinary diversion. *J Urol* **147**: 1343–1347.
86. Koch MO, McDougal WS, Reddy PK *et al.* (1991) Metabolic alterations following continent urinary diversion through colonic segments. *J Urol* **145**: 270–273.
87. Kock NG, Ghoneim MA, Lycke KG *et al.* (1989) Replacement of the bladder by the urethral Kock pouch: functional results, urodynamics and radiological features. *J Urol* **141**: 1111–1116.
88. Kock NG, Nilson AE, Nilsson LO *et al.* (1982) Urinary diversion via a continent ileal reservoir: clinical results in 12 patients. *J Urol* **128**: 469–475.
89. Koff SA (1988) Guidelines to determine the size and shape of intestinal segments used for reconstruction. *J Urol* **140** (5 pt 2): 1150–1151.
90. Kramolowsky EV, Clayman RV and Weyman PJ (1988) Management of ureterointestinal anastomotic strictures: comparison of open surgical and endourological repair. *J Urol* **139**: 1195–1198.
91. Kreder K, Das AK and Webster GD (1992) The hemi-Kock ileocystoplasty: a versatile procedure in reconstructive urology. *J Urol* **147**: 1248–1251.
92. Lerner SP, Skinner DG, Lieskovsky G *et al.* (1993) The rationale for en bloc pelvic lymph node dissection for bladder cancer patients with nodal metastases: long-term results. *J Urol* **149**: 758–764.
93. Levinson AK, Johnson DE and Wishnow KI (1990) Indications for urethrectomy in an era of continent urinary diversion. *J Urol* **144**: 73–75.
94. Lieskovsky G and Skinner DG (1984) Role of lymphadenectomy in the treatment of bladder cancer. *Urol Clin N Am* **11**: 709–716.
95. Lieskovsky G, Skinner DG and Boyd SD (1988) Complications of the Kock pouch. *Urol Clin N Am* **15**: 195–205.
96. Light JK (1991) Continence mechanisms following continent urinary diversion and orthotopic bladder replacement. *Prog Clin Biol Res* **370**: 83–92.
97. Light JK and Marks JL (1990) Total bladder replacement in the male and female using the ileocolonic segment (LeBag). *Br J Urol* **65**: 467–472.
98. Linker DG and Whitmore WF (1975) Ureteral carcinoma in situ. *J Urol* **113**: 777–780.

99. Lipponen PK, Nordling S, Eskelinen MJ *et al.* (1993) Flow cytometry in comparison with mitotic index in predicting disease outcome in transitional-cell bladder cancer. *Int J Cancer* **53**: 42–47.
100. Liu BC and Liotta LA (1992) Biochemistry of bladder cancer invasion and metastasis. Clinical implications. *Urol Clin N Am* **19**: 621–627.
101. Lockhart JL, Pow SJ, Persky L *et al.* (1991) Results, complications and surgical indications of the Florida pouch. *Surg Gynecol Obstet,* **173**: 289–296.
102. Lue TF (1991) Impotence after radical pelvic surgery: physiology and management. *Urol Int* **46**: 259–265.
103. Malkowicz SB and Skinner DG (1990) Development of upper tract carcinoma after cystectomy for bladder carcinoma. *Urology* **36**: 20–22.
104. Malkowicz SB, Nichols P, Lieskovsky G *et al.* (1990) The role of radical cystectomy in the management of high grade superficial bladder cancer (PA, P1, PIS and P2). *J Urol* **144**: 641–645.
105. Månsson W, Ahlgren G and White T (1989) Glomerular filtration rate up to 10 years after urinary diversion of different types. A comparative study of ileal and colonic conduit, refluxing and antirefluxing ureteral anastomosis and continent caecal reservoir. *Scand J Urol Nephrol* **23**: 195–200.
106. Månsson W, Colleen S, Forsberg L *et al.* (1984) Renal function after urinary diversion. A study of continent caecal reservoir, ileal conduit and colonic conduit. *Scand J Urol Nephrol* **18**: 307–315.
107. Månsson W, Colleen S and Mardh PA (1989) Urine from continent caecal reservoirs. Studies on chemical composition and bacterial growth. *Eur Urol* **16**: 18–22.
108. Månsson W, Colleen S and Stigsson L (1979) Four methods of uretero-intestinal anastomosis in urinary conduit diversion. A comparative study of early and late complications and the influence of radiotherapy. *Scand J Urol Nephrol* **13**: 191–199.
109. Månsson W, Colleen S and Sundin T (1984) Continent caecal reservoir in urinary diversion. *Br J Urol* **56**: 359–365.
110. Månsson A Johnson G and Månsson W (1988) Quality of life after cystectomy. Comparison between patients with conduit and those with continent caecal reservoir urinary diversion. *Br J Urol* **62**: 240–245.
111. Månsson A, Johnson G and Månsson W (1991) Psychosocial adjustment to cystectomy for bladder carcinoma and effects on interpersonal relationships. *Scand J Caring Sci* **5**: 129–134.
112. Månsson W and Lindstedt E (1978) Electrolyte distrubances after jejunal conduit urinary diversion. *Scand J Urol Nephrol* **12**: 17–21.
113. Månsson W and Willen R (1988) Mucosal morphology and histochemistry of the continent cecal reservoir for urine. *J Urol* **139**: 1199–1201.
114. Marshall FF (1991) Ileocolic neobladder after cystectomy. *Urol Clin N Am* **18**: 631–639.

115. Marshall FF and Treiger BF (1991) Radical cystectomy (anterior exenteration) in the female patient. *Urol Clin N Am* **18**: 765–775.
116. McDougal WS (1992) Metabolic complications of urinary intestinal diversion. *J Urol* **147**: 1199–1208.
117. McDougal WS and Koch MO (1986) Accurate determination of renal function in patients with intestinal urinary diversions. *J Urol* **135**: 1175–1178.
118. McDougal WS, Koch MO, Shands C, III *et al.* (1988) Bony demineralization following urinary intestinal diversion. *J Urol* **140**: 853–855.
119. Melchior H, Spehr C, Knop WI *et al.* (1988) The continent ileal bladder for urinary tract reconstruction after cystectomy: a survey of 44 patients. *J Urol* **139**: 714–718.
120. Merz VW, Marth D, Kraft R *et al.* (1995) Analysis of early failures after intravesical instillation therapy with Bacille Calmette–Guerin for carcinoma in situ of the bladder. *Br J Urol* **75**: 180–184.
121. Miller K, Wenderoth UK, de Petriconi R *et al.* (1991) The ileal neobladder. Operative technique and results. *Urol Clin N Am* **18**: 623–630.
122. Mitrofanoff P (1980) Trans-appendicular continent cystostomy in the management of the neurogenic bladder. *Chir Pediatr* **21**: 297–305.
123. Montie JE, Straffon RA and Stewart BH (1984) Radical cystectomy without radiation therapy for carcinoma of the bladder. *J Urol* **131**: 477–482.
124. Morabito RA, Kandzari SJ and Milam DF (1979) Invasive bladder carcinoma treated by radical cystectomy: survival of patients. *Urology* **14**: 478–481.
125. Morales P and Golimbu M (1975) Colonic urinary diversion: 10 years of experience. *J Urol* **113**: 302–307.
126. Neal DE (1985) Complications of ileal conduit diversion in adults with cancer followed up for at least five years. *BMJ* **290**: 1695–1697.
127. Neal DE (1989) Urodynamic investigation of the ileal conduit: upper tract dilatation and the effects of revision of the conduit. *J Urol* **142**: 97–100.
128. Nesbit RM (1949) Ureterosigmoid anastomosis by direct elliptical connection: a preliminary report. *J Urol* **61**: 728–734.
129. Ngan JH, Lau JL, Lim ST *et al.* (1993) Long-term results of antral gastrocystoplasty. *J Urol* **149**: 731–734.
130. Nguyen DH and Mitchell ME (1991) Gastric bladder reconstruction. *Urol Clin N Am* **18**: 649–657.
131. Nordström GM and Nyman CR (1991) Living with a urostomy. A follow up with special regard to the peristomal-skin complications, psychosocial and sexual life. *Scand J Urol Nephrol (Suppl.)* **138**: 247–251.
132. Nordström GM and Nyman CR (1992) Male and female sexual function and activity following ileal conduit urinary diversion. *Br J Urol* **70**: 33–39.
133. Nordström G, Nyman CR and Theorell T (1992) Psychosocial adjustment and general state of health in patients with ileal conduit urinary diversion. *Scand J Urol Nephrol* **26**: 139–147.

134. Nurse DE and Mundy AR (1989) Metabolic complications of cystoplasty. *Br J Urol* **63**: 165–170.
135. Office of Population Censuses and Surveys (1994) *Mortality Statistics: General 1992*, HMSO, London.
136. Ortlip SA and Fraley EE (1982) Indications for palliative urinary diversion in patients with cancer. *Urol Clin N Am* **9**: 79–84.
137. Pagano F, Artibani W, Villi G *et al.* (1992) The vesica ileale Padovana, in *Continent Urinary Diversion*, (eds R Wammack and R Hohenfellner), Churchill Livingstone, Edinburgh, pp. 117–125.
138. Pagano F, Bassi P, Galetti TP *et al.* (1991) Results of contemporary radical cystectomy for invasive bladder cancer: a clinicopathological study with an emphasis on the inadequacy of the tumor, nodes and metastases classification. *J Urol* **145**: 45–50.
139. Pernet FPPM. and Jonas U (1985) Ileal conduit urinary diversion: early and late results of 132 cases in a 25 year period. *World J Urol* **3**: 140–144.
140. Philipson BM, Kock NG, Höckenström T *et al.* (1986) Ultrastructural and histochemical changes in ileal reservoir mucosa after long-term exposure to urine. A study in patients with continent urostomy (Kock pouch). *Scand J Gastroenterol* **21**: 1235–1244.
141. Philp NH, Williams JL and Byers CE (1980) Ileal conduit urinary diversion: long-term follow-up in adults. *Br J Urol* **52**: 515–519.
142. Pitts WR Jr and Muecke EC (1979) A 20-year experience with ileal conduits: the fate of the kidneys. *J Urol* **122**: 154–157.
143. Pritchett TR, Schiff WM, Klatt E *et al.* (1988) The potency-sparing radical cystectomy: does it compromise the completeness of the cancer resection? *J Urol* **140**: 1400–1403.
144. Reddy PK (1991) The colonic neobladder. *Urol Clin N Am* **18**: 609–614.
145. Remigailo RV, Lewis EL, Woodard JR *et al.* (1976) Ileal conduit urinary diversion. Ten-year review. *Urology* **7**: 343–348.
146. Richie JP and Skinner DG (1975) Urinary diversion: the physiological rationale for non-refluxing colonic conduits. *Br J Urol* **47**: 269–275.
147. Richie JP and Skinner DG (1978) Carcinoma in situ of the urethra associated with bladder carcinoma: the role of urethrectomy. *J Urol* **119**: 80–81.
148. Roehrborn CG, Sagalowsky AI and Peters PC (1991) Long-term patient survival after cystectomy for regional metastatic transitional cell carcinoma of the bladder. *J Urol* **146**: 36–39.
149. Rowland RG (1992) The plicated or tapered ileal outlet – 'Indiana pouch'. *Scand J Urol Nephrol (Suppl.)* **142**: 70–72.
150. Rowland RG, Mitchell ME, Bihrle R *et al.* (1987) Indiana continent urinary reservoir. *J Urol* **137**: 1136–1139.
151. Sandberg AA and Berger CS (1994) Review of chromosome studies in urological tumors. II. Cytogenetics and molecular genetics of bladder cancer. *J Urol* **151**: 545–560.

152. Sarkis AS, Dalbagni G, Cordon Cardo C *et al.* (1993) Nuclear overexpression of p53 protein in transitional cell bladder carcinoma: a marker for disease progression. *J Natl Cancer Inst* **85**: 53–59.
153. Sarosdy MF (1992) Management of the male urethra after cystectomy for bladder cancer. *Urol Clin N Am* **19**: 391–396.
154. Sawczuk IS, deVere White R, Gold RP *et al.* (1983) Sensitivity of computed tomography in evaluation of pelvic lymph node metastases from carcinoma of bladder and prostate. *Urology* **21**: 81–84.
155. Schellhammer PF and Whitmore WJ (1976) Transitional cell carcinoma of the urethra in men having cystectomy for bladder cancer. *J Urol* **115**: 56–60.
156. Scher HI, Yagoda A, Herr HW *et al.* (1988) Neoadjuvant M-VAC (methotrexate, vinblastine, doxorubicin and cisplatin) effect on the primary bladder lesion. *J Urol* **139**: 470–474.
157. Schiff SF and Lytton B (1991) Incontinence after augmentation cystoplasty and internal diversion. *Urol Clin N Am* **18**: 383–392.
158. Schlegel PN and Walsh PC (1987) Neuroanatomical approach to radical cystoprostatectomy with preservation of sexual function. *J Urol* **138**: 1402–1406.
159. Schmidt JD, Hawtrey CE, Flocks RH *et al.* (1973) Complications, results and problems of ileal conduit diversions. *J Urol* **109**: 210–216.
160. Schreiter F and Noll F (1989) Kock pouch and S bladder: 2 different ways of lower urinary tract reconstruction. *J Urol* **142**: 1197–1200.
161. Sidransky D, Frost P, von Eschenbach A *et al.* (1992) Clonal origin bladder cancer. *N Engl J Med* **326**: 737–740.
162. Sidransky D and Messing E (1992) Molecular genetics and biochemical mechanisms in bladder cancer. Oncogenes, tumor suppressor genes, and growth factors. *Urol Clin N Am* **19**: 629–639.
163. Simon J (1852) Ectopia vesicae (absence of the anterior walls of the bladder and pubic abdominal parietes): operation for directing the orifices of the ureters into the rectum. Temporary success: subsequent death: autopsy. *Lancet* **2**: 568.
164. Simoneau AR and Jones PA (1994) Bladder cancer: the molecular progression to invasive disease. *World J Urol* **12**: 89–95.
165. Skinner DG (1980) Current perspectives in the management of high-grade invasive bladder cancer. *Cancer* **45** (Suppl. 7): 1866–1874.
166. Skinner DG (1982) In search of the ideal method of urinary diversion. *J Urol* **128**: 476
167. Skinner DG (1992) The Kock pouch for continent urinary reconstruction focusing on the afferent segment and the reservoir. *Scand J Urol Nephrol Suppl.* **142**: 77–78.
168. Skinner DG, Boyd SD and Lieskovsky G (1984) Clinical experience with the Kock continent ileal reservoir for urinary diversion. *J Urol* **132**: 1101–1107.

169. Skinner DG, Boyd SD, Lieskovsky G *et al.* (1991) Lower urinary tract reconstruction following cystectomy: experience and results in 126 patients using the Kock ileal reservoir with bilateral ureteroileal urethrostomy. *J Urol* **146**: 756–760.
170. Skinner DG and Lieskovsky G (1984) Contemporary cystectomy with pelvic node dissection compared to preoperative radiation therapy plus cystectomy in management of invasive bladder cancer. *J Urol* **131**: 1069–1072.
171. Skinner DG, Lieskovsky G and Boyd S (1989) Continent urinary diversion. *J Urol* **141**: 1323–1327.
172. Smith JAJ and Whitmore WFJ (1981) Regional lymph node metastasis from bladder cancer. *J Urol* **126**: 591–593.
173. Spencer JR and Filmer RB (1992) Malignancy associated with urinary tract reconstruction using enteric segments. *Cancer Treat Res* **59**: 75–87.
174. Splinter TA, Pavone Macaluso M, Jacqmin D *et al.* (1992) A European Organization for Research and Treatment of Cancer–Genitourinary Group phase 2 study of chemotherapy in stage T3–4N0-XM0 transitional cell cancer of the bladder: evaluation of clinical response. *J Urol* **148**: 1793–1796.
175. Splinter TA, Scher HI, Denis L *et al.* (1992) The prognostic value of the pathological response to combination chemotherapy before cystectomy in patients with invasive bladder cancer. European Organization for Research on Treatment of Cancer–Genitourinary Group. *J Urol* **147**: 606–608.
176. Stanley P, Craven JD, Skinner DG *et al.* (1975) The natural history of the upper renal tracts in adults following ureteroileal diversion (Bricker procedure). *Am J Roentgenol Radium Ther Nucl Med* **125**: 804–811.
177. Stein JP, Cote R, Freeman JA *et al.* (1994) Lower urinary tract reconstruction in women following cystectomy for pelvic malignancy: a pathological review of female cystectomy specimens. *J Urol* **151**: 304A.
178. Stein JP, Huffman JL, Freeman JA *et al.* (1994) Stenosis of the afferent antireflux valve in the Kock pouch continent urinary diversion: diagnosis and management. *J Urol* **151**: 338–340.
179. Stein JP, Stenzl A, Grossfeld GD *et al.* (1996) The use of orthotopic neobladders in women undergoing cystectomy for pelvic malignancy. *World J Urol* **14**: 9–14.
180. Steiner MS and Morton RA (1991) Nutritional and gastrointestinal complications of the use of bowel segments in the lower urinary tract. *Urol Clin N Am* **18**: 743–754.
181. Steiner MS, Morton RA and Marshall FF (1993) Vitamin B12 deficiency in patients with ileocolic neobladders. *J Urol* **149**: 255–257.
182. Stenzl A, Colleselli K and Poisel S *et al.* (1995) Rationale and technique of nerve-sparing radical cystectomy before an orthotopic neobladder procedure in women. *J Urol* **154**: 2044–2049.

183. Stenzl A, Draxl H, Hernegger B *et al.* (1994) Localization study of bladder cancer in the female: can urethral segments safely be spared at radical cystectomy? *J Urol* **151**: 475A.
184. Stewart M, Macrae FA and Williams CB (1982) Neoplasia and ureterosigmoidostomy: a colonoscopy survey. *Br J Surg* **69**: 414–416.
185. Stockle M, Gokcebay E, Riedmiller H *et al.* (1990) Urethral tumor recurrences after radical cystoprostatectomy: the case for primary cystoprostatourethrectomy? *J Urol* **143**: 41–42.
186. Strohmeyer TG and Slamon DJ (1994) Proto-oncogenes and tumor suppressor genes in human urological malignancies. *J Urol* **151**: 1479–1497.
187. Studer UE (1992) The role of ureteral implantation in continent urinary reservoirs, in *Continent Urinary Diversion*, (eds R Hohenfellner and R Wammack), Churchill Livingstone, Edinburgh, pp. 209–223.
188. Studer UE, Ackermann D, Casanova GA *et al.* (1989) Three years' experience with an ileal low pressure bladder substitute. *Br J Urol* **63**: 43–52.
189. Studer UE, Bacchi M, Biedermann C *et al.* (1994) Adjuvant cisplatin chemotherapy following cystectomy for bladder cancer: results of a prospective randomized trial. *J Urol* **152**: 81–84.
190. Studer UE, Casanova GA and Zingg EJ (1991) Historical aspects of continent urinary diversion, in *Problems in Urology*, (ed. RG Rowland), Lippincott, Philadelphia, Vol. 5, pp. 197–202.
191. Studer UE, Gerber E, Springer J *et al.* (1992) Bladder reconstruction with bowel after radical cystectomy. *World J Urol* **10**: 11–19.
192. Studer UE, Ruchti E, Greiner RM *et al.* (1983) Important factors determining survival after radical cystectomy for bladder cancer. *Akt Urol* **14**: 70–77.
193. Studer UE, Spiegel T, Casanova GA *et al.* (1991) Ileal bladder substitute: antireflux nipple or afferent tubular segment? *Eur Urol* **20**: 315–326.
194. Studer UE and Turner WH (1996) Ileal low pressure bladder substitute with an afferent tubular isoperistaltic segment, in *Comprehensive Textbook of Genitourinary Oncology*, (eds NJ Vogelzang, PT Scardino, WU Shipley *et al.*), Williams and Wilkins, Baltimore, pp. 495–508.
195. Studer UE, Wallace DMA, Ruchti E *et al.* (1985) The role of pelvic lymph node metastases in bladder cancer. *World J Urol* **3**: 98–103.
196. Sweeney P, Kursh ED and Resnick MI (1992) Partial cystectomy. *Urol Clin N Am* **19**: 701–711.
197. Thomas PJ, Nurse DE, Deliveliotis C *et al.* (1992) Cystoprostatectomy and substitution cystoplasty for locally invasive bladder cancer. *Br J Urol* **70**: 40–42.
198. Thrasher JB and Crawford ED (1993) Current management of invasive and metastatic transitional cell carcinoma of the bladder. *J Urol* **149**: 957–972.
199. Thüroff JW, Alken P, Riedmiller H *et al.* (1988) 100 cases of Mainz pouch: continuing experience and evolution. *J Urol* **140:** 283–288.

200. Treiger BF and Marshall FF (1991) Carcinogenesis and the use of intestinal segments in the urinary tract. *Urol Clin N Am* **18**: 737–742.
201. Turnberg LA, Bieberdorf FA, Morawski SG *et al.* (1970) Interrelationships of chloride, bicarbonate, sodium, and hydrogen transport in the human ileum. *J Clin Invest* **49**: 557–567.
202. Turner WH, Bitton A and Studer UE (1997) Reconstruction of the urinary tract after cystectomy: the case for continent urinary diversion. *Urology* **49**: 663–667.
203. Turner Warwick R and Ashken MH (1967) The functional results of partial, subtotal and total cystoplasty with special reference to uretero-caecocystoplasty, selective sphincterotomy and cystocystoplasty. *Br J Urol* **39**: 3–12.
204. Velagapudi SRC, Ruckle H, Timm P *et al.* (1994) Nodal positive bladder carcinoma (T_4 N_1M_0); 21 year experience at a single institution. *J Urol* **151**: 451A.
205. Vet JAM, Debruyne FMJ and Schalken JA (1994) Molecular prognostic factors in bladder cancer. *World J Urol* **12**: 84–88.
206. Voges GE, Tauschke E, Stöckle M *et al.* (1989) Computerized tomography: an unreliable method for accurate staging of bladder tumors in patients who are candidates for radical cystectomy. *J Urol* **142**: 972–974.
207. Walther PJ (1993) Adjunctive adjuvant or neoadjuvant chemotherapy for locally advanced bladder cancer: a critical appraisal of the present status. *Sem Urol* **11**: 227–234.
208. Wammack R, Fisch M, Thüroff JW *et al.* (1992) The MAINZ pouch, in *Continent Urinary Diversion*, (eds R Hohenfellner and R Wammack), Churchill Livingstone, Edinburgh, pp. 127–143.
209. Wear JBJ and Barquin OP (1973) Ureterosigmoidostomy. Long-term results. *Urology* **1**: 192–200.
210. Webster GD and Bertram RA (1986) Continent catheterizable urinary diversion using the ileocecal segment with stapled intussusception of the ileocecal valve. *J Urol* **135**: 465–469.
211. Williams DF, Burkholder GV and Goodwin WE (1969) Ureterosigmoidostomy: a 15-year experience. *J Urol* **101**: 168–170.
212. Wilson TG, Moreno JG, Weinberg A *et al.* (1994) Late complications of the modified Indiana pouch. *J Urol* **151**: 331–334.
213. Wishnow KI, Johnson DE, Ro JY *et al.* (1987) Incidence, extent and location of unsuspected pelvic lymph node metastasis in patients undergoing radical cystectomy for bladder cancer. *J Urol* **137**: 408–410.
214. Wolinska WH, Melamed MR, Schellhammer PF *et al.* (1977) Urethral cytology following cystectomy for bladder carcinoma. *Am J Surg Pathol* **1**: 225–234.
215. Wood DJ, Spencer M, Hocevar BJ *et al.* (1988) Office management of urinary stomas. *Urol Clin N Am* **15**: 753–767.

216. Woodhouse CR and Gordon EM (1994) The Mitrofanoff principle for urethral failure. *Br J Urol* **73**: 55–60.
217. Woodhouse CR and MacNeily AE (1994) The Mitrofanoff principle: expansions upon a versatile technique. Presented at the meeting of British Association of Urological Surgeons, Birmingham, p. 110.
218. Zabbo A and Montie JE (1984) Management of the urethra in men undergoing radical cystectomy for bladder cancer. *J Urol* **131**: 267–268.
219. Zincke H and Segura JW (1975) Ureterosigmoidostomy: critical review of 173 cases. *J Urol* **113**: 324–327.
220. Zincke H, Garbeff PJ and Beahrs JR (1984) Upper urinary tract transitional cell cancer after radical cystectomy for bladder cancer. *J Urol* **131**: 50–52.

Commentary

Donald G. Skinner

This chapter is an outstanding review of the indications for cystectomy, surgical technique and the options for urinary diversion, including late and early complications. Several points are worthy of emphasis.

INDICATIONS FOR CYSTECTOMY

These should be based on a careful assessment of the initial tumour grade, evidence of muscle invasion, presence or absence of lymphovascular invasion, and presence or absence of coexisting carcinoma in situ. In addition, genetic markers using immunohistochemical techniques are now available in many major centres that allow selection of patients with adverse prognoses for cystectomy at an early stage when cure is still possible. Recent data strongly suggest that nuclear p53 alteration by deletion or mutation, as detected by immunohistochemical staining, can predict adverse prognosis[1,6]. It is clearly established that patients destined for cystectomy should do so when their cancer is still organ-confined, when cure is likely, rather than deferring the decision until the cancer extends outside the bladder, invades contiguous structures such as the prostatic stroma, or metastasizes to pelvic lymph nodes, when cure is less likely[2]. The availability of orthotopic urinary diversion removes much of the psychological barrier to early cystectomy because of improved quality of life compared to that provided by ileal conduit diversion.

CYSTECTOMY: STANDARD TREATMENT FOR INVASIVE BLADDER CANCER

Five fundamental observations support the view that radical cystectomy is optimal therapy for muscle invasive bladder cancer. First, the morbidity and mortality associated with the procedure have declined in the past three decades. Second, the best long-term survival to date is achieved following surgical resection of invasive tumours. Wide, *en bloc* resection of the anterior pelvic viscera, perivesical fat and pelvic lymph nodes provides negative surgical margins and low local recurrence rates. Third, transitional cell carcinoma is resistant to even high doses of radiation therapy. Fourth, chemotherapy alone or in combination with bladder sparing surgery has not yielded equivalent long-term survival comparable to radical cystectomy. Finally, surgical reconstruction of the genitourinary tract can return the patient to a satisfactory and functional life style following radical cystectomy. The majority of patients are able to micturate via the urethra, preserve renal function, and resume functional sexual activity.

PELVIC LYMPHADENECTOMY

The importance of a meticulous pelvic lymph node dissection should be emphasized. Two large series have shown that an *en bloc* pelvic lymphadenectomy can cure up to 35% of patients with pelvic lymph node metastases[5,12]. In addition, it reduces the risk of pelvic recurrence. In our experience, only 11% of 132 patients with extensive primary tumours associated with pelvic nodal metastases developed recurrences[5]. Most long-term survivors harbour microscopic metastases, not detectable by preoperative imaging (CT, MRI) or palpation at the time of surgery. Even some patients with grossly positive nodes are potential long-term survivors but most will also require adjuvant chemotherapy. The rationale for an *en bloc* dissection is that the surgeon does not know from preoperative imaging or surgical palpation who will or will not have nodal metastases; those that have benefited the most from the lymphadenectomy were those with microscopic disease. Performance of a lymphadenectomy greatly facilitates the operation, makes the performance more anatomic, decreases blood loss and is not associated with any increase in operative morbidity[5]. I routinely initiate the procedure at the aortic bifurcation and sweep all lymphatic tissue *en bloc* toward the bladder, leaving the lymphatic channels between the bladder and the hypogastric, obturator and perivesical nodes intact. After the bladder is removed I submit the presciatic nodes from each side and the presacral nodes as individual specimens to pathology. We have reported a number of long-term cures in patients with multiple positive and grossly positive metastases but most have also received adjuvant chemotherapy. Lymphadenectomy is important for staging as well as therapy and indicates those patients who might benefit from chemotherapy.

MOLECULAR AND GENETIC STUDIES

Any individual urologist involved in the management of invasive bladder cancer needs to become versed in important molecular and genetic studies that have an impact on bladder tumour progression and prognosis. In addition to routine histopathological evaluation of the cystectomy specimen, recent studies have shown that p53 alteration, as assessed by immunohistochemical nuclear accumulation of p53, is an important predictor of bladder cancer recurrence and death. In a study of 243 patients undergoing radical cystectomy between 1983 and 1988, Esrig and associates[1] determined that those patients whose tumour showed a mutated or deleted p53 gene had a very significant probability of tumour recurrence and death following radical cystectomy compared to those patients whose tumour showed a normal or wild-type p53 expression. Furthermore, in patients with organ-confined bladder cancer, p53 was the only predictor of progression, independent of stage and presence or absence of lymphovascular invasion. It also appears that patients with p53 alteration benefit from chemotherapy.

BLADDER RECONSTRUCTION

Reconstructive techniques have now been available for over 10 years[8]. These techniques, well summarized by Drs Turner and Studer, allow for the construction of a urinary reservoir with many of the characteristics of the native bladder: large capacity, low pressure, non-refluxing, continent, permitting volitional voiding per urethra with no abdominal stoma necessary. The patients are able to return to a near normal life style, including the resumption of sexual activity. Lower urinary tract reconstruction is now routine in between 80 and 90% of both men and women. Fear of urethral recurrences are unfounded, with two large series reporting less than a 4% late recurrence in men[3,4]. I agree that it is premature to determine the long-term risk in women, but the urethra is rarely involved by primary bladder cancer and many centres are now routinely performing lower urinary tract reconstruction in women[9,10,11]. Current informed consent should make every candidate for cystectomy aware of the option of lower urinary tract reconstruction. If a urologic surgeon feels uncomfortable with newer techniques, or is unwilling to spend the time learning these techniques or performs that surgery infrequently, the patient should be referred for a second opinion or to have the procedure performed at centres with the necessary experience.

CONTINENCE

The key to continence is preparation of the urethra with minimal dissection anteriorly. There is no evidence that preservation of the sympathetic nerves improves continence – we routinely remove them as part of the pelvic node dissection and radical cystectomy. In fact, preservation of the sympathetic neurovascular bundle in females may result in hypercontinence with the need for intermittent catheterization[7]. Continence in both males and females following cystectomy with lower urinary tract reconstruction is a function of the pudendal nerves which lie anatomically under the endopelvic fascia within the levator muscles and innervate the so-called rhabdoid sphincter. The majority of this sphincter is anterior and the surgeon must be very careful not to place large sutures, right-angled clamps, or do significant dissection in the area of the dorsovenous complex for fear of injuring the control mechanism. We prefer to divide the dorsovenous complex sharply in both men and women and use eight 2–0 Vicryl® sutures to incorporate a small portion of the urethra including the mucosa with the edge of the rhabdoid sphincter which will control the venous plexus. Those sutures are used to anastomose the intestinal reservoir to the urethra.

CONCLUSION

The past three decades have seen tremendous progress in the surgery of invasive bladder cancer. It is possible to perform radical cystectomy with acceptable morbidity in patients of all ages. The option of continent orthotopic urinary diversion has increased both patient and physician acceptance of the operation. The operation provides important staging information which can select a subset patients who are appropriate candidates for adjuvant chemotherapy. Newer molecular and immuno-histochemical techniques will help identify those patients at greatest risk for progression and death. Hopefully, these techniques will identify patients requiring more aggressive additional therapy and spare others not needing adjuvant chemotherapy and its toxicity. Finally, these advances will encourage both patients and physicians to seek early, appropriate treatment of invasive bladder cancer.

COMMENTARY REFERENCES

1. Esrig D, Elmagian D, Groshen S, Freeman JA *et al.* (1994) Accumulation of nuclear p53 and tumor progression in bladder cancer. *N Engl J Med* **331:** 1259–1264.
2. Freeman JA, Esrig D, Simoneau AR *et al.* (1995) Radical cystectomy for high risk patients with superficial bladder cancer in the era of orthotopic urinary retention. *Cancer* **76**: 833–839.
3. Freeman J, Tarter TA, Esrig D *et al.* (1996) Urethral recurrence in patients with ileal neo-bladder. *J Urol* **156**: 1615–1610.
4. Hautmann RE, Miller K, Steiner H *et al.* (1993) The ileal neobladder; 6 years of experience with more than 200 patients. *J Urol* **150**: 40–45.
5. Lerner SP, Skinner DG, Lieskovsky G *et al.* (1993) The rationale for en bloc pelviclymph node dissection for bladder cancer patients with nodal metastases: long term results. *J Urol* **149**: 758–764.
6. Sarkis AS, Dalbangni G, Gordon-Cardo C *et al.* (1993) Nuclear over expression of p53 protein in transitional cell bladder carcinoma. A marker for disease progression. *J Natl Can Inst* **85**: 53–59.
7. Skinner DG (1996) Editorial commentary on: Hautman RE, Paiso T, DePetriconi R. Ileal neobladder in women: 9 years of experience in 18 patients. *J Urol* **155**: 76.
8. Skinner DG, Studer U, Okada K *et al.* (1995) Which patients are suitable for continent diversion or bladder substitution following cystectomy or other definitive local treatment? *Int J Urol* **2** (Suppl. 2): 105–112.

9. Stein JP, Cote RJ, Freeman JA *et al.* (1995) Indications for lower urinary tract reconstruction in women after cystectomy for bladder cancer: pathological review of female cystectomy specimens. *J Urol* **154**: 1329–1333
10. Stein JP, Stenzl A, Esrig D *et al.* (1995) Lower urinary tract reconstruction in women following cystectomy using the Kock ileal reservoir with bilateral ureteroileal urethrostomy: initial clinical experience. *J Urol* **152**: 1404–1408.
11. Stenzl A, Draxl H, Posch B *et al.* (1995) The risk of urethral tumors in female bladder cancer: can the urethra be used for reconstruction of the lower urinary tract? *J Urol* **153**: 950–955.
12. Vieweg J, Whitmore WF Jr, Herr HW *et al.* (1994) The role of pelvic lymphadenectomy and radical cystectomy for lymph node positive bladder cancer. *Cancer* **73**: 3020–3028.

8

Radiotherapy for bladder cancer

Mary K. Gospodarowicz, Padraig Warde and Rob G. Bristow

INTRODUCTION

Preservation of a functional organ without compromising survival is the optimal treatment approach to the management of cancer and has been used successfully in the treatment of larynx cancer, breast cancer and limb sarcoma. Radiation therapy (RT) is often used either alone or in combination for organ preservation. Such an approach is particularly successful in cancers amenable to a complete surgical resection. When a potentially curative salvage treatment is available in the event of local failure following RT, overall survival has not been adversely affected by attempts at organ preservation. Definitive RT has been used for muscle invasive bladder cancer since the early 1900s and there is evidence that patients can achieve durable local control and maintain a functional bladder without compromising survival[4,14,27,28,30,57]. However, in bladder cancer, there has been considerable resistance to organ preservation with primary radiotherapy and selective cystectomy for RT failure[52] particularly since the availability of continent urinary diversion and orthotopic bladder substitution.

The primary reason for the lack of acceptance of bladder preservation is the paucity of well-executed trials comparing radical radiotherapy with radical cystectomy. Other reasons include the possibility of local disease progression prior to salvage surgery, possible compromise of survival by the inevitable delay in cystectomy, concerns that post-radiation cystectomy is associated with greater morbidity and that the total amount of treatment would be greater than with immediate cystectomy. The latter point is certainly valid and emphasizes that bladder preservation should not be attempted in patients who are likely to relapse and require an early salvage cystectomy. There are additional concerns that a radical course of radiation may damage bladder function to the extent that a patient's quality of life is compromised even if the cancer is controlled. These quality of life issues have been addressed rarely but Lynch and colleagues have documented excellent quality of life in patients following complete response to radical radiation for bladder cancer, similar to a control group with no prior bladder malignancy[48]. In this study, 72 patients who showed an initial response to radiotherapy for muscle invasive bladder cancer, were compared to 55 control patients with no history of bladder cancer. Using a patient-based and validated quality of life questionnaire (Nottingham Health Profile), and physician-administered bladder symptom score, no difference in patient quality of life or urinary symptoms were found between the two groups.

A randomized trial comparing definitive radiation therapy with cystectomy alone has not been performed. Although radical radiation therapy has been used successfully in the treatment of muscle invasive bladder cancer, the reported overall 5-year survival rates range from only 20% to 40% (Table 8.2)[4,14,27,29,61,79]. The 5-year survival for patients with clinical T2/T3a disease has been reported to range from 26% to 59%. While these results seem disappointing for patients with T2 disease, it must be appreciated that the criteria used for clinical staging were inconsistent. While the TMN classification requires a transurethral resection to be performed to distinguish T2 from T3a tumours, most reports do not indicate whether a TUR was performed. Also, the majority of reports quote the results for all T3 tumours, ignoring the well-documented difference in outcome for patients with T3a tumours as compared with those with T3b tumours, the latter by definition having an extravesical mass. The lack of uniformity in interpreting and reporting clinical stage and advances in imaging of bladder cancer in the past two decades make historical series difficult to interpret.

Comparisons of the results of institutional experience and non-randomized studies have consistently shown higher local control and survival for surgically staged patients treated with cystectomy compared with clinically staged patients treated with RT. However, selection bias, stage migration, clinical versus pathological staging, and differences in prognostic factors in patients selected for RT (i.e. performance status and co-morbid illnesses) have not been considered in these comparisons. Not surprisingly, younger, fitter patients are usually managed with surgery, and their outcome is better than the older, less fit population of patients treated with RT. Furthermore, in many reviews of RT the use of early and aggressive salvage cystectomy has not been consistent. Indirect evidence suggests that the survival of patients treated with cystectomy is not significantly different from those treated with definitive RT and selective salvage cystectomy. This evidence comes from prospective randomized trials showing no significant difference in survival for cystectomy as compared to preoperative RT plus cystectomy[12,80], or preoperative RT plus cystectomy versus definitive RT[2,5,76]. Table 8.1 summarizes the most relevant factors for comparison and Table 8.2 illustrates the results that can be obtained by radiotherapy as definitive treatment. Bladder preservation utilizing definitive RT is a well-studied and feasible approach aimed to preserve the long-term quality of life for patients with muscle invasive bladder cancer.

Although radical radiation is an important modality in bladder preservation, local control rates and survival have not yet been optimized. In this chapter we will focus

Table 8.1 Definitive radiation therapy versus cystectomy

	Definitive RT	Radical cystectomy
Staging	Clinical	Surgical/pathological
Applicability	T2–T3b disease	pT2–pT3b disease
Local control	Optimal in T2, T3a T2–T3a 60–70% T3b 25–35%	Optimal in pT2, pT3a pT2–pT3a 80–90% pT3b 60–65%
Survival (5 years)	T2–T3a 40–50% T3b 25–35%	pT2–pT3a 60–70% pT3b 25–40%
Benefits	Normal bladder function Improved quality of life	Immediate local control
Relative contra-indications	Previous pelvic RT incontinence Inflammmatory bowel disease Chronic pelvic infection	Anaesthetic risk Age
Acute complications	Rectal and bladder irritation Mortality < 1%	Mortality 1–3% Post-operative infection, fistulae
Chronic complications	Radiation cystitis <5% Radiation proctitis <5% Bowel obstruction <3%	Urinary diversion 100% Calculi, stenosis, infection Self-catheterization

Table 8.2 Results of radical radiation therapy in patients with muscle invasive bladder cancer

Author	Centre	Number of patients	5-Year survival (%)			
			T2	T3 T3a/T3b	T4 T4a/T4b	Overall
Duncan	Edinburgh	699	40.2	25.9	11.6	30.0
Gospodarowicz*	PMH	121	59.0	52/29.7	50/16	44.8
Pollack	MD Anderson	135	42.0	20.0	0.0	26.0
Vale	St Barts	60	38.0	12.0		
Blandy	London	614	27.0	38.0	9.0	
Smaaland	Bergen	146	25.7	9.9**		
Davison*	Glasgow	675	49.1	27.7	2.3	

*Cause specific survival.
**Results for T3/T4 combined.
See references [4,13,14,27,60,79,84].

on strategies to improve the results of radiation therapy in muscle invasive bladder. These include selection of patients who are well suited for bladder conservation, selection of the overall treatment plan (RT alone, concurrent chemotherapy and RT and neoadjuvant chemotherapy), discussion of issues concerning RT treatment planning and delivery, post-RT follow-up management, and finally possible novel treatment strategies based on the molecular genetics of radiation response in bladder tumours.

PATIENT SELECTION

The patient

Not all patients with bladder cancer are suitable candidates for organ preservation. Detailed assessment of patients who are being considered for treatment with RT is essential. Patients who have had multiple transurethral resections for recurrent superficial bladder tumours, or multiple courses of intravesical chemotherapy or BCG are often not suitable for definitive RT. Experience suggests that these patients are at higher risk for RT-induced complications, and have a lower probability of long-term bladder preservation. Patients with diffuse malignant involvement of bladder mucosa and pre-existing suboptimal bladder function are also at risk for increased acute radiation toxicity and rarely recover useful bladder function. For example, definitive RT is not appropriate for patients with incontinence or severe urinary symptoms and a small bladder capacity. In these patients, early cystectomy is usually the most appropriate treatment, both to produce a rapid improvement in symptoms and achieve local tumour control. Patients with a history of pelvic inflammatory disease, symptomatic adhesions from previous pelvic surgery, or inflammatory bowel disease, are not suitable candidates because of the increased risk of radiation toxicity[3,9]. Also, patients with large atonic bladders, and those patients with large bladder diverticula, are not optimal candidates for definitive RT alone as variation in bladder volume makes daily RT treatment set-up difficult to reproduce.

The tumour

Favourable prognostic factors for local control with radiotherapy include solitary tumours, categories T2 and T3a, no ureteric obstruction and complete transurethral resection of visible tumour[24,27,30,32]. Poor local control after RT alone is observed in patients with multiple tumours, categories T3b, T4b, large tumours (extravesical mass greater than 5 cm) and ureteric obstruction. Also, the presence of carcinoma in situ (CIS), squamous differentiation, DNA aneuploidy, high percentage of tumour cells in S-phase, β-HCG (human chorionic gonadotrophin) expression by the tumour and high levels of neurone-specific enolase have all been implicated as adverse prognostic factors[20,40,54]. However, at present, these latter factors are not helpful in selecting patients for radiotherapy and their value in determining clinical radiocurability is unclear.

To assure consistent treatment delivery to the target volume, a meticulous assessment of disease extent with cystoscopy, bimanual palpation under anaesthesia, and imaging of the primary tumour are essential. The documentation of the depth of bladder wall invasion by the tumour is important and, as a minimum, the presence of muscle invasion should be confirmed histologically. Patients with superficial bladder cancer (Ta, T1) do not require and do not benefit from definitive RT. The TNM classification for bladder cancer has recently been revised to reflect the importance of documenting extravesical tumour extension (Chapter 2).

Pelvic lymph node involvement is a common finding at cystectomy in patients with locally advanced bladder cancer. Conversely, the absence of lymph node metastases is difficult to establish clinically. Frequently, CT or MRI imaging is negative but lymph node dissection is positive. Although laparoscopic lymph node dissection has been used in staging prostate cancer, its use has not been explored in patients with bladder cancer. Similarly to prostate cancer, laparoscopic lymph node dissection may provide the information required to select the most appropriate treatment approach for bladder cancer patients with a high risk of nodal involvement.

Patients with regional lymph node involvement are at very high risk for occult metastatic disease. In the past, no effective systemic treatment was available for such patients, and therefore the detection of lymph node involvement was not essential. However, some recent chemotherapy trials of MVAC (methotrexate, vinblastine, Adriamycin®, cisplatin), CMV (cisplatin, methotrexate and vinblastine), or CISCA regimens have suggested benefit of adjuvant chemotherapy in these patients, making detection of lymphatic spread an important factor in bladder cancer management. The value of systemic chemotherapy in this setting has been tested by several investigators[19,21,75,81] and is reviewed in Chapter 9. In the absence of conclusive randomized trials, it is reasonable to treat patients with regional lymph node involvement with chemotherapy and a bladder conserving approach with RT[64].

SELECTION OF THE OVERALL TREATMENT PLAN

Having established the extent of disease and selected the patient for bladder preservation, the next decision is the choice between RT alone or RT combined with concurrent cisplatin chemotherapy, or neoadjuvant combination chemotherapy. The use of altered fractionation regimens or brachytherapy may be considered but is still experimental. If the probability of durable local control with radiation therapy alone is high and the likelihood of distant metastatic disease is low, then definitive RT alone is recommended.

Radiation therapy as a single modality

The overall complete response rate to radiotherapy approximates 50%, but significantly better results can be achieved in selected patients using the prognostic factors discussed above. Full-dose external-beam RT as single modality therapy is appropriate for patients with tumours that are T2, T3a, solitary, with no associated CIS and no ureteric obstruction.

Patients with a gross **complete transurethral resection of bladder tumour** prior to RT have been reported to have improved local control after radiotherapy[24,71]. In a series of 50 patients treated with radiotherapy and concurrent cisplatin, 20 of 24 patients (83%) who underwent macroscopically complete transurethral resection achieved a complete remission, as compared to 21 of 30 patients (70%) who had an incomplete resection[71]. It seems rational to expect better local control after debulking of the primary tumour. However, in the absence of prospective randomized studies, some of this benefit may be attributed to case selection rather than the transurethral resection itself.

The presence of **coexistent carcinoma in situ** (CIS) contributes to a lower long-term local tumour control following RT, with a higher rate of new tumour recurrence and less likelihood of long-term bladder preservation[24,27,85]. The contrary view has been published by Quilty *et al.*[63] where the presence of mucosal abnormalities, including carincoma in situ, was not an independent prognostic factor for local recurrence following radiotherapy. In recent years, there have been encouraging reports of the early use of intravesical BCG therapy in patients with either persistent, or recurrent, CIS following RT, suggesting excellent response rates. For example, Pisters and colleagues reported a 70% response rate for patients with CIS after radiation therapy treated with intravesical BCG[58]. Long-term outcome data on these patients were not available. Further studies are necessary to clarify to what extent transitional cell CIS can be controlled with external-beam radiotherapy alone, and to what extent post-radiotherapy intravesical BCG can increase long-term local disease control.

Loco-regional failure in the pelvis remains a significant problem in patients treated either with cystectomy or radiation. Local control in these patients is clearly suboptimal and alternative strategies should be explored. In the past decade attempts to increase the local control of bladder cancer have included the use of concurrent chemotherapy and RT, or a combined modality therapy approach consisting of systemic chemotherapy followed by definitive RT. Patients with a large extravesical mass or a tumour involving the pelvic side wall (T4b) or other poor prognostic factors rarely benefit from definitive RT alone. Any symptomatic improvement obtained is usually short lasting. As local control is uncommon, such patients should be considered for phase I–II studies of innovative treatment approaches, or offered palliative treatment with an emphasis on improving quality of life rather than survival.

Concurrent chemotherapy and radiation therapy

To improve the therapeutic ratio of RT, concurrent chemotherapy has been investigated as a radiation sensitizer, cisplatin as a single agent being the drug of choice for this purpose. In addition to its cytotoxic activity, it is also a radiation sensitizer. Enhancement of radiation effect has been demonstrated *in vivo*, *in vitro*, and under hypoxic conditions. The major toxicities of cisplatin and radiation therapy do not overlap, there is no evidence that concurrent cisplatin increases the acute radiation toxicity, but the toxicity is additive. A large number of phase II studies of concurrent cisplatin and radiation therapy have been conducted (Table 8.3) and the results are encouraging. High complete response rates, bladder preservation in over 60% of patients and no apparent detrimental effect on survival have been reported. These results suggest significant selection of favourable cases, which appears to be confirmed by the low distant failure rates observed in these trials. Toxicity in these studies was limited, and treatments appeared to be tolerable for the majority of patients.

To determine whether the addition of concurrent cisplatin to preoperative or definitive radiation therapy improved local control and survival, the National Cancer Institute of Canada has conducted a prospective randomized trial[10]. Ninety-nine patients were treated with either definitive radiotherapy or cystectomy with preoperative radiotherapy (40 Gy) and were randomized to receive either intravenous cisplatin (100 mg/m^2 at 2-week intervals during radiation therapy), or no concurrent chemotherapy. The pelvic recurrence-free survival was significantly better for the cisplatin than the non-cisplatin arm (67% versus 47% ($P = 0.038$)). No difference in overall survival was detected, although the power of the study to detect a small survival benefit was limited by the sample size[10].

Table 8.3 Phase II studies of concurrent chemotherapy and RT

Author	Centre	Number of patients	Treatment	Results
Shipley	NBCCGA	57	Cisplatin + RT	79% CR
Jakse	Innsbruck	22	Cisplatin + VM-26 + RT	77% CR
Coppin	Vancouver	29	Cisplatin + RT	76% CR
Sauer	Erlangen	67	Cisplatin + RT	85% CR
Eapen	Ottawa	26	Cisplatin (i.a.) + RT	92% CR
Rotman	SUNY Brooklyn	18	5FU ± Mit C + RT	61% CR
Russell	Seattle	16	5FU + RT	60% CR
Housset	Paris	54	5FU, cisplatin + RT	74%

CR, complete response; 5FU, 5-fluorouracil; i.a., intra-arterial; Mit C, mitomycin-C.
See references[10,16,38,41,68,69,71,77].

In spite of the numerous phase II studies, the optimal dose schedule for administering concurrent cisplatin has not been determined. The optimal treatment probably would require exposure to cisplatin every day during the radiation therapy but this is limited by the toxicity of daily cisplatin. In non-small cell lung cancer, different schedules of concurrent cisplatin and radiotherapy have been investigated and differences in response rates in favour of daily cisplatin administration have been demonstrated[72]. Eapen and colleagues have investigated the use of intra-arterial cisplatin and concurrent definitive radiation therapy in bladder cancer reporting excellent results with 89% complete response rate and 80% cause-specific survival[15,16]. The toxicity of this approach can be significant in inexperienced hands, and no randomized trials have been conducted to show superiority of this approach over the intravenous route. The strategy of concurrent cisplatin and RT and other radiation sensitizing drugs is promising. Currently, concurrent cisplatin with RT is standard therapy in the United States and Canada. However, since there is no consensus as to the optimal route for cisplatin adminstration and the dose prescription, this issue deserves further study.

Neoadjuvant chemotherapy followed by radical radiotherapy

Treatment failure in bladder cancer occurs most often from the growth of occult systemic metastases which seeded prior to local therapy. This suggests that if effective adjuvant chemotherapy is given before local treatment, loco-regional relapse and distant metastatic spread may be reduced. However, this strategy is effective only if chemotherapy eradicates all subclinical metastases and if the primary tumour can be treated effectively by local therapy. A phase II study of neoadjuvant CMV chemotherapy followed by definitive radiation with concurrent cisplatin has been conducted in 91 patients with T2/T4a tumours by the Radiation Oncology Therapy Group. The 4-year survival was 62% and 4-year actuarial survival with bladder preservation was 44%[83]. While these results were encouraging, the hope that neoadjuvant chemotherapy would improve survival in patients with bladder cancer has been dashed by the results of the MRC/EORTC trial[33] . These results suggest that a short course of currently available chemotherapy is unlikely to produce major survival differences in this group of patients. Also, there are no definitive data to suggest that neoadjuvant therapy improves local control. The search for more effective chemotherapy should be continued.

RADIOTHERAPY – TREATMENT DELIVERY

Radiation therapy for bladder cancer is usually administered by the use of external-beam radiotherapy using a linear accelerator. Less frequently, brachytherapy is used for small tumours, in which the radiation source is implanted in the bladder wall. With both techniques, the aim is to obtain a homogeneous dose of radiation within the target volume, while minimizing the radiation dose received by the surrounding normal tissue.

External-beam radiation therapy

There is surprisingly little information available on the influence of radiation dose, target volume, treatment planning and technique on the outcome of radiotherapy in bladder cancer. Most centres recommend treating the pelvic lymph nodes to a microscopic tumour dose and then delivering a further dose to a boost volume to a total bladder dose of 60–68 Gy given in 30–40 fractions over 6–8 weeks. Whether the regional lymph nodes should be included in the radiotherapy field and whether the whole bladder should be irradiated has never been established. Shipley *et al.*[77] have recommended that the final boost dose be given to the primary tumour with a margin, rather than the whole bladder, as is the practice in many other sites.

There are several potential logistic problems in defining the target volumes in patients with bladder cancer. First, the size of the bladder varies depending on the volume of the urine stored within it, but the influence of patient hydration on the reproducibility of the target volume has not been studied. Patients are asked to void prior to receiving radiation therapy as a manoeuvre to standardize the size of the target volume. However, depending on the interval between the patient emptying the bladder and the time of actual treatment delivery, there could be a considerable accumulation of urine. This could be affected further by the use of diuretics and the ingestion of diuretic beverages such as coffee or soft drinks immediately prior to radiation treatment. Second, the bladder is a very mobile organ. Changes in its exact location during treatment have clear potential to affect the accurate delivery of radiation therapy[82]. In prostate cancer, significant changes in the location of the prostate gland during a course of external-beam RT have been well documented. The use of fiducial markers and portal imaging to ensure the organ position in the treatment field on a daily basis is currently being tested. Similar strategies in bladder cancer may be important. Another issue in the planning process of radiation therapy is the use of the CT planning to help in delineating the target volume. CT planning has been shown to be superior to the use of planning cystograms as the latter delineate the inside of the bladder only and do not allow imaging of the thickness of the tumour, or the bladder wall[67]. Although the use of CT allows for visualization of the bladder wall thickness, the imaging of patients with an empty bladder compromises the ability to delineate the bladder contour accurately. Therefore, the use of both a CT scan to plan the volume, and a cystogram to confirm the location of the superior bladder wall are recommended.

Studies conducted almost 30 years ago have shown that higher RT doses improved local control in bladder cancer[53]. However, the use of higher RT doses are associated with higher rates of late complications. The use of conformal planning techniques and limiting higher radiation doses to patients with small bladder tumours, thus protecting at least a part of the bladder from the effect of high radiation doses may improve local control and minimize late toxicity. Experience with brachytherapy has taught us that small areas of the bladder are able to tolerate much higher radiation doses without the compromise of future bladder function. It is important to limit the amount of radiation that the normal or uninvolved bladder receives to prevent late radiation damage, as this would necessitate cystectomy for radiation complications, negating the benefit of the organ-preservation approach.

Alternate fractionation

Analysis of radiation dose–response data for cells derived from normal and tumour tissues suggests that relative differences exist in the repair of radiation-induced damage following doses in the clinically relevant range of 1–3 Gy per fraction. Both experimental and clinical data support the use of a decreased dose per RT fraction (less than 2 Gy) as a means to deliver tumouricidal radiation treatment while sparing late-reacting normal tissues and increasing the overall therapeutic ratio. This use of an increased number of small radiation fractions delivered over the same overall time is termed 'hyperfractionated radiotherapy', and is currently being explored for its efficacy in a number of tumours. In squamous cell cancers of the head and neck, the initial reports are encouraging. Randomized clinical trials have indicated improved results in a subset of patients[37]. In bladder cancer, Naslund *et al.* have reported results from 168 patients with muscle invasive tumours (T2/T4) randomized to receive 84 Gy in 84 fractions given three times a day, or 64 Gy in 32 fractions using daily treatments[55]. Both groups of patients were treated over a period of 8 weeks using a two-week rest interval in the standard fractionation group. There was a statistically significant improvement in local control and survival in the hyperfractionated treatment group, maintained with 10-year follow-up. The actuarial risk of late bowel complications was not reported in this study, but there did not appear to be a major risk of significant late toxicity in normal tissues. The split-course treatment approach used in the standard arm of the study has largely been abandoned in modern radiotherapy practice because of the potential for tumour regrowth during the 2-week rest interval.

Clinical observations have also suggested that an increased overall radiation treatment time is associated with a decreased probability of tumour control, and have highlighted the potential deleterious effects of tumour cell proliferation during treatment. In 830 patients with carcinoma of the cervix treated with RT at the Princess Margaret Hospital, a significant effect of treatment on pelvic control was observed[25]. The loss of control was approximately 1% per day. Patients receiving postoperative RT for head and neck cancers in recently reported randomized trials have shown improved loco-regional control when treatment was given over 5 weeks, rather than over 7 weeks[1]. These clinical observations have been confirmed experimentally using flow cytometric end points in tumours treated with chemo-radiotherapy[62]. A number of cell proliferation assays, i.e. the potential doubling time of Tpot, and p53, KI-57, PCNA, are being evaluated as to their ability to predict rates of tumour cell proliferation. In the future, these assays may identify patients with tumours who would benefit from accelerated fractionation protocols which are designed to deliver RT over a shorter overall treatment time. The feasibility of a partially accelerated approach using 60 Gy in 30 fractions over 5 weeks has been reported by Plataniotis *et al.* in 39 patients with T2/T3 bladder cancer[59]. It is unclear, however, whether any therapeutic gain will be obtained from this approach and clinical trials are currently addressing this hypothesis.

Brachytherapy

With interstitial therapy, it is possible to deliver a high radiation dose to a limited area of the bladder in a relatively short period of time. Various approaches, including the permanent implantation of radon seeds and gold grains, and temporary implants of radium, caesium needles, tantalum or iridium wires have been used. In order to decrease the potential hazard of radiation exposure to paramedical and medical staff, afterloading techniques using iridium or tantalum wires are recommended. Treatment is usually confined to patients with solitary, small (less than 5 cm in diameter) tumours, and in many cases a short course of external-beam radiation is given prior to the implant. In a recent series reported from The Netherlands Cancer Institute, 28 patients with T2/T3a disease were treated with a combination of external beam and implant using iridium[51]. With a minimum follow-up of 2 years, only six patients (21%) have relapsed in the bladder, and two of these patients relapsed with a superficial disease only. Other centres have reported equally encouraging results. However, the application of interstitial radiotherapy has remained confined to only a few centres in France and The Netherlands. In the absence of prospective randomized trials comparing external-beam radiotherapy with interstitial treatment, it is impossible to conclude whether a brachytherapy-based approach offers a significant advantage to that provided by external-beam therapy.

POST-TREATMENT MANAGEMENT

Organ-preservation strategy in bladder cancer must include the early identification of patients with residual or recurrent disease to ensure prompt salvage cystectomy. In our institution, the first follow-up cystoscopy, with a biopsy of any suspicious or residual lesions, is carried out within 6–10 weeks of completion of treatment. In patients with initial complete remission, follow-up cystoscopy with assessment of urine cytology should be carried out on a regular basis. The optimal follow-up schedule for patients who achieve complete response has not been determined, although cystoscopy at intervals of 3 or 4 months for 2 years, and less frequent intervals thereafter, is the usual practice. Biopsy confirmation of recurrence is mandatory prior to salvage cystectomy as radiation necrosis and inflammatory changes in the bladder may mimic recurrent or persistent tumour.

Patients with preserved bladders following treatment of invasive transitional cell carcinoma may be at a lifelong risk of developing new bladder cancers as well as possible local recurrence of the original tumour. Superficial recurrences may be managed successfully with transurethral resection and intravesical chemotherapy. Patients with residual or recurrent carcinoma in situ should be treated with intravesical BCG. In patients with positive urine cytology, but no evidence of tumour in the bladder, it is important to consider the presence of transitional cell carcinoma in the upper tracts, the lower end of the ureter and the urethra, as secondary urothelial tumours are common[35]. Possible chemopreventive approaches, such as that observed with retinoids or retinoid analogues, could be evaluated as a novel approach to preclude post-RT tumour recurrence[36].

POSSIBLE CELLULAR AND MOLECULAR APPROACHES TO IMPROVING THE RESULTS OF BLADDER RADIATION TREATMENT

The curative potential of radiation therapy is limited by a number of cellular and microenvironmental factors including intrinsic tumour cell radioresistance, the number of tumour stem cells or ‘clonogens’, the radiation tolerance of normal tissues, and the presence of coexisting metastases. Current advances in the understanding of tumour and molecular biology will no doubt have a major influence on future treatment strategies, as the past decade has seen an explosive increase in our knowledge of the molecular effects of ionizing radiation and other DNA-damaging agents.

Intratumoural hypoxia as a cause of radioresistance and tumour progression

Hypoxic cells typically require two to three times the radiation dose to produce a similar degree of cell killing compared to cells with normal oxygen status. The presence of both transient and chronic hypoxic cell populations has been demonstrated and may contribute significantly to clinical radioresistance. Furthermore, the presence of intratumoural hypoxia may provide a microenvironment for the clonal selection of genetically unstable mutant cells which are resistant to apoptosis and subsequently lead to the generation of aggressive tumour cell phenotypes[31,65].

Much effort has gone into the design of agents which could potentially reverse the level of hypoxia found in tumours as a means to improve the efficacy of radiation treatment and to decrease the potential mutagenic properties of the hypoxic environment. For example, compounds are available which act to mimic the radiosensitization effect of oxygen with increased electron-affinity (i.e. Nimorazole, SR4233, etc.) and have been shown to augment radiation damage among hypoxic tumour cells. Radiotherapy can also be combined with agents that are selectively toxic to hypoxic cells, such as mitomycin C, E09, and Tiropazamine. Ongoing development of predictive assays for hypoxic cells is paramount to the appropriate selection of patients as candidates for hypoxia-targetting protocols as a recent meta-analysis has suggested that there is benefit in local tumour control using these approaches[56].

Advances in molecular radiobiology that may refine the selection of patients for bladder-preservation radiotherapy protocols

Traditionally, radiation cell survival was thought to be determined by the relative repair of single-strand and double-strand DNA breaks. However, more recent data suggest that proteins contained in the cellular membrane may also be an important target for cells that preferentially undergo radiation-induced apoptosis[23]. The progression of non-invasive to invasive bladder cancer requires a series of molecular changes including the activation of certain oncogenes, and the inactivation of tumour suppressor genes (i.e. p53 and pRB-retinoblastoma genes) which can alter the cellular mechanisms that normally control the cell cycle and the stability of the genome following DNA damage (Figure 8.1). Other changes may involve the bax and bcl-2 proteins, whose relative ratio of expression may determine the ability of cells to undergo spontaneous or radiation-induced apoptosis[43]. These changes may impact on the relative aggressiveness of the tumour cells and their relative survival following radio- and chemotherapy. As such, pretreatment determination of gene expression and intrinsic radiosensitivity may be an important factor in radio-therapeutic prognostication and in the selection of patients for bladder preservation protocols.

Some, but not all, retrospective clinico-pathological studies have suggested that alterations in p53, pRB or bcl-2 protein expression may predict for decreased survival in patients undergoing cystectomy[17,26,43]. Furthermore, tumours that have lost both normal p53 and pRB protein function may be more lethal than those tumours that have lost the cellular function of only one of these proteins (reviewed by Liebert and Seigne[45]). Unfortunately, these studies suffer from a number of methodological problems, including low numbers of patients with muscle invasive disease, the use of immunohistochemistry rather than DNA sequencing to document alterations in gene function[17], results which are stage-specific[18], and the

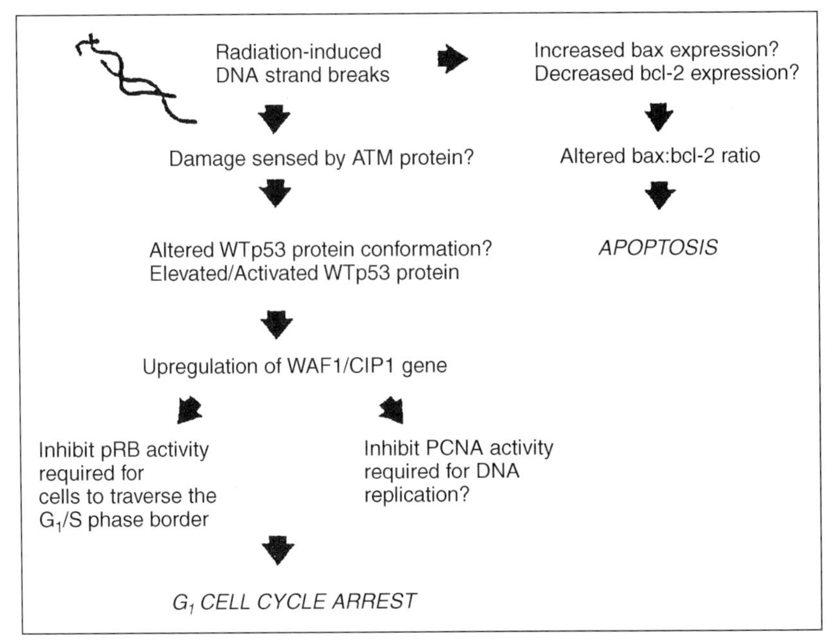

Figure 8.1 In normal cells, ionizing radiation can activate a number of proteins involved in controlling DNA damage cell-cycle checkpoints in the G_1 and G_2 phases of the cell cycle, including modifying the activities of the p53 and pRB (retinoblastoma) tumour suppressor proteins. One current model suggests that the DNA strand breaks caused by ionizing radiation activate the cellular functions of the p53 protein. One of these p53-mediated activities leads to an increase in the p21WAF1 protein which can inhibit the activity of other proteins involved in cell-cycle control, including the pRB-retinoblastoma and cyclin protein. Depending on the cell type, ionizing radiation can lead to either a radiation-induced G_1 cell-cycle arrest, or, as shown, radiation-induced apoptosis.

lack of information regarding co-expressed oncogenic sequences which may modify tumour suppressor gene function in the transformed cell (reviewed in Bristow *et al.*[6]).

Tumour-cell radioresistance may be acquired following loss of normal p53 protein function[6], a decreased bax:bcl-2 protein ratio[8], increases in DNA repair capability[46] and altered intracellular signalling cascades that utilize the mitogen-activated and stress-activated kinase pathways (i.e. MAP- and SAP-kinase pathways, respectively). Studies conducted on panels of cultured tumour cell lines representing transitional cell carcinomas (TCCs) of the bladder or other human or rodent histologies, have shown that cells that lack normal p53 function can acquire relative radioresistance and have decreased rates of radiation-induced apoptosis *in vitro* and *in vivo*[6,50,78].

Pollack and colleagues[86] have published data suggesting that the radiation response of bladder tumours treated with pre-cystectomy radiotherapy may be correlated with an increased level of pretreatment tumour apoptosis (i.e. spontaneous) and a loss of pRB protein expression. Increased p53 protein expression was not correlated with radioresponse, but was correlated with overall survival in T3b tumours. Although the authors contend that experimental data suggest that a loss of pRB function leads to increased radiosensitivity, other studies have actually reported the opposite result[6]. Additionally, two other recent studies[11,70] reported conflicting results concerning the role that the p53-mediated pathways have in determining the relative chemosensitivity of bladder cancer to doxorubicin and platinum-containing clinical regimens. Given the inherent problems with p53 immunohistochemical analyses[17,39], it will be important to determine the role of the p53 and pRB genes in radioresponse based on more sophisticated 'functional' end points (i.e. flow cytometry) among tumour stem cell populations[6]. The determination of tumour and host genetic factors may one day allow for the determination of a 'molecular therapeutic ratio' for each patient undergoing radiotherapy [42].

Gene therapy as an adjunct to radiotherapy as a means to improve the therapeutic ratio

Pretreatment determination of altered cell-cycle arrests or apoptosis following radiation may afford the design of novel radiotherapy strategies or select patients for other modalities (Figure 8.2). For example, preclinical studies have observed a sensitization of bladder cancer cells to *cis*-platinum and radiation-induced DNA damage with the concomitant use of p53- and pRB-mediated gene therapy; a result recently supported by the results of a phase I clinical trial[22,66,87]. Other novel genetic approaches utilize radiation-inducible genetic sequences to express cytotoxic and radiosensitizing agents (tumour necrosis factor (TNF)) selectively within a irradiated treatment volume. By design, this form of 'genetic radiotherapy' leads to an increased therapeutic ratio by increasing the concentration of agents within tumour-bearing tissues to levels that may have been unattainable if the same agents were delivered systemically[34]. Bladder tumours may be amenable to both types of molecular therapies, given their potential accessibility to cystoscopic instillation of genetic constructs.

It should be noted that standard or novel strategies designed to reverse primary tumour radioresistance would be less useful if the radioresistant phenotype was correlated with the metastatic phenotype. If true, gains in local tumour control would be offset by distant spread of disease without attendant gains in survival. However, both experimental and clinical studies support the concept that these two aggressive clinical phenotypes do not correlate with each other and, particularly, that genetic-mediated radioresistance does not always co-segregate with an increased ability for spontaneous metastasis[7]. As such, current molecular efforts to improve bladder preservation could theoretically have an impact on patient survival in radioresistant

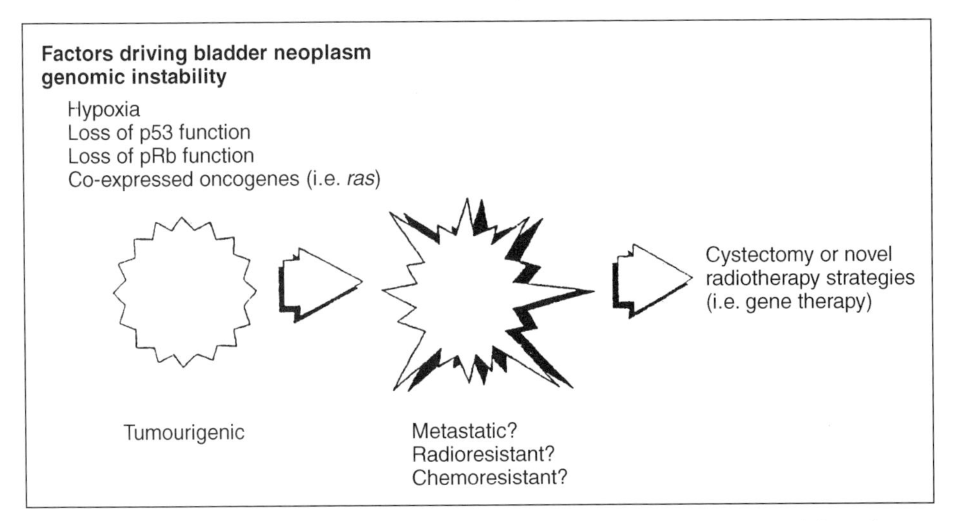

Figure 8.2 Acquired genetic changes secondary to the inherent genetic instability of tumour cells can lead to tumour progression and the subsequent acquisition of aggressive tumour phenotypes, as described in the text. Cellular and molecular assays that can determine these genetic and epigenetic changes prior to treatment will be required to triage patients between standard versus experimental therapies, and select the best candidates for bladder-preservation protocols.

T2 and T3 tumours. The results of phase II and phase III clinical trials that use these novel strategies are eagerly awaited to usher in a new area of combined modality treatments.

SUMMARY

The role of definitive radiation therapy as a means of organ-preserving curative treatment for muscle invasive bladder cancer has not been utilized optimally. The paucity of data from prospective randomized trials combined with clinicians' rigidity in their approach towards this disease has led to the abandonment of prospective investigation of radiation therapy in this disease. Definitive RT is not considered as a standard treatment option for bladder cancer in North America[73]. In contrast, RT with salvage cystectomy remains a common approach in Great Britain[47]. This diversity of opinion further inhibits progress in this area. Without renewed interest on the part of urologists with skills for accurate diagnosis, assessment, supportive and salvage therapy of patients with bladder cancer, progress in RT-based bladder preservation will not occur. As in many adult cancers, combined assessment and a multimodality treatment approach is required to optimize success of organ preservation.

Patients with localized bladder cancer should be informed of the available management options. There are excellent data to support the use of organ preservation and little suggestion that survival is compromised by this approach. By applying treatment concepts currently in use in other malignancies and optimal use of currently available treatment methods, we can improve the outcome for patients with bladder cancer. Major improvements in survival will not be achieved until better systemic management becomes available. Future research and discussion between all disciplines will determine the best strategy for the integration of RT into the overall treatment strategy for this disease. The role of RT as an adjuvant approach to either partial or total cystectomy needs to be re-evaluated as local control remains a problem in patients with locally advanced extravesical disease. New approaches of gene therapy, a new generation of hypoxic cell sensitizers, and dose escalation with conformal RT will add to the radiation oncologist's armamentarium of clinical investigation initiatives. The challenge for clinicians involved in the management of these patients is to evaluate these new therapeutic avenues prospectively and identify patients who are most likely to benefit from an organ-preservation approach.

REFERENCES

1. Ang K, Trotti A, Garden A *et al.* (1996) Overall time factor in postoperative radiation: Results of a prospective randomized trial. *Radiother Oncol* **40**: S30.
2. Barlebo H, Steven K, Sprensen BL *et al.* (1990) Preoperative irradiation (40 Gy) and cystectomy versus radiotherapy (60 Gy) followed by salvage cystectomy in the treatment of advanced bladder cancer (T2–T4a): a randomized study. (DAVECA 8201). *J Urol (Suppl.)* **143**: 291A.
3. Bentzen S and Overgaard J (1994) Patient-to-patient variability in the expression of radiation-induced normal tissue injury. *Sem Radiat Oncol* **4**: 68–80.
4. Blandy JP, Jenkins BJ, Fowler CG *et al.* (1988) Radical radiotherapy and salvage cystectomy for T2/3 cancer of the bladder. *Prog Clin Biol Res* **260**: 447–451.
5. Bloom HJ, Hendry WF, Wallace DM *et al.* (1982) Treatment of T3 bladder cancer: controlled trial of pre-operative radiotherapy and radical cystectomy versus radical radiotherapy. *Br J Urol* **54**: 136–151.
6. Bristow R, Benchimol S and Hill R (1996) The p53 gene as a modifier of intrinsic radiosensitivity: implications for radiotherapy. *Radiother Oncol* **40**: 197–223.
7. Bristow R, Brail L, Jang A *et al.* (1996) p53-mediated radioresistance does not correlate with metastatic potential in tumorigenic rat embryo cell lines following oncogene transfection. *Int J Rad Oncol, Biol Phys* **34**: 341–355.

8. Chresta C, Masters J and Hickman J (1996) Hypersensitivity of human testicular tumors to etoposide-induced apoptosis is associated with functional p53 and a high Bax–Bcl-2 ratio. *Cancer Res* **56**: 1834–1841.
9. Coia L, Won M, Lanciano R *et al.* (1990) The Patterns of Care outcome study for cancer of the uterine cervix: results of the Second National Practice Survey. *Cancer* **66**: 2451–2456.
10. Coppin C, Gospodarowicz M *et al.* (1996) The NCI-Canada trial of concurrent cisplatin and radiotherapy for muscle invasive bladder cancer. *J Clin Oncol* **14**: 2901–2907.
11. Cote R, Esrig D, Groshen S *et al.* (1997) p53 and treatment of bladder cancer. *Nature* **385**: 123–124.
12. Crawford ED, Das S, Smith JA (1987) Preoperative radiation therapy in the treatment of bladder cancer. *Urol Clin N Am* **14**: 781–787.
13. Davidson SE, Symonds RP, Snee MP *et al.* (1990) Assessment of factors influencing the outcome of radiotherapy for bladder cancer. *Br J Urol* **66**: 288–293.
14. Duncan W and Quilty PM (1986) The results of a series of 963 patients with transitional cell carcinoma of the urinary bladder primarily treated by radical megavoltage x-ray therapy. *Radiother Oncol* **7**: 299–310.
15. Eapen L, Stewart D, Danjoux C *et al.* (1989) Intraarterial Cisplatin and concurrent radiation for locally advanced bladder cancer. *J Clin Oncol* **7**: 230–235.
16. Eapen L, Stewart D, Crook J *et al.* (1992) Intra-arterial Cisplatin and concurrent pelvic radiation in the management of muscle invasive bladder cancer. *Int J Rad Oncol, Biol Phys* **24** (Suppl. 1): 211.
17. Elledge R (1996) Assessing p53 status in breast cancer prognosis: where should you put the thermometer if you think your p53 is sick? *J Natl Cancer Inst* **88**: 141–143.
18. Esrig D, Elmajian D, Groshen S *et al.* (1994) Accumulation of nuclear p53 and tumor progression in bladder cancer. *N Engl J Med* **331**: 1259–1263.
19. Evans PE and Swanson DA (1996) What to do if the lymph nodes are positive. *Sem Urol Oncol* **14**: 96–102.
20. Fossa SD, Berner AA, Jacobsen AB *et al.* (1993) Clinical significance of DNA ploidy and S-phase fraction and their relation to p53 protein, c-erbB-2 protein and HCG in operable muscle-invasive bladder cancer. *Br J Cancer* **68**: 572–578.
21. Freiha F, Reese J, Torti FM (1996) A randomized trial of radical cystectomy versus radical cystectomy plus cisplatin, vinblastine and methotrexate chemotherapy for muscle invasive bladder cancer. *J Urol* **155**: 495–499. [See comments]
22. Fujiwara T, Cai D, Georges R *et al.* (1994) Therapeutic effect of a retroviral wild-type p53 expression vector in an orthotopic lung cancer model. *J Natl Cancer Inst* **86:** 1458–1462.

23. Fuks Z, Haimovitz-Friedman A and Kolesnick R (1995) The role of the sphingomyelin pathway and protein kinase C in radiation-induced cell kill. *Imp Adv Oncol* 19–31.
24. Fung CY, Shipley WU, Young RH *et al.* (1991) Prognostic factors in invasive bladder carcinoma in a prospective trial of preoperative adjuvant chemotherapy and radiotherapy. *J Clin Oncol* **9**: 1533–1542. [See comments]
25. Fyles A, Keane T, Barton M *et al.* (1992) The effect of treatment duration in the local control of cervix cancer. *Radiother Oncol* **25**: 273–279.
26. Glick S, Howell L and White R (1996) Relationship of p53 and bcl-2 to prognosis in muscle-invasive transitional cell carcinoma of the bladder. *J Urol* **155**: 1754–1757.
27. Gospodarowicz MK, Hawkins NV, Rawlings GA *et al.* (1989) Radical radiotherapy for the muscle invasive transitional cell carcinoma of the bladder: failure analysis. *J Urol* **142**: 1448–1454.
28. Gospodarowicz MK, Quilty PM, Scalliet P *et al.* (1995) The place of radiation therapy as definitive treatment of bladder cancer. *Int J Urol* **2**: 41–48. [Review]
29. Gospodarowicz MK, Rider WD, Keen CW *et al.* (1991) Bladder cancer: long term follow-up results of patients treated with radical radiation. *Clin Oncol* **3**: 155–161.
30. Gospodarowicz MK and Warde PR (1993) A critical review of the role of definitive radiation therapy in bladder cancer. *Semin Urol* **11**: 214–226. [Review]
31. Graeber T, Osmanian C, Jacks T *et al.* (1996) Hypoxia-mediated selection of cells with diminished apoptotic potential in solid tumours. *Nature* **379**: 88–91.
32. Greven KM, Solin LJ and Hanks GE (1990) Prognostic factors in patients with bladder carcinoma treated with definitive irradiation. *Cancer* **65**: 908–912.
33. Hall R (1996) Neo-adjuvant CMV chemotherapy and cystectomy or radiotherapy in muscle invasive bladder cancer. First analysis of MRC/EORTC intercontinental trial. *Proc Am Soc Clin Oncol* **15**: 244, Abstract 612.
34. Hallahan DE, Maucera HJ, Seug LP *et al.* (1995) Spatial and temporal control of gene therapy using ionising radiation. *Nat Med* **1**: 786–791.
35. Herr HW, Cookson MS and Soloway SM (1996) Upper tract tumors in patients with primary bladder cancer followed for 15 years. *J Urol* **156**: 1286–1287.
36. Hong W and Lippman S (1995) Cancer chemoprevention. *J Natl Canc Inst Monographs* **17**: 49–53.
37. Horiot J, Bontemps P, Le Fur R *et al.* (1996) An overview of the EORTC accelerated and hyperfractionated radiotherapy trials in head and neck cancers. *Radiother Oncol* **40**: S30.
38. Housset M, Maulard C, Chretien Y *et al.* (1993) Combined radiation and chemotherapy for invasive transitional cell carcinoma of the bladder: a prospective study. *J Clin Oncol* **11**: 2150–2157.

39. Jacobs T, Prioleau J, Stillman I *et al.* (1996) Loss of tumor marker-immunostaining intensity on stored paraffin slides of breast cancer. *J Natl Cancer Inst* **88**: 1054–1059.
40. Jacobsen AB, Nesland JM, Fossa SD *et al.* (1990) Human chorionic gonadotropin, neuron specific enolase and deoxyribonucleic acid flow cytometry in patients with high grade bladder carcinoma. *J Urol* **143**: 706–709.
41. Jakse G and Frommhold H (1985) Radiotherapy and chemotherapy in locally advanced bladder cancer. *Eur Urol* **14** (Suppl. 1): 45.
42. Jorgensen T and Shiloh Y (1996) The ATM gene and the radiobiology of ataxia-telangiectasia. *Int J Rad Oncol, Biol Phys* **69**: 527–537.
43. King E, Matteson J, Jacobs S *et al.* (1996) Incidence of apoptosis, cell proliferation and bcl-2 expression in transitional cell carcinoma of the bladder: association with tumor progression. *J Urol* **155**: 316–320.
44. Koraitim M, Kamal B, Metwalli N *et al.* (1995) Transurethral ultrasonographic assessment of bladder carcinoma: its value and limitation. *J Urol* **154**: 375–378.
45. Liebert M and Seigne J (1996) Characteristics of invasive bladder cancers: histological and molecular markers. *Sem Urol Oncol* **14**: 62–72.
46. Lynch T, Anderson P, Wallace D *et al.* (1991) A correlation between nuclear supercoiling and the response of patients with bladder cancer to radiotherapy. *Br J Cancer* **64**: 867–871.
47. Lynch TH, Waymont B, Dunn JA *et al.* (1992) Urologists' attitudes to the management of bladder cancer. *Br J Urol* **70**: 522–525.
48. Lynch WJ, Jenkins BJ, Fowler CG *et al.* (1992) The quality of life after radical radiotherapy for bladder cancer. *Br J Urol* **70**: 519–521.
49. Malmstrom PU, Lonnemark M, Busch C *et al.* (1993) Staging of bladder carcinoma by computer tomography-guided transmural core biopsy. *Scand J Urol Nephrol* **27**: 193–198.
50. McIlwrath A, Vasey P, Ross G *et al.* (1994) Cell cycle arrests and radiosensitivity of human tumor cell lines: dependence on wild-type p53 for radiosensitivity. *Cancer Res* **54**: 3718–3722.
51. Moonen LM, van Horenblas S *et al.* (1994) Bladder conservation in selected T1G3 and muscle-invasive T2–T3a bladder carcinoma using combination therapy of surgery and iridium-192 implantation. *Br J Urol* **74**: 322–327.
52. Moore MJ, O'Sullivan B and Tannock IF (1988) How expert physicians would wish to be treated if they had genitourinary cancer. *J Clin Oncol* **6**: 1736–1745.
53. Morrison R (1975) The results of treatment of cancer of the bladder – a clinical contribution to radiobiology. *Clin Radiol* **26**: 67–75.
54. Moutzouris G, Yannopoulos D, Barbatis C *et al.* (1993) Is beta-human chorionic gonadotrophin production by transitional cell carcinoma of the bladder a marker of aggressive disease and resistance to radiotherapy? *Br J Urol* **72**: 907–909.

55. Naslund I, Nilsson B and Littbrand B (1994) Hyperfractionated radiotherapy of bladder cancer. A ten-year follow-up of a randomized clinical trial. *Acta Oncol* **33**: 397–402.
56. Overgaard J (1994) Clinical evaluation of nitroimidazoles as modifiers of hypoxia in solid tumors. *Oncol Res* **6**: 509–518.
57. Paschkis R (1911) Radiumbehandling von Blasengeschwulsten. *Wienner Klinische Wochenschrift* **24:** 1962.
58. Pisters LL, Tykochinsky G and Wajsman Z (1991) Intravesical bacillus Calmette–Guerin or mitomycin C in the treatment of carcinoma in situ of the bladder following prior pelvic radiation therapy. *J Urol* **146**: 1514–1517.
59. Plataniotis G, Michalopoulos E, Kouvaris J *et al.* (1994) A feasibility study of partially accelerated radiotherapy for invasive bladder cancer. *Radiother Oncol* **33**: 84–87.
60. Pollack A, Zagars GK, Cole CJ *et al.* (1995) The relationship of local control to distant metastasis in muscle invasive bladder cancer. *J Urol* **109**: 2059–2064.
61. Pollack A, Zagars GK (1996) Radiotherapy for stage T3b transitional cell carcinoma of the bladder. *Sem Urol Oncol* **14:** 86–95. [Review]
62. Priesler H, Kotelnikov V, LaFollette S *et al.* (1996) Continued malignant cell proliferation in head and neck tumours during cytoxic therapy. *Clin Canc Res* **2**: 1453–1460.
63. Quilty PM, Hargreave TB, Smith G and DuncanW (1987) Do normal mucosal biopsies predict prognosis in patients with transitional cell carcinoma of bladder treated by radical radiotherapy. *Br J Urol* **59**: 242–247.
64. Raghavan D and Huben R (1995) Management of bladder cancer. *Curr Probl Cancer* **19**: 1–64. [Review]
65. Reynolds T, Rockwell S and Glazer PM (1996) Genetic instability induced by the tumour microenviroment. *Cancer Res* **56**: 5754–5757.
66. Roth JA *et al.* (1996) Retrovirus-mediated wild type p53 gene transfer to tumours of patients with lung cancer. *Nat Med* **2**: 985–987.
67. Rothwell R, Ash D and Jones W (1983) Radiation treatment planning for bladder cancer: a comparison of cystogram localisation with computed tomography. *Clin Radiol* **34**: 103–111.
68. Rotman M, Aziz H, Porrazzo M *et al.* (1990) Treatment of advanced transitional cell carcinoma of the bladder with irrigation and concomitant 5-fluorouracil infusion. *Int J Rad Oncol, Biol Phys* **18**: 1131–1137.
69. Russel KJ, Boileau MA, Higano C *et al.* (1990) Combined 5-flourouracil and irradiation for transitional cell carcinoma of the urinary bladder. *Int J Rad Oncol, Biol Phys* **19**: 693–699.
70. Sarkis A, Bajorin D, Reuter V *et al.* (1995) Prognostic value of p53 nuclear overexpression in patients with invasive bladder cancer treated with neoadjuvant MVAC. *J Clin Oncol* **13**: 1384–1390.

71. Sauer R, Dunst J, Altendorf-Hofmann A *et al.* (1990) Radiotherapy with and without cisplatin in bladder cancer. *Int J Rad Oncol, Biol Phys* **19**: 687–691.
72. Schaake-Koning C, van den Bogaert W, Dalesio O *et al.* (1992) Effects of concomitant Cisplatin and radiotherapy on inoperable non-small cell lung cancer. *N Eng J Med* **326**: 524–530.
73. Scher H, Shipley W and Herr H (1997) Cancer of the bladder, in *Cancer – Principles and Practice of Oncology*, 5th edn, (eds VJ DeVita, S Hellman and S Rosenberg), Lippincott-Raven, Philadelphia.
74. See WA and Fuller JR (1992) Staging of advanced bladder cancer. Current concepts and pitfalls. *Urol Clin N Am* **19**: 663–683.
75. Seidman AD and Scher HI. (1991) The evolving role of chemotherapy for muscle infiltrating bladder cancer. *Sem Oncol* **18**: 585–595.
76. Sell A, Jakobsen A, Nerstrom B *et al.* (1991) Treatment of advanced bladder cancer category T2, T3 and T4a. *Scand J Urol Nephrol (Suppl.)* **138**: 193–201.
77. Shipley WU, Van der Schueren E, Kitigawa T *et al.* (eds) (1986) Guidelines for radiation therapy in clinical research on bladder cancer, in *Developments in Bladder Cancer,* Alan R Liss, New York, pp. 109–121.
78. Siles E, Villalobos M, Valenzuela M *et al.* (1996) Relationship between p53 status and radiosensitivity in human tumour cell lines. *Br J Cancer* **73**: 581–588.
79. Smaaland R, Akslen L, Tonder B *et al.* (1991) Radical radiation treatment of invasive and locally advanced bladder cancer in elderly patients. *Br J Urol* **67**: 61–69.
80. Smith J, Crawford E, Paradelo J *et al.* (1997) Treatment of advanced bladder cancer with combined preoperative irradiation and radical cystectomy versus radical cystectomy alone: a Phase III Intergroup study. *J Urol* **157**: 805–808.
81. Stockle M, Meyenburg W, Wellek S *et al.* (1995) Adjuvant polychemotherapy of nonorgan-confined bladder cancer after radical cystectomy revisited: long-term results of a controlled prospective study and further clinical experience. *J Urol* **153**: 47–52.
82. Sur RK, Clinkard J, Jones WG *et al.* (1993) Changes in target volume during radiotherapy treatment of invasive bladder carcinoma. *Clin Oncol* **5**: 30–33.
83. Tester W, Porter A, Asbell S *et al.* (1993) Combined modality program with possible organ preservation for invasive bladder carcinoma: results of RTOG protocol 85–12. *Int J Radiat Oncol, Biol Phys* **25**: 783–790.
84. Vale JA, A'Hern RP, Liu K *et al.* (1993) Predicting the outcome of radical radiotherapy for invasive bladder cancer. *Eur Urol* **24**: 48–51.
85. Wolf H, Olsen PR and Hojgaard K (1985) Urothelial dysplasia concomitant with bladder tumours: A determinant for future new occurrences in patients treated by full-course radiotherapy. *Lancet* **1**: 1005–1008.

86. Wu CS, Pollack A, Czerniak B *et al.* (1996) Prognostic value of p53 in muscle-invasive bladder cancer treated with preoperative radiotherapy. *Urology* **47**: 305–310.
87. Xu H, Zhou Y, Seigne J *et al.* (1996) Enhanced tumor suppressor gene therapy via replication-deficient adenovirus vectors expressing an N-terminal truncated retinoblasto protein. *Cancer Res* **56**: 2245–2249.

Commentary 1

Alan Horwich

This excellent review makes clear that a precept of the use of radical radiotherapy in the treatment of localized muscle invasive bladder cancer is that organ conservation would be preferable to cystectomy if these alternative approaches were equally effective and if bladder conservation was not associated with a high risk of late complication or of new tumour formation. The Institute of Urology Trial sought to address this question in T3 tumours by prospective randomized comparison of radical radiotherapy alone versus preoperative radiotherapy and cystectomy[1]. A long-term analysis of data from this trial has recently been conducted and the results are shown in Table 8.4. This analysis has been conducted on an intention-to-treat basis and it can be seen that cause-specific survival appears to be higher in patients who had initial cystectomy. However, the trial was modest in size and survival differences were not statistically significant. A similar trial was conducted in Denmark subsequently, based on patients with T2, T3 and T4a tumours, again randomized between preoperative radiotherapy and cystectomy compared to radical radiotherapy. In this study, the radical radiotherapy was given with a treatment gap and this would now be considered an inferior technique. The trial result again indicated higher survival in patients treated by cystectomy; the difference between cystectomy and radical radiotherapy was not significant[7].

Given these results, it is appropriate that radiation oncologists be cautious about a recommendation for radical radiotherapy in the treatment of muscle invasive bladder cancer for a patient who is fit for cystectomy and willing to accept the side-effects. At the same time, the oncologist will recognize that there have been marked improvements in the radiotherapy of bladder cancer since these

Table 8.4 Long-term cause-specific survival of Institute of Urology trial[1] comparing radical radiotherapy alone with preoperative radiotherapy and cystectomy in T3 bladder cancer

Randomised treatment allocations	*N*	% 5-year survival	(95% CI)	% 10-year survival	(95% CI)
RT	91	27.9	(19.1–37.8)	19.6	(11.9–29.1)
RT → Cx	98	39.8	(30.3–29.8)	30.4	(21.4–40.5)

RT, Radical radiotherapy; RT→Cx, preoperative radiotherapy and cystectomy; CI, confidence interval.

trials were conducted. In particular, advances include improved accuracy of staging and treatment planning based on axial imaging techniques, and improved tumour control associated with concomitant treatments. Additionally, advances in our understanding of the biology of bladder cancer[2] allows improved selection for radical radiotherapy, and radiobiological studies have led to investigations of altered fractionation. In particular, there has been evidence that bladder cancer has the capacity for rapid stem cell repopulation during a protracted course of treatment[8]. Maciejewski and Majewski[6] analysed the local control of 77 transitional carcinomas of the bladder. Protraction of the overall treatment time from 40 to 55 days led to a decrease in local control rate from 50% to about 5% and these results, together with data from the literature, suggested that tumour clonogen repopulation accelerates after a lag period of about 5 weeks, requiring a dose increment of about 0.36 Gy/day to compensate.

A pilot study of accelerated fractionation in the treatment of localized muscle invasive bladder cancer has been completed at the Royal Marsden Hospital[3]. The study was based on 85 patients treated with radiotherapy twice per day to a total dose of between 57.6 and 64.0 Gy in 32 fractions during an overall time period of 28 days. The treatment schedule incorporated a brief gap after the first 10 fractions. In view of the range of tumours treated, the high local control rate (80% at 3 months) should be interpreted with caution, but the pilot study did demonstrate good tolerance (Table 8.5) and has led to a prospective randomized trial of accelerated versus conventional fractionation which will complete accrual during 1997 with 250 patients registered.

In conclusion, the previous randomized comparisons of radical radiotherapy and surgery, while failing to demonstrate a significant difference in outcome, have not been sufficiently robust to demonstrate with security that these treatment approaches are equivalent. In any event, there have been considerable advances in staging and treatment techniques since these trials were carried out. With respect to radiotherapy, there have been advances in radiation biology, radiation treatment planning and also encouraging results from concomitant therapies and altered fractionation regimens which will form the basis for future clinical trials. Furthermore, evidence is accumulating that patients may be selected for an organ-conserving approach, based on their initial treatment response[4,5].

Table 8.5 Toxicity of accelerated radiotherapy in a pilot study of bladder cancer (Royal Marsden Hospital 1987–1992)

	Bowel	Bladder
Acute[2] RTOG Grade 3	9%	7%
Late[2] RTOG Grade 3	9%	18%

COMMENTARY REFERENCES

1. Bloom HJ, Hendry WF, Wallace DM and Skeet RG (1982) Treatment of T3 bladder cancer: controlled trial of preoperative radiotherapy and radical cystectomy versus radical radiotherapy, second report and review. *Br J Urol* **54**: 136–151.
2. Esrig D, Elmajian D, Groshen S *et al.* (1994) Accumulation of nuclear p53 and tumour progression in bladder cancer. *N Engl J Med* **331**: 1259.
3. Horwich A, Pendlebury S and Dearnaley DP (1995) Organ conservation in bladder cancer. *Eur J Cancer* **31A**: S208, Abstract 1000.
4. Housset M, Dufour B, Maulard-Durdux C *et al.* (1997) Concomitant fluorouracil (5-FU)–cisplatin (CDDP) and bifractionated split course radiation therapy (BSCRT) for invasive bladder cancer. *Proc Am Soc Clin Oncol* **16**: 319A, Abstract 1139.
5. Kachnic LA, Kaufman DS, Heney NM *et al.* (1997) Bladder preservation by combined modality therapy for invasive bladder cancer. *J Clin Oncol* **15**: 1022.
6. Maciejewski B and Majewski S (1991) Dose fractionation and tumour repopulation in radiotherapy for bladder cancer. *Radiother Oncol* **21**: 163.
7. Sell A, Jakobsen A, Nerstrom B *et al.* (1991) Treatment of advanced bladder cancer category T2, T3 and T4a. A randomized multicenter study of preoperative irradiation cystectomy versus radical irradiation and early salvage cystectomy for residual tumour. DAVECA protocol 8201. Danish Vesical Cancer Group. *Scand J Urol Nephrol (Suppl.)* **138**: 193.
8. Trott KR and Kummermehr J (1985) What is known about tumour proliferation rates to choose between accelerated fractionation or hyperfractionation? *Radiother Oncol* **3**: 1.

Commentary 2

Luc M. F. Moonen

BRACHYTHERAPY

A brachytherapy-based approach has been used successfully in selected bladder carcinoma for many years. The theoretical advantage of such an approach is the possibility to deliver a very high radiation dose to a limited area of the bladder, thereby increasing the chance of obtaining tumour control without increasing the incidence of severe complications. A theoretical risk of treating only part of the bladder is the development of secondary tumours in the untreated part of the bladder.

The treatment is usually confined to poorly differentiated T1 tumours and to all grades of muscle-infiltrating cancer, provided they are solitary and smaller than 5 cm. Lesions extending outside the bladder (T3b) are generally considered unsuitable for implantation because the tumour coverage using a single-plane implant would be insufficient. Treatment of larger areas to high dose would result in a high incidence of badly healing ulcers and even necrosis of the bladder wall. In general, the implant procedure is preceded by a short course of external-beam radiation which is intended to prevent tumour implantation due to perioperative tumour spill.

Various forms of brachytherapy have been used, including the permanent implantation of radon seeds and gold grains, the temporary implantation of radium or caesium needles, and tantalum or iridium wires. In order to decrease the potential hazard of radiation exposure to paramedical and medical staff, remote controlled after-loading techniques using iridium wires are used most commonly at the present time. In some centres the implantation is preceded by a partial cystectomy, while in other centres only implantation is performed.

Despite the variety of isotopes and techniques employed, the results reported by different centres are comparable. In a report from The Netherlands Cancer Institute, 53 patients with T2/T3a bladder cancer were treated with a combination of external-beam treatment and iridium implant[3,4]. Nine of these patients developed an isolated bladder relapse of whom four could be salvaged with a transurethral resection, and three with cystectomy. An additional seven patients developed bladder recurrence and distant metastases while a further eight patients developed distant metastases only. Of the 16 bladder relapses, only seven were true recurrences at the primary site as nine were new tumours located elsewhere in the bladder. Six of the recurrences were superficial tumours managed by bladder conservative treatment and the rest were muscle invasive.

Others have reported equally encouraging results. Van der Werf-Messing reported local control rates of 91% for T1 lesions, 84% for T2 and 72% for T3 cancers after radium and caesium implantation[5]. Batterman and Boon, also using radium and caesium implants, obtained a local control rate of 82% for T1 tumours and 74% for T2/T3a tumours [1]. Mazeron *et al.* reported a 77% local control rate in T1 tumours and 93% in T2 tumours after partial cystectomy plus iridium implantation[2]. Typically, in all these series, roughly half the recurrences were true recurrences at the primary tumour site while the other half were secondary tumours elsewhere in the bladder.

Despite these apparently excellent results with respect to local tumour control and bladder conservation, the application of interstitial radiotherapy has remained restricted to only a few centres, mainly in France and The Netherlands. Some critics of brachytherapy believe that the good results reflect the natural behaviour of the selected tumours rather than the superiority of the treatment. In the absence of prospective randomized trials comparing external-beam radiotherapy with interstitial treatment, it is impossible to conclude whether brachytherapy offers a significant advantage to that provided by external-beam radiotherapy.

COMMENTARY REFERENCES

1. Batterman J and Boon T (1988) Interstitial therapy in the management of T2 bladder tumours. *Endocuriether/hyperther* **4**: 1–6.
2. Mazeron JJ, Crook J, Chopin D *et al.* (1988) Conservative treatment of bladder carcinoma by partial cystectomy and interstitial irridium-192. *Int J Rad Oncol, Biol Phys* **15**: 1323–1330.
3. Moonen LMF, Horenblas S, Bos F, Meinhardt W and Bartolink H (1996) Good results of bladder-conserving treatment of poorly differentiated and muscle invasive bladder carcinomas with the aid of interstitial Irridium-192 radiotherapy. *Ned Tijdschr Geneesk* **140**: 1406–1410.
4. Moonen LMF, Horenblas S, van der Voet JCM, Nuyten MJC and Bartolink H (1994) Bladder conservation in selected T1, G3 and muscle invasive T2-T3a bladder carcinoma using combined therapy of surgery and irridium 192 implantation. *Br J Urol* **74**: 333–327.
5. Van der Werf-Messing BHP (1986) Interstitial radiation therapy of carcinoma of the urinary bladder. *Endocuriether/hyperther* **2**: 67–76.

9

Neoadjuvant and adjuvant chemotherapy in the treatment of muscle invasive bladder cancer

Christopher J. Logothetis

The use of chemotherapy in metastatic transitional cell carcinoma has evolved from the use of single agents to combination chemotherapy. The evolution of chemotherapy has been such that it has clearly shown an impact on survival relative to that of single-agent therapy[2,4] and that long-term disease-free survival can be achieved in selective series[5]. In view of this demonstrable effect on metastatic disease, it is logical to assume that chemotherapy given earlier in transitional cell carcinoma in combination with local modalities would result in improved impact. The logic for combining early therapy with systemic therapy is based on the reasoning that the combination treatment of small volume disease would result in a more effective outcome. In addition, there are significant data that suggest that cancers are in a vulnerable state immediately in the perioperative setting.

Despite the apparent logic for combining chemotherapy with local modalities, interpretation of the existing data is not simple. First, the end points of clinical trials for locally advanced bladder carcinoma are more complex and are not as simple as those for metastatic disease. When treating patients with metastatic disease, the end points of response rate and long-term disease-free survival are self-evident, as patients would be expected to die of their cancer. However, in the treatment of early disease a portion of patients will survive without the use of cytotoxic chemotherapy and will not be benefited. Such patients, and the result of such trials, are therefore under the influence of the paradox that a smaller portion of patients benefit from therapy the earlier the disease is treated. This is based on three assumptions:

1. Only those patients destined to metastasize benefit from the cytotoxic therapy.
2. The earlier the patients are treated, the higher the portion of those who will receive unnecessary chemotherapy.
3. The proportion of patients cured is not increased by the earlier introduction of systemic therapy (Table 9.1).

Table 9.1 Theoretical calculations of benefit based on risk of recurrence with systemic local therapy*

	Cured by primary surgery	
	50%	80%
NED from surgery	50	80
Number at risk	50	20
NED from chemotherapy	25	10
Relapse	25	10
Total	100	100

*Calculation based on 50% survival of patients at risk, treated with chemotherapy.
NED, no evidence of disease.

In addition to the paradox associated with the treatment of early disease are the complications that arise from the use of different end points. Like metastatic disease, the end point of long-term survival is valid, but the use of chemotherapy in localized disease may have other values. For example, the ability to preserve organ function, to reduce the problems of local relapse and the ability to palliate and perhaps even reduce the morbidity of surgery. All of these have been evoked as possible end points and benefits of cytotoxic chemotherapy[3]. In this chapter I will outline the clinical experience of The University of Texas M. D. Anderson Cancer Center (UTMDACC) of the use of chemotherapy and the basic premise used to guide trial design. The focus at the UTMDACC is driven by an intent to increase survival. Trial design is based on the evidence that chemotherapy has an impact on the survival of patients with metastatic disease and is therefore likely to have at least an equal impact on localized disease. Although intuitive, the assumption remains to be proved.

EVIDENCE FOR THE USE OF ADJUNCTIVE COMBINATION CHEMOTHERAPY IN UROTHELIAL TUMOURS

The CISCA chemotherapy combination (cyclophosphamide, Adriamycin® and cisplatin) was studied in the Department of Genitourinary Medical Oncology at UTMDACC between 1979 and 1986. The first clinical trials focused on the use of CISCA in metastatic bladder carcinoma. The randomized trial[1] failed to reveal a survival advantage for the combination over single-agent cisplatin, which supported the widely held view at that time that combination therapy was not superior to single-agent therapy[4]. Despite the negative results of the randomized trial, we interpreted our experience as revealing evidence in favour of combination chemotherapy not detected by the randomized trial. This belief was based on the small, long-term, tumour-free survival rate seen with the use of CISCA chemotherapy in a previous study, in patients with regionally unresectable bladder cancer, which had not been reported previously with the single agent. In that single-arm trial, the policy of giving chemotherapy to patients undergoing cystectomy was tested and its feasibility confirmed[3]. The failure of the randomized trial to detect the existence of the putative benefit was considered to be due to the small number of patients treated.

Of equal importance was that patients with regionally advanced disease (T4b, N+) had a higher likelihood of achieving a gratifying response to chemotherapy. We interpreted this as evidence that treatment of disease with a lesser degree of tumour dissemination (T4b, N+ but no distant metastases) was more likely to result in a survival benefit from chemotherapy.

ADJUVANT CISCA CHEMOTHERAPY IN PATIENTS WITH RESECTED UROTHELIAL CANCER

Based on the evidence that chemotherapy of metastatic disease appeared to permit long-term disease-free survival in a portion of patients, we felt that there was sufficient evidence that chemotherapy may increase the long-term disease-free survival rate when combined with surgery, and further study was warranted[5]. A prospective non-randomized study was performed in which patients with localized bladder cancer were treated with chemotherapy following surgery. We concluded that adjuvant chemotherapy was feasible and could be delivered immediately postoperatively without excess morbidity in patients with resected regional disease. In addition, there was a suggestion of a long-term disease-free survival advantage with the addition of adjuvant chemotherapy. Based on the conclusions of that trial, we felt confident that adjuvant chemotherapy was feasible in 'unselected patients' with bladder cancer following a cystectomy.

Based on the assumption of a likely survival advantage of chemotherapy when combined with surgery, the trials we performed thereafter focused on the optimal integration of chemotherapy and surgery, and were meant to be complimentary to trials conducted by others whose purpose was to confirm the existence and to assess the degree of the survival advantage. The principles that guided study developments at UTMDACC were:

1. Optimal local care was defined as a cystectomy and pelvic lymph node dissection.
2. Optimal chemotherapy was five courses of MVAC (methotrexate, vinblastine, Adriamycin®, cisplatin).
3. Selection of therapy (surgery only versus surgery and chemotherapy) was to be determined by the risk of relapse.

OPTIMAL INTEGRATION OF CHEMOTHERAPY AND SURGERY FOR LOCALLY ADVANCED BLADDER CANCER

Localized invasive bladder cancer has a wide range of clinical behaviour. Some patients with invasive bladder cancer may enjoy a high rate of tumour-free survival in the era of a modern cystectomy[7,8]. Others have a low rate of survival, with the development of locally recurrent cancers and metastases which are unresectable at their initial clinical presentation. Selection for adjuvant therapy should be determined by the degree of local tumour extent and probability of cure by surgery only. We therefore studied patients at high risk for developing recurrent tumours after an optimal cystectomy. The patient selection criteria for our next randomized trial permitted the inclusion of only patients at high risk for relapse. The basic assumptions were that chemotherapy improved the survival of patients with bladder cancer; surgery is the optimal local therapy for patients with an invasive and locally advanced bladder cancer, and clinical stages T2 with vascular invasion, T3b and T4a have a low cure rate with local therapy only.

The purpose of the trial was to compare the relative therapeutic advantage of preoperative (neoadjuvant) chemotherapy versus postoperative (adjuvant) chemotherapy in patients with biopsy-confirmed bladder carcinoma categories T3a (with vascular or lymphatic invasion), T3b or T4a. The rationale for this comparison was based on the reported advantages of preoperative tumour control in other tumour types and the potential benefits of initial chemotherapy compared to that of postoperative chemotherapy. Patients were randomized to receive either two cycles of pre-cystectomy MVAC + three cycles postoperatively, or five cycles postoperatively. The trial was designed to detect a 20% difference in disease-free survival between the two treatment arms, and plans to recruit 148 patients. Secondary aims of the trial were to detect differences in morbidity of therapy and to examine the validity of the clinical staging. The initial results of an interim analysis of the trial based on the first 100 patients recruited have been reported[6].

The clinical stage of the patients was found to be a reliable predictor of adverse pathology. The validity of clinical staging was demonstrated by the finding that T category was a reliable predictor of adverse pathology at cystectomy and correctly selected patients to receive chemotherapy. In the 48 patients who received no preoperative MVAC adverse prognostic factors justifying chemotherapy were confirmed in 46 of the 48 patients (Table 9.2). By comparison of the 52 patients receiving initial (neoadjuvant) chemotherapy, 14 had no pathological tumour in the pathology specimen, i.e. pT0 (Table 9.3). This represents considerable downstaging, amounting to a pathological complete response rate of 28% from two cycles of MVAC. Furthermore, unresectable cancer was found in 3/51 patients who received neoadjuvant MVAC and 8/48 who underwent immediate cystectomy. Unresectable disease was defined as cancer that could not be removed surgically or viable carcinoma extending to the margin of the cystectomy pathology specimen. These criteria were adopted to ensure accurate

Table 9.2 Correlation of clinical and pathological stage in 48 patients treated by immediate cystectomy

Clinical T category	Pathological stage						
	pT0	pTIS/pTa	pT1–3	pT3b	pT4a	N+	Unresectable
T3 + vascular invasion	0	1	3	2	2	2	2
T3b	1	0	6	12	1	7	3
T4a	0	0	0	1	1	1	3
Total	1	1	9	15	4	10	8

Table 9.3 Correlation of clinical and pathological stage in 51 patients who received two cyles of neoadjuvant MVAC

Clinical T category	Pathological stage						
	pT0	pTIS/pTa	pT1–3	pT3b	pT4a	pN+	Unresectable
T3 + vascular invasion	2	2	2	1	0	3	0
T3b	10	2	4	7	1	6	1
T4a	2	1	1	2	1	1	2
Total	14	5	7	10	2	10	3

assessment of the true rate of resectability, and to exclude the possibility that resectability rates were affected by surgeon behaviour, influenced by knowledge of the use of preoperative chemotherapy.

Of the total trial population of 100 patients, 70% received at least four courses of MVAC. Seven of the 48 randomized did not receive chemotherapy following cystectomy. Tolerability of MVAC was favourable. Increase in serum creatinine (6–29%) and episodes of infection (6–19%) were the two most significant unwanted effects. One patient died of surgical complications and two patients died of chemotherapy toxicity, one in each arm. After 31.7 months of follow-up, overall disease-free survival was 49% in the neoadjuvant group and 52% in those treated by immediate cystectomy and adjuvant MVAC.

Based on these preliminary data, we feel that the clinical stage is a reliable tool to assess prognosis and, hence, select therapy. In addition, we see no advantage in delaying chemotherapy to be delivered postoperatively. Long-term follow-up of the total trial population will be required to assess the relative impact of each of these therapies on outcome, and to evaluate further the appropriate integration of chemotherapy and surgery and the selection of patients for future study.

EARLY STAGE INVASIVE BLADDER CANCER

The focus of the studies performed at the UTMDACC has been to assess the impact of therapy on survival. Recent studies indicate that some series of properly staged patients with T2 and T3a bladder cancer treated with a modern cystectomy have excellent survival rates[7,8]. Based on this data we believe that patients with stage T2 or T3a bladder cancer are clinical categories in which survival end points are not the only worthy focus of study. The rate of organ preservation in these patients who have an excellent tumour-specific survival with frequently localized disease is appropriate. Hence, patients in these clinical categories may derive a benefit from chemotherapy by increasing the probability of achieving organ preservation. The range of therapeutic options for these patients includes transurethral resection of the

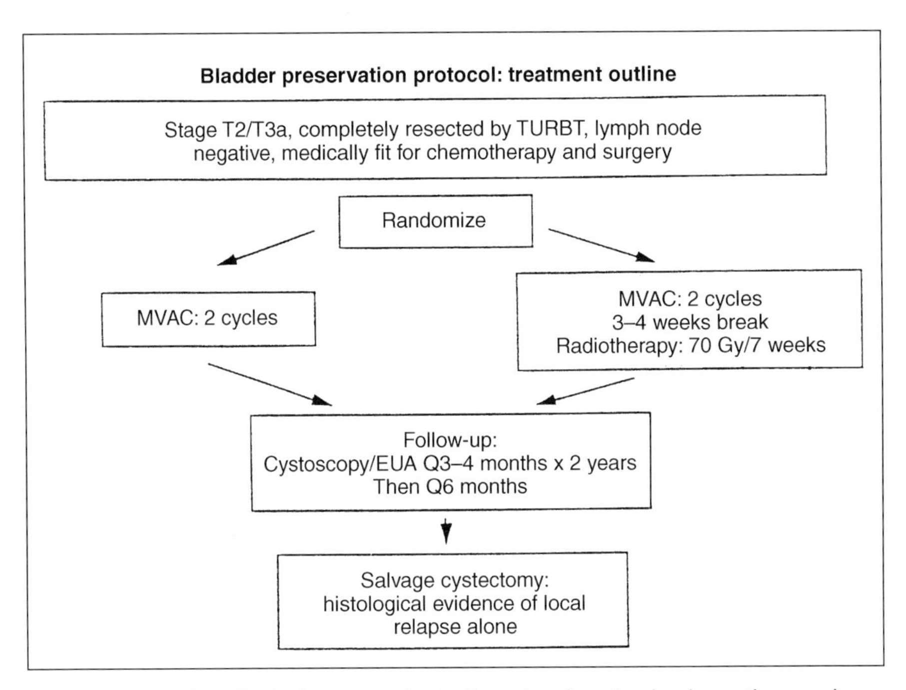

Figure 9.1 Proposed study design to evaluate the role of systemic chemotherapy in organ-preserving treatment of good-prognosis muscle invasive bladder cancer (University of Texas M. D. Anderson Cancer Center). EUA, examination under anaesthetic; MVAC, methotrexate, vinblastine, Adiamycin®, cisplatin; TURBT, transurethral resection of the bladder tumour.

bladder tumour alone or in combination, although evidence to date is not convincing that the rate of organ preservation has been expanded by the use of preoperative chemotherapy in combination with reduced surgery[3]. In an attempt to answer this question, we have designed a clinical trial for patients who meet the criteria for a cystectomy but who may be candidates for the benefit of chemotherapy in reducing the need for cystectomy (Figure 9.1).

SUMMARY

We believe that sufficient evidence exists that selected patients with advanced and unresectable bladder cancer may achieve long-term disease-free survival with neoadjuvant or adjuvant systemic combination chemotherapy. There is also evidence that chemotherapy can be tolerated without excessive morbidity in therapeutically significant doses immediately before or after cystectomy. However, there is no convincing evidence that cytotoxic chemotherapy impacts either on survival or on the rate of organ preservation in patients with localized but resectable advanced urothelial malignancies of the bladder. Ongoing trials will answer the question and tell us how to combine local therapy with systemic therapy appropriately.

REFERENCES

1. Khandekar JD, Elson PJ, DeWys WD and Slayton RE (1981) Comparative activity and toxicity of *cis*-diaminedichloroplatinum (DDP) vs. cyclophosphamide (CTX), adriamycin (ADR) and DDP (CAD) in disseminated transitional cell carcinoma of the urinary tract (DTCUT). *Proc Am Ass Can Res* **22**: 461, Abstract C-500.
2. Loehrer PJ, Einhorn LH, Elson PJ *et al.* (1992) A randomised comparison of cisplatin alone or in combination with methotrexate, vinblastine and doxorubicin in patients with metastatic urothelial carcinoma: a co-operative group study. *J Clin Oncol* **10**: 1066–1072.
3. Logothetis CJ (1991) Organ preservation in bladder carcinoma: a matter of selection. *J Clin Oncol* **9**: 1525–1526.
4. Logothetis CJ, Dcxeus FH, Chong C *et al.* (1989) Cisplatin, cyclophosphamide and doxorubicin chemotherapy for unresectable urothelial tumors: The M D Anderson Experience. *J Urol* **141**: 33–37.
5. Logothetis CJ, Johnson DE, Chong C *et al.* (1988) Adjuvant cyclophosamide, doxorubicin and cisplatin chemotherapy for bladder cancer: an update. *J Clin Oncol* **6**: 1590–1596.
6. Logothetis CK, Swanson D, Amato R *et al.* (1996) Optimal delivery of perioperative chemotherapy: preliminary results of randomised propective comparative trial of pre-operative and post-operative chemotherapy for invasive bladder carcinoma. *J Urol* **155**: 1241–1245.

7. Schoenberg MP, Walsh PC, Breazeale DR *et al.* (1996) Local recurrence and survival following nerve sparing radical cystoprostatectomy for bladder cancer: ten year follow up *J Urol* **155**: 490–494.
8. Studer UE, Bacchi M, Biedermann N *et al.* (1994) Adjuvant cisplatin chemotherapy following cystectomy for bladder cancer: results of a prospective randomised trial *J Urol* **152**: 81–84.

Commentary

Mahesh K. B. Parmar and Sarah Burdett

Dr Logothetis has provided a full review of the experience of the University of Texas M. D. Anderson Texas Cancer Center in the use of cisplatin-based chemotherapy with surgery in patients with locally advanced (muscle invasive) bladder cancer. As he points out, the main reason for using such chemotherapy in locally advanced disease is to improve survival. This clear account of the practice of a single, highly regarded institution is enlightening, but must be considered in a wider context of evidence and experience elsewhere. In an effort to summarize all such evidence from completed randomized controlled trials, in 1995 Ghersi *et al.*[6] performed a systematic review and meta-analysis of all randomized trials of neoadjuvant cisplatin-based chemotherapy. For the purposes of this discussion, we have updated this meta-analysis and systematically reviewed the evidence from reported randomized trials on the basis of a single end point of overall survival using cisplatin-based chemotherapy in either an adjuvant or neoadjuvant setting. In undertaking this review the approach to systematic reviews specified by the Cochrane Collaboration[3] has been used. In particular, a recommended search strategy has been employed (Appendix 1) and this has been augmented by inspection of the reference lists of relevant articles to identify as many published randomized trials as possible. It should be stressed that although the previous review published by Ghersi *et al.*[6] endeavoured to collect individual patient data and tried to identify both published and unpublished trials, the present update is much more limited. Our review has concentrated on identifying trials published in the English language, indexed by MEDLINE, and, when available, we have extracted relevant data on overall survival from each publication, rather than analysing source data for individual patients. Well-established methods, using the stratified logrank test, were used to combine the results of trials, when appropriate [11].

The rationale for giving chemotherapy before local treatment (neoadjuvant or primary chemotherapy) is to allow introduction of chemotherapy as early as possible to impact immediately on any possible micro-metastases and also to reduce the size of the primary tumour to make local treatment more effective. In this context local treatment could be either radical radiotherapy or surgery. The alternative is to give chemotherapy after the primary local treatment (adjuvant chemotherapy). There may be some advantages for this compared with neoadjuvant treatment. First, a decision to treat with chemotherapy can be made on the basis of pathological rather than clinical stage, perhaps more accurately identifying those at higher risk of micro-metastases and subsequent relapse. Second, chemotherapy is given when there is minimal tumour bulk, perhaps maximizing the opportunity for complete eradication of the disease.

NEOADJUVANT CHEMOTHERAPY

By the end of 1994 Ghersi *et al.* had identified four published randomized trials[10,13,21] , testing the value of adding neoadjuvant cisplatin-based chemotherapy to local therapy (surgery or radiotherapy). In our updated review a further three trials have been found, one performed by an Egyptian group[1], one performed by a number of groups in collaboration led by the Medical Research Council and the European Organization for Research and Treatment of Cancer[7], and one performed by a collaborative Nordic group[9]. These three trials have reported in abstract form only. A publication updating the results of one of the trials included in the previous collaboration was also found[8]. A summary of the design of these trials and the numbers recruited is given in Table 9.4.

Cisplatin alone

Three trials have used cisplatin alone, from Australia[21], the UK[21] and Spain[10]. In the Australian trial 137 patients were randomized to receive radiotherapy alone or two cycles of cisplatin at 100 mg/m^2 followed by radiotherapy. The UK trial randomized 159 patients to radiotherapy alone or to three cycles of cisplatin at 100 mg/m^2 also

Table 9.4 Randomized trials comparing neoadjuvant cisplatin-based chemotherapy plus local treatment against local treatment alone in patients with locally advanced bladder cancer

Trial	Chemotherapy dose and schedule	Local	Opened	Closed	Patients (*n*)
Australia[21]	C 100 mg/m^2	Radiotherapy	Feb. 85	Feb. 88	137
UK[21]	C 100 mg/m^2	Radiotherapy	Jun. 84	Jun. 88	159
Spain[10]	C 100 mg/m^2	Cystectomy	Oct. 84	Aug. 89	122
Nordic 1[13]	C 100 mg/m^2 A 30 mg/m^2	Radiotherapy + cystectomy	Dec. 85	May 89	325
Nordic 2[9]	C Not available M Not available	Radiotherapy + cystectomy	91	Not available	325
MRC/EORTC[7]	C 100 mg/m^2 M 30 mg/m^2 V 4 mg/m^2	Radiotherapy or cystecomy	Nov. 89	July 95	976
SWOG[16]	M 30 mg/m^2 V 3 mg/m^2 A 30 mg/m^2 C 70 mg/m^2	Cystectomy	Aug 87	Not applicable since ongoing	Target accrual 298

C, cisplatin; A, Adriamycin® (doxorubicin); M, methotrexate; V, vinblastine.

followed by radiotherapy. Finally, the Spanish trial randomized 122 patients to cystectomy alone or three cycles of the cisplatin (100 mg/m^2) followed by cystectomy. The results of these individual trials and the overall results combining them are shown in Figure 9.2. It can be seen by examining the confidence intervals around the hazard ratio estimate of each trial that none of the trials shows clear evidence of a benefit for single-agent cisplatin chemotherapy. In fact, the estimates for two of the trials (Australian and UK) favour radiotherapy alone, although these results are not conventionally significant. The combined overall estimated effect of the three trials slightly favours local treatment alone, with an estimated overall hazard ratio of 1.11 ($P = 0.41$; 95% confidence interval = 0.86–1.43). There is therefore no clear evidence of an effect of chemotherapy. However, the total number of patients randomized in the combination of these trials is 379, sufficient only to detect relatively large differences in survival reliably. Smaller differences are likely to remain undetected. The level of uncertainty is reflected by the width of the 95% confidence interval for the overall hazard ratio, stretching from 0.86 to 1.43. These limits translate into possible absolute differences of 5% (from 50% 3-year survival to 55%) in favour of chemotherapy plus local treatment, to 13% (from 50% to 37%) in favour of local treatment alone.

Cisplatin in combination with other drugs

Four trials – the Nordic trials 1[8] and 2[9], the MRC/EORTC Intergroup trial[7] and the Egyptian trial[1] – have reported results assessing the value of cisplatin in combination with other drugs. However, only two provide sufficient data to be included in this review. The Egyptian trial did not provide any overall survival information, and the Nordic 2 trial has reported preliminary results with insufficient detail to extract the information necessary to include quantitative information in this review. The Nordic 1 trial used two cycles of cisplatin and doxorubicin and randomized 325 patients (analysed 311 patients). The Nordic 2 trial, was conducted with the aim of confirming the exploratory results of the Nordic 1 trial (a survival benefit in patients with T3/T4 disease, but no benefit in patients with T1/T2 disease) but changed the chemotherapy regimen to two cycles of neoadjuvant cisplatin plus methotrexate. In both Nordic 1 and 2 the local treatment used was short-term (flash) radiotherapy followed by cystectomy. The MRC/EORTC Intergroup trial used three cycles of cisplatin, methotrexate and vinblastine and randomized and analysed 976 patients. Local treatment in this trial was left to institutional choice of radiotherapy, cystectomy or preoperative radiotherapy plus cystectomy. This choice was made before randomization, and thus was balanced across the chemotherapy and no-chemotherapy arms.

The updated results of the Nordic 1 trial indicate a possible benefit for chemotherapy with a hazard ratio of 0.69 with a 95% confidence interval of 0.49–0.98, just excluding the equivalence value 1. This trial included patients with

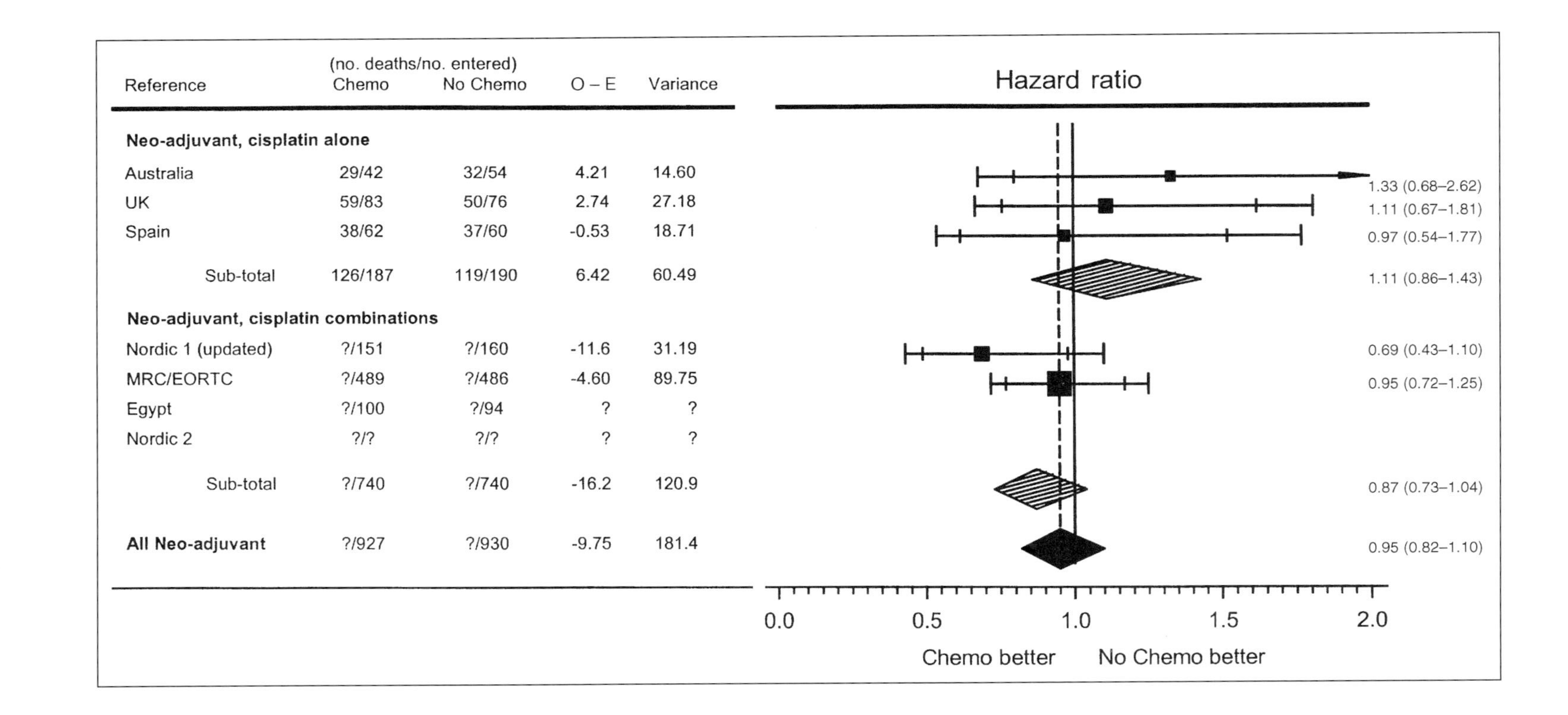

Reference
(no. deaths/no. entered)
Chemo
No Chemo
O – E
Variance
Hazard ratio
Neo-adjuvant, cisplatin alone
Australia 29/42 32/54 4.21 14.60 1.33 (0.68–2.62)
UK 59/83 50/76 2.74 27.18 1.11 (0.67–1.81)
Spain 38/62 37/60 -0.53 18.71 0.97 (0.54–1.77)
Sub-total 126/187 119/190 6.42 60.49 1.11 (0.86–1.43)
Neo-adjuvant, cisplatin combinations
Nordic 1 (updated) ?/151 ?/160 -11.6 31.19 0.69 (0.43–1.10)
MRC/EORTC ?/489 ?/486 -4.60 89.75 0.95 (0.72–1.25)
Egypt ?/100 ?/94 ? ?
Nordic 2 ?/? ?/? ? ?
Sub-total ?/740 ?/740 -16.2 120.9 0.87 (0.73–1.04)
All Neo-adjuvant ?/927 ?/930 -9.75 181.4 0.95 (0.82–1.10)
0.0
0.5
1.0
1.5
2.0
Chemo better
No Chemo better

T1G3 disease but, from the information given in the publication, it was not possible to 'extract' these patients from the overall analysis. The authors did perform subgroup analyses by T category and commented in their conclusions that 'neoadjuvant chemotherapy seems to improve long-term survival after cystectomy in patients with stages T3 to T4a bladder carcinoma, while no survival benefit was found for stages T1 to T2 disease'. This analysis of subgroups, although interesting, should be considered no more than in generating hypotheses for testing in other trials (Chapter 12). In an endeavour to confirm the apparent survival benefit seen in patients with T3–T4 disease, the Nordic 2 trial started in 1991. The design was the same as for the Nordic 1 trial except that only patients with T3 or T4 were entered. A total of 325 patients were entered. Preliminary results suggested no evidence of a difference in overall survival between the two randomized groups. The report does not contain sufficient detail of the survival results to be able to include quantitative estimates within this review.

Figure 9.2 Hazard ratio plot for trials comparing neoadjuvant cisplatin-based chemotherapy plus local treatment against local treatment alone in patients with locally advanced bladder cancer. For each trial the number of deaths (events) and the number of patients randomized are given. The statistic $O - E$ is the difference between the observed (O) and expected (E) deaths for chemotherapy. The expected deaths are calculated using a logrank-type analysis for survival types data. Under the assumption that there is no difference between the two arms, $O - E$ should differ only randomly from zero. (Similarly the total ($O - E$) should also differ only randomly from zero.) The statistic V, the variance, is a measure of the information contained in the trial. So, the MRC/EORTC trial with $V = 89.75$ has almost three times more information than Nordic 1 trial with $V = 31.19$. The right-hand side of the figure shows the hazard ratio (HR) for each trial, calculated using the expression: $\exp[(O - E)/V]$, and represented by the centre of the open square. The hazard ratio given by the centre of each square and also at the end of each line is a measure of the relative death rate in the two arms of the trial: a value of 1 represents no difference in death rates. The value of 1.33 for the Australian trial represents an increase in the death rate of 33% as a result of the use of chemotherapy. A value less than 1 represents a reduction in the risk of death, e.g. 0.97 for the Spanish trial represents a 3% reduction $[(1-0.97) \times 100\%]$ in the risk of death with the use of chemotherapy. Confidence intervals for the hazard ratio estimate for each trial are given by the lines either side of the square. The inner ticks represent the 95% interval, while the outer ticks represent the 99% interval. The 99% interval is also given numerically in brackets at the end of the line. If the line between the inner ticks does not cross the equivalence line 1, then the trial has produced a result significant at the 5% level (that is $P < 0.05$). Similarly, if the line enclosed in the outer ticks does not cross the equivalence line, the trial has produced a result significant at the 1% level (that is $P < 0.01$). The hazard ratio combined across groups of trials is given by the diamond. The first two shaded diamonds represent the results combined across trials of cisplatin alone and cisplatin in combination, respectively. The filled diamond represents the results across both groups of trials. The ends of each diamond represent the 95% confidence interval for the combined result. If the diamond does not cross the equivalence line 1, then the result is significant at the 5% level ($P < 0.05$).

The largest trial of all, the MRC/EORTC trial, gave a hazard ratio of 0.95 with no evidence of a difference between the chemotherapy and no-chemotherapy groups ($P = 0.63$; 95% confidence interval = 0.77–1.17). Nevertheless, even in such a large trial the confidence intervals are relatively wide, indicating that we cannot reliably exclude differences of the order of 5–7% absolute improvement or detriment in 3-year survival.

Combining the results of the Nordic and MRC/EORTC trials gives a hazard ratio of 0.87 ($P = 0.14$; 95% confidence interval 0.73–1.04), representing a 13% reduction in the risk of death with combination chemotherapy. The result is, however, not significant by conventional statistical criteria. Thus, there is no clear evidence of an effect of chemotherapy. The estimate translates into an absolute difference of 5% (from 50% 3-year survival to 55%) and the 95% intervals translate into an absolute difference of 10% (from 50% 3-year survival to 60%) in favour of chemotherapy to an absolute difference of 1% (from 50% to 49%) in favour of local treatment alone. It may be thought that addition of the Nordic 2 data would reduce the overall effect as it was a 'negative' trial. However, Nordic 2 was not large enough to detect a difference of the order of 5%, and thus it may be entirely consistent with the combined estimate from the Nordic 1 and MRC/EORTC trials. It is difficult, therefore, to predict the likely impact of this trial on the overall result.

Ongoing trials

There is one ongoing randomized trial[16] of neoadjuvant chemotherapy which will contribute more information in the future. The South-West Oncology Group (SWOG) initiated this trial in 1987, and is comparing three cycles of the MVAC regimen (methotrexate, vinblastine, doxorubicin and cisplatin) given before surgery with surgery alone. The aim is to enter 298 patients and first results are awaited.

ADJUVANT CHEMOTHERAPY

Five randomized trials have been performed to assess the value of adjuvant cisplatin-based chemotherapy (Table 9.5).

Cisplatin alone

In the only trial[19] testing single-agent cisplatin chemotherapy, the SAKK group from Switzerland randomized 80 (77 eligible) patients to cystectomy plus three cycles of cisplatin or cystectomy alone. With 5-year survival figures of 54% and 57% in the chemotherapy and no-chemotherapy groups, respectively, the authors

Table 9.5 Randomized trials comparing adjuvant cisplatin-based chemotherapy plus local treatment against local treatment alone in patients with locally advanced bladder cancer

Trial	Chemotherapy	Local treatment	Opened	Closed	Patients
Studer *et al.*[19]	C 90 mg/m^2	Cystectomy	April 84	May 89	80
Skinner *et al.*[15]	Varied, most patients C 100 mg/m^2 A 60 mg/m^2 CY 600 mg/m^2	Cystectomy	July 80	Dec. 88	91
Stockle *et al.*[18]	M, V, A or M, V, E, C (doses not given)	Cystectomy	May 87	Dec. 90	49
Freiha *et al.*[5]	C 100 mg/m^2 M 30 mg/m^2 V 4 mg/m^2	Cystectomy	April 86	April 93	55
Bono *et al.*[2]	C 70 mg/m^2 M 40 mg/m^2	Cystectomy	Dec. 84	Dec. 87	125

C, cisplatin; A, Adriamycin® (doxorubicin); M, methotrexate; V, vinblastine; CY, cyclophosphamide.

reported no evidence of a difference between the two groups (P = 0.65). The authors, however, do point out that this is a very small trial which, with only 35 deaths, could have detected only very large differences in survival.

Cisplatin in combination with other drugs

Four randomized trials[2,5,15,18] have reported results comparing local treatment plus adjuvant combination chemotherapy against local treatment alone. In all four trials the local treatment was cystectomy.

In the first trial, Skinner *et al.*[15] randomized 91 patients to radical cystectomy plus chemotherapy versus radical cystectomy alone. A large but not conventionally statistically significant difference (P = 0.099) in survival between the chemotherapy and control groups was reported, with median survivals of 51 months for the chemotherapy group and 29 months for the no-chemotherapy group. A larger difference was observed in time to progression, with 70% of patients free of disease at 3 years in the chemotherapy group and 46% free of disease at the same point in the no-chemotherapy group (P = 0.011). The survival and time to progression from this trial are presented in Figure 9.3. However, this trial had a number of flaws,

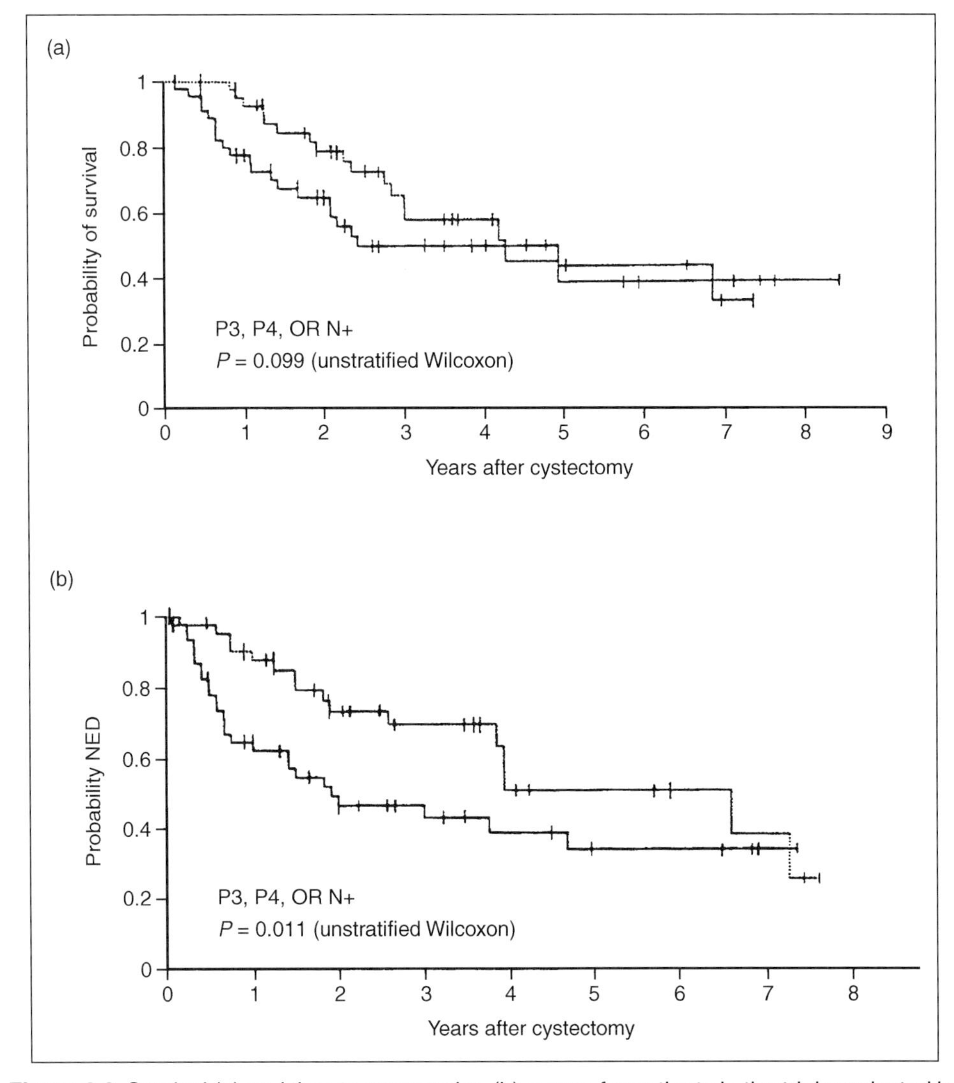

Figure 9.3 Survival (a) and time to progression (b) curves for patients in the trial conducted by Skinner *et al.*[15]. NED, no evidence of disease. (Reproduced from Skinner *et al.* (1990) Adjuvant chemotherapy following cystectomy benefits patients with deeply invasive bladder cancer. *Sem Urol* **8**: 279–84, by permission of W. B. Saunders Co. Ltd, London.)

which have been highlighted elsewhere[20]. The chemotherapy was of variable type and duration for different patients and a large proportion (25%) of patients allocated chemotherapy did not actually receive it. It is therefore difficult to believe that the observed effect on survival is real and not caused by an imbalance of important but unknown prognostic factors. Such an imbalance is likely, because such a small number of patients were randomized and analysed.

Stockle *et al.*[18] randomized 49 patients to three cycles of MVAC or MVEC (methotrexate, vinblastine, epirubicin and cisplatin) versus no chemotherapy. The trial reported a large improvement in disease-free survival, with median times to progression of 19.7 and 11.6 months in the chemotherapy and no-chemotherapy groups, respectively ($P = 0.0015$). In this trial, again, a large proportion (31%) of patients allocated chemotherapy did not receive it. The trial was stopped before the pre-planned number of patients had been recruited because of the large difference observed in progression-free survival. Such early stopping is likely to have provided an inflated estimate of the effect on progression-free survival[17]. No data on overall survival were presented but the authors state that 'overall and progression-free survival curves, however, ran almost parallel and, thus, would not have added

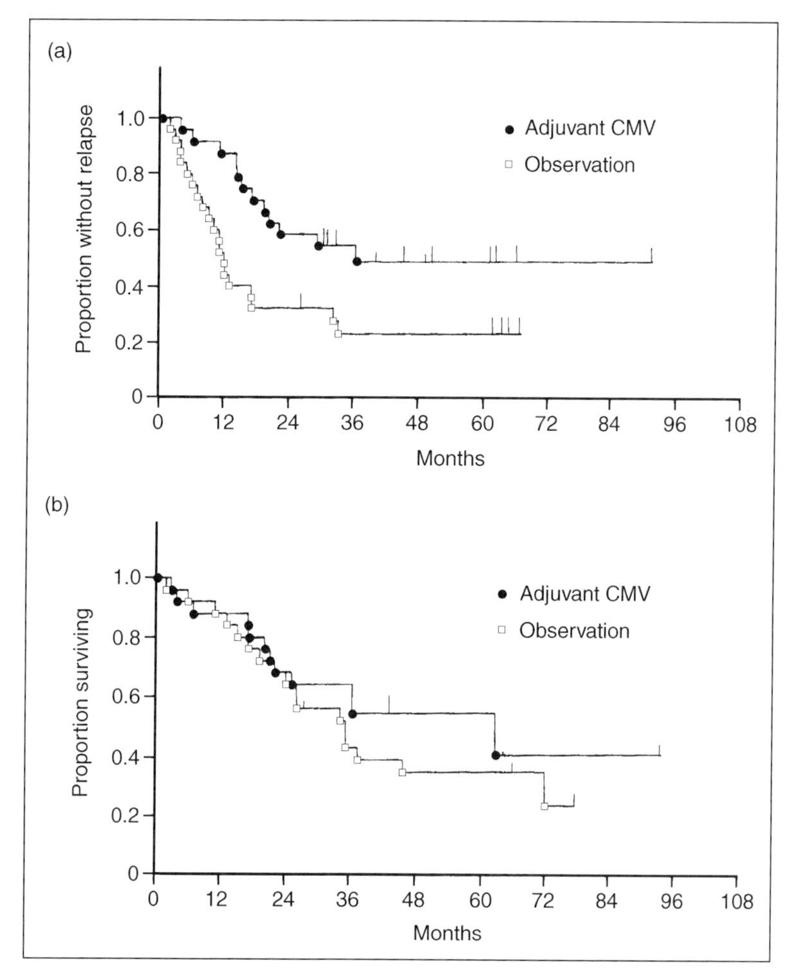

Figure 9.4 Time to progression (a) and survival curves (b) for patients in the trial conducted by Freiha *et al*[5]. CMV, cisplatin, vinblastine and methotrexate. (Reproduced from Freiha *et al.* (1996) A randomised trial of radical cystectomy versus radical cystectomy plus cisplatin, vinblastine and methotrexate chemotherapy for muscle invasive bladder cancer. *Journal of Urology,* **155**, 495–9, by permission of Williams & Wilkins.)

relevant additional information because our protocol did not include general therapeutic recommendations for the patients with tumour progression'. As a consequence very few patients on the no-chemotherapy arm received salvage chemotherapy at the time of tumour progression. This would probably be considered unacceptable in present-day clinical practice where most patients, if fit enough, would be offered salvage chemotherapy. This is reflected in the design of more recent adjuvant trials which aim to compare the policy of adjuvant chemotherapy with the policy of delayed chemotherapy given at disease progression.

Freiha *et al.*[5] randomized 55 patients to four cycles of CMV (cisplatin, methotrexate and vinblastine) versus no chemotherapy after cystectomy. There was a large improvement in disease-free survival in favour of the chemotherapy group, with a median time to relapse of 37 months in the chemotherapy group and 12 months in the no-chemotherapy group ($P = 0.01$). There was no evidence of a difference in survival ($P = 0.32$). The time to progression and overall survival curves are shown in Figure 9.4. This trial, like the Stockle trial, was terminated early because of an observed difference in disease-free survival and thus is likely to have provided an inflated estimate of the effect on this end point. However, in contrast to the Stockle trial, salvage chemotherapy was given to patients in the no adjuvant chemotherapy arm when disease progression occurred. This may, in part at least, explain why a difference in time to progression was observed, but no evidence of a difference in overall survival was observed.

The fifth trial of adjuvant chemotherapy was conducted by Bono and colleagues[2] in Italy. They randomized 125 patients to receive cystectomy plus four cycles of cisplatin and methotrexate or cystectomy alone. Although details of the comparisons are not given, the investigators found no evidence of a difference between the two arms in terms of time to progression or overall survival.

CONCLUSION

Randomized trials which have compared adjuvant or neoadjuvant cisplatin-based chemotherapy plus local (definitive) treatment against local treatment alone have been reviewed systematically, concentrating on the end point of overall survival. Although the basic tenets set down for all systematic reviews[3] have been followed, it must be stressed that this review is based on identifying and retrieving information from only those trials that have been published in the English language. Of the 10 reported trials identified, it was possible to extract sufficient summary information to include the data in a meta-analysis from only five of them. This is a direct consequence of poor reporting of trials. Thus our review may be incomplete and biased because of the exclusion of unpublished trials, exclusion of trials not published in the English language, exclusion of patients in individual trials and exclusion of trials with insufficient information to be included in the quantitative synthesis of results.

The most common and serious flaw raised in this review is the size of the individual trials. Adjuvant chemotherapy in breast and colon cancer have improved survival by only modest amounts, of the order of an absolute improvement in survival of 5–10%, that is from 50% to 55 or 60%[4,12]. The chemotherapy employed in breast and colon cancer has similar levels of activity in metastatic disease as has cisplatin-based chemotherapy in metastatic bladder cancer. Thus it is unrealistic to expect larger improvements in survival from the use of adjuvant (or neoadjuvant) chemotherapy in bladder cancer. However, to detect reliably an improvement in 3-year survival from 50% to 60% requires that nearly 1000 patients are randomized, while to detect an improvement from 50% to 55% requires that nearly 4000 patients are randomized. With these caveats, a number of conclusions can be reached from the present review.

1. There is no good evidence that **cisplatin used by itself** as neoadjuvant therapy or as adjuvant therapy improves survival. With only 379 patients randomized into trials of neoadjuvant cisplatin and 80 patients randomized into trials of adjuvant cisplatin, this conclusion is not surprising. A total of 379 patients is only sufficient to detect reliably improvements in survival of the order of 15%, from 50% to 65%.
2. There is no good evidence that **neoadjuvant cisplatin-based combination chemotherapy** improves survival. This is the best studied of the questions, as 1625 patients have already been randomized into three trials. However, it was not possible to include the data from the Nordic 2 trial into the current meta-analysis and the MRC/EORTC trial has reported in abstract form only and longer-term results are awaited. When the SWOG trial completes accrual this will expand the number of patients to at least 1923. When these trials have reported their full results, a meta-analysis of individual patient data of these trials together with Nordic 1 will provide more reliable information.
3. There is no good evidence that **adjuvant cisplatin-based combination chemotherapy** improves survival. The four trials of adjuvant cisplatin-based chemotherapy present intriguing results, with at least three suggesting large benefits in time to progression. However, each of these trials has a number of flaws which have been outlined in the discussion above. The total numbers randomized in these trials is 273, sufficient to detect reliably absolute benefits of the order of only 20% (from 50% to 70%). As discussed above, such differences seem inherently implausible.

The question of whether cisplatin-based chemotherapy improves the survival of patients with locally advanced bladder cancer has been with us for more than 15 years. Unfortunately, we still do not have a reliable answer to this question. Most trials have considered the role of neoadjuvant chemotherapy, and we can conclude that if there is a survival benefit for patients in this setting then it is modest – of the order of a 5% improvement (from 50% to 55%). When the trials of the MRC/EORTC and SWOG are published and an individual patient based meta-analysis is performed we are likely to have a clearer answer to the neoadjuvant cisplatin-based combination chemotherapy question. The few trials of adjuvant chemotherapy are very provocative. Unfortunately, these trials had methodological flaws and the number of patients randomized was less than 300, when at least four to ten times this number are probably required to answer this question reliably. Some have asked whether it is worthwhile performing such trials[14], suggesting that if at best neoadjuvant chemotherapy improves survival by only a modest amount, is adjuvant chemotherapy likely to be any more successful? As a consequence it is argued that no such trials should be launched until more effective chemotherapy becomes available. However, it is possible that chemotherapy may in fact be more effective when given after local therapy, when the tumour burden is minimal. It should be also be recognized that many patients with locally advanced, node-positive disease are currently given adjuvant chemotherapy because they are considered to have a poor prognosis. There is a need to know whether such management is justified. Unfortunately, unless some really large trials – randomizing at least 1000 patients – of adjuvant cisplatin-based or new chemotherapy are performed, it is likely that this question will remain unanswered.

COMMENTARY REFERENCES

1. Abol-Enein H, El-Makresh M, El-Baz M *et al.* (1997) Neoadjuvant chemotherapy in the treatment of invasive transitional bladder cancer. A controlled, prospective randomised study. *Br J Urol* **79** (Suppl. 4): 43.
2. Bono AV, Benvenuti C, Gibba A *et al.* (1997) Adjuvant chemotherapy in locally advanced bladder cancer. Final analysis of a controlled multicentre study *Acta Urol Ital* **11**: 5–8.
3. Dickersin K, Scherer R and Levebvre C (1995) Identifying relevant studies for systematic reviews, in *Systemic Reviews,* (eds I Chalmers and DG Altman), BMJ Publishing Group, London, pp.17–36.
4. Early Breast Cancer Trialists' Collaborative Group (1992) Systematic treatment of early breast cancer by hormonal, cytotoxic or immune therapy. *Lancet* **339**(a): 1–15.
5. Freiha F, Reese J and Torti FM (1996) A randomised trial of radical cystectomy versus radical cystectomy plus cisplatin, vinblastine and methotrexate chemotherapy for muscle invasive bladder cancer (see comments). *J Urol* **155**: 495–499.
6. Ghersi D, Stewart LA, Parmar MKB *et al.* (1995) Does neoadjuvant cisplatin-based chemotherapy improve the survival of patients with locally advanced bladder cancer: a meta-analysis of individual patient data from randomized clinical trials. *Br J Urol* **75**: 206–213.

7. Hall RR for MRC Advanced Bladder Cancer Working Party, EORTC GU Group, NCI Canada, Norwegian Bladder Cancer Group, Australian Bladder Cancer Study Group, Club Urologia Espanol de Tratamiento and FinBladder (1996) Neoadjuvant CMV chemotherapy and cystectomy or radiotherapy in muscle invasive bladder cancer. First analysis of MRC/EORTC Intercontinental Trial. *Proc Am Soc Clin Oncol* **15**: 244, Abstract 612.
8. Malmström PU, Rintala E, Wahlqvist R and members of the Nordic Cooperative Bladder Cancer Study Group (1996) Five year follow up of a prospective trial of radical cystectomy and neoadjuvant chemotherapy: Nordic Cystectomy Trial 1. *Urol* **155**: 1903–1906.
9. Malmström PU, Rintala E, Wahlqvist R *et al.* (1997) Prospective trials of radical cystectomy and neoadjuvant chemotherapy. *Br J Urol* **80** (Suppl. 2): 50, Abstract 193.
10. Martinez-Pineiro JA, Gonzalez Martin IM, Arocena F *et al.* (1995) Neoadjuvant cisplatinum chemotherapy before radical cystectomy in invasive transitional cell carcinoma of the bladder: Advanced Bladder Cancer Overview Collaboration. *J Urol* **153**: 964–970.
11. Parmar MKB, Stewart LA and Altman DG (1996) Meta-analysis of randomised trials: when the whole is more than just the sum of the parts. *Br J Cancer* **74**: 496–501.
12. Parmar MKB, Ungerleider RS and Simon R (1996) Assessing whether to perform a confirmatory randomised clinical trial. *J Natl Cancer Inst* **88**: 1645–1651.
13. Rintala E, Hannisdal E, Fossa SD *et al.* (1993) Neoadjuvant chemotherapy in bladder cancer: a randomized Study. Nordic Cystectomy Trial 1. *Scand J Urol Nephrol* **27**: 355–362.
14. Roth BJ and Bajorin DF (1995) Advanced bladder cancer: the need to identify new agents in the post-M-VAV (methotrexate, vinblastine, doxorubicin and cisplatin) world. *J Urol* **153**: 894–900.
15. Skinner D, Daniels JR, Russell CA *et al.* (1990) Adjuvant chemotherapy following cystectomy benefits patients with deeply invasive bladder cancer. *Sem Urol* **8**: 279–284.
16. South West Oncology Group (SWOG). Phase III randomised comparison of cystectomy alone vs neoadjuvant MVAC (MTX, VBL, ADR, CACP) plus cystectomy in patients with locally advanced transitional cell carcinoma of the bladder. Trial protocol available from the South Western Oncology Group, USA.
17. Stewart LA and Parmar MKB (1996) Bias in the analysis and reporting of randomised controlled trials. *Int J Tech Assess in Health Care* **12**: 264–275.
18. Stockle M, Meyyenburg W, Wellek S *et al.* (1995) Adjuvant polychemotherapy of nonorgan-confined bladder cancer after radical cystectomy revisited: long-term results of a controlled prospective study and further clinical experience. *J Urol* **153**: 47–52.

19. Studer UE, Bacchi M, Biedermann C *et al.* (1994). Adjuvant cisplatin chemotherapy following cystectomy for bladder cancer: results of a prospective randomized trial. *J Urol* **152**: 81–84.
20. Tannock IF (1990) The current status of adjuvant chemotherapy for bladder cancer. *Sem Urol* **8**: 291–297.
21. Wallace D, Raghavan K, Kelly K *et al.* (1991) Neoadjuvant (pre-emptive) cisplatin therapy in invasive transitional cell carcinoma of the bladder. *Br J Urol* **67**: 608–615.

Appendix 1

Strategy used to search MEDLINE for randomized trials of neoadjuvant and adjuvant cisplatin-based chemotherapy in locally advanced bladder cancer (as of September 1997)

Label	Items	Search statement
1	89 358	PT=Randomized-Controlled-Trial
2	9690	Randomized ADJ Controlled ADJ Trial.DE.
3	36 276	Random ADJ Allocation.DE.
4	49 160	Double ADJ Blind ADJ Method.DE.
5	3273	Single ADJ Blind ADJ Method.DE.
6	133 508	1 OR 2 OR 3 OR 5
7	2 677 554	Animal.DE.
8	5 973 927	Human.DE.
9	2 150 339	7 NOT (7 AND 8)
10	126 677	6 NOT 9
11	205 239	PT=Clinical-Trials
12	97 714	Clinical-Trials#.DE.
13	42 522	(Clin$ WITH Trial$).AB,TI.
14	47 655	((Sing$ OR Doub$ OR Treb$ OR Trip$) ADJ (Blind$ OR Mask$)).AB,TI.
15	17 978	Placebos.DE.
16	49 175	Placebo$.AB,TI.
17	41 617	Random.AB,TI.
18	20 163	Research ADJ Design.DE.
19	312 225	11 OR 12 OR 13 OR 14 OR 15 OR 16 OR 17 OR 18
20	295 915	19 NOT 9
21	177 295	20 NOT 10
22	126 677	10 NOT 21
23	20 935	Bladder-Neoplasms#.DE.
24	110 762	Drug-Therapy#.DE.
25	557	23 AND 24
26	70	22 AND 25
27	20 951	Bladder ADJ Neoplasm$
28	6477	Bladder ADJ Cancer
29	21 643	27 OR 28
30	81 303	Chemotherapy
31	1921	29 AND 30
32	1936	26 OR 31
33	25 341	Superficial
34	1578	32 NOT 33

10

Bladder preservation in muscle invasive bladder cancer

J. Trevor Roberts

Muscle invasive bladder cancer is fatal for the majority of patients. Approximately 50% of patients with muscle invasive bladder cancer have occult metastatic disease and cannot be cured by eradication of the primary tumour, however this is achieved. The median age at presentation of bladder cancer in Britain is 67 years. In this elderly population a proportion will die, with or without their bladder cancer, of other diseases. As anaesthetic and surgical techniques have improved, the number who die as a result of their surgery has decreased. Nevertheless, radical surgery carries a risk of death and a numerically greater risk of significant morbidity. Those with cancers curable only by surgery are the only beneficiaries of a radical surgical approach.

Only a minority of patients with muscle invasive disease may therefore benefit from radical surgical extirpation of their primary tumour. It may be argued that those destined to die of their metastatic disease are not well served by being allowed to live out the remainder of their lives with a urinary stoma. Of those cured by radical surgery there is undoubtedly a proportion whose tumours were, potentially, equally curable by methods that allowed preservation of bladder function.

In a number of centres ingenious and elaborate techniques for the construction of neobladders have been developed and improved. While they provide a (mainly) continent, non-prosthetic alternative to a urinary diversion, they cannot function as well as a normal, or perhaps even a damaged, preserved urinary bladder.

For cancer arising in other anatomical sites, radiotherapy allows cure of selected patients with organ preservation and normal, or near normal, function. The first series of patients with early laryngeal cancer cured by radiotherapy, without serious long-term sequelae, was reported more than 70 years ago[44]. Similarly, the role of radiotherapy in allowing organ preservation in selected patients with cancers of the head and neck, breast, rectum and anus is established. In the case of breast cancer, not only may acceptable or even excellent cosmesis be achieved, but preservation of function to the extent of allowing breast feeding, has been reported[61]. The addition of neoadjuvant, concurrent or conventional adjuvant chemotherapy to radiotherapy may permit a broadening of the selection criteria for suitability for organ-preserving radiotherapy or increase the likelihood of organ preservation or cure[26].

The role of radical radiotherapy in allowing cure of bladder cancer with preservation of organ function is established in, for example, Britain and Canada, and is recognized as a potentially curative alternative to cystectomy. Combinations of radiotherapy and chemotherapy are also being actively examined and early results suggest the possibility of greater organ preservation rates, if not greater survival, with such an approach[7,9,24,27,60].

Radiotherapy is, however, not without its morbidity, and the possibility that other organ-preserving approaches may allow cure of those whose disease is capable of cure with preservation of good, or acceptable, function is worthy of exploration. Other potential beneficiaries of this approach are those whose disease is, *ab initio*, incurable and who may thus be allowed to die of, or with, their metastases with intact bladder function and with a bladder free of tumour while escaping the morbidities of radical surgery or full-dose radiotherapy.

NON-RADIOTHERAPEUTIC APPROACHES TO BLADDER PRESERVATION

Conservative surgery, either TUR or partial or segmental cystectomy, and systemic cytotoxic chemotherapy are treatments that are conventionally regarded as adjunctive to other radical approaches to the cure of bladder cancer. However, there exists a literature on the sole use of each and on the combination of a conservative surgical approach with cytotoxic chemotherapy in the treatment of muscle invasive bladder cancer.

TRANSURETHRAL RESECTION

Transurethral resection (TUR) is established as a diagnostic measure, to determine the presence of invasive carcinoma by providing biopsy material for histological examination. Staging of bladder cancer depends upon the depth of muscle invasion documented in TUR biopsy specimens and upon the findings of bimanual examination performed before and after the procedure[20].

Although conventionally regarded as capable of dealing with low-grade, low-stage tumours, TUR may contribute to the success of bladder-conserving strategies for high-grade, muscle invasive tumours[2]. As the preliminary step in an integrated, non-surgical, bladder-conserving treatment strategy, TUR may serve several purposes. First, it may allow the identification of those patients with smaller (generally less than 5 cm diameter) tumours for whom a bladder conserving approach may be thought possible. The fact that a 'thorough' or complete TUR can be performed may itself identify a group of good-prognosis tumours, with an inherently favourable natural history, for which a conservative approach may be appropriate. The development of the means to identify, through the use of molecular markers of tumour behaviour, other good-prognostic tumours may allow further refinement of this process. TUR provides tumour material for such analyses.

The removal of a bulk of tumour tissue (a process akin to teilectomy in breast cancer) may be therapeutic, leaving subclinical or microscopic disease which chemotherapy, radiotherapy, chemoradiotherapy and/or the host immune response are capable of eliminating. In this context, 'thorough' TUR contributes to increased local control rates after radiotherapy, at least in some series[50]. It has also been integrated into other chemotherapy-based strategies aimed at bladder conservation, which are discussed in more detail later in this chapter[3,14,16]. Its importance as part of an integrated combination of 'aggressive' TUR with chemoradiotherapy has been emphasized[9,24,27,50].

TUR may also contribute to the therapeutic success of other treatments, in that it removes tumour bulk and hence produces prompt palliation of symptoms, such as haematuria, prior to the instigation of cytostatic treatment[29]. The release of cytokines as a result of the trauma of TUR may also contribute a cytostatic effect which may result in killing of carcinoma cells. However, concern has been expressed elsewhere that surgical debulking by TUR may change the growth kinetics of metastasis adversely.

TUR alone

Although the conventional view is that TUR cannot be regarded as more than a debulking procedure for T2–4 tumours[64], for a minority of patients with muscle invasive bladder cancer the therapeutic benefits of TUR may extend beyond palliation; in selected patients muscle invasive bladder cancer may be cured by TUR[20]. There are two bodies of evidence for this:

1. Approximately 10% of patients undergoing total cystectomy for T2 or T3 bladder tumour will have no tumour found within the cystectomy specimen (p0), suggesting that the TUR, performed with diagnostic intent, had completely removed the invasive tumour (Table 10.1).
2. A number of authors have reported 5-year survival rates for patients with muscle invasive bladder cancer (T2, T3) treated by TUR (Table 10.2). Furthermore, the survival figures in these studies approximate those achievable with more radical therapies. These results suggest that selected patients with muscle invasive bladder cancer may be treated adequately by TUR alone. The possibility of identifying such patients has been considered by Herr[19]. He restaged 217 consecutive patients, referred to him after TUR performed by other urologists, and found 172 (79%) to have invasive carcinoma. These patients underwent partial or total cystectomy. Forty-five patients (21%) were selected to undergo

Table 10.1 Number of patients without tumour in the bladder at radical cystectomy

Author	No. with cystectomy	No. p0	%
Slack *et al.* (1977)[53]	158	14	9
Mathur *et al.* (1991)[34]	58	4	7
Montie *et al.* (1984)[38]	99	10	10
Brendler *et al.* (1990)[6]	76	13	17
Skinner *et al.* (1990)[51]	80	12	15
Pagano *et al.* (1991)[42]	270	25	9
Totals	741	78	10.5

Table 10.2 Five-year survival rates after transurethral resection for muscle invasive bladder cancer

Author	No. of patients	T category (%)		
		T2	T3a	T2 + T3a
Flocks (1951)[11]	142	56	43	47
Milner (1954)[37]	88	57	23	53
Barnes *et al.* (1967)[4]	114	–	–	40
O'Flynn *et al.* (1975)[40]	123	59	20	52
Barnes *et al.* (1977)[5]	75	–	–	31
Herr (1987)[19]	45	70	57	68
Henry *et al.* (1988)[18]	43	63	38	52
Kondás and Szentogyörgyi (1992)[29]	27	32*	20*	30*

*Relapse-free survival.

conservative management by TUR with intravesical therapy as necessary. Twenty had no tumour in the bladder (T0); 17 had carcinoma in situ; four had T1 disease; and four, T2. Of these 45 patients 30 (67%) did not undergo cystectomy and remained free of disease. Nine of the 30 had no further therapy, while 21 required repeat TUR, with or without intravesical chemotherapy, for recurrent superficial bladder cancer. Four of the remaining 15 patients were alive with metastatic disease and two of these four remained free of tumour in the bladder at the time of his last report[20]. Eleven patients underwent cystectomy between 9 and 30 months after restaging. In eight cases cystectomy was performed because of muscle invasive bladder cancer. The remaining three patients developed rapidly recurrent superficial bladder cancer which was not amenable to conservative measures and underwent cystectomy. Herr advocated transurethral resection for local control of T2 and T3a tumours

> if bladder function is good, there are no more than 2 papillary tumours, neither greater than 2 cm diameter at the base, well differentiated and not associated with a mass palpable bimanually; positioned for 'safe and sure resection' (i.e. located on the base or lateral walls rather than the posterior wall or dome or involving the bladder neck); not of high grade (Grade 3) and invading muscle on a broad front, rather than in a tentacular fashion.

However, his subsequent editorial comment[21] on a more recently reported prospective series would suggest the development of a more cautious approach; perhaps as a result of later, unreported, adverse experiences? In this brief editorial comment he suggests that TUR be reserved for solitary papillary tumours.

In the series that prompts this comment, Solsona *et al.*[55] report their experience with 95 patients, in two consecutive series, with muscle invasive transitional cell carcinoma. The proportion of patients with T2 tumours is not stated, but 64% had G3 and 36%, G2 tumours. Following complete TUR, 52.5% of the first series (with a median follow-up of 58 months) and 66.6% of the second series (median follow-up 26 months) remained free of recurrence at the time of the report. Of the first series 83% remained alive and 72.8% retained their bladder. Henry and her colleagues[18] reported on a retrospective series of patients, treated at the university of Iowa between 1974 and 1973 for stage B transitional cell carcinoma of the bladder. Of 114 patients treated in this period, 43 had been treated by TUR alone. The remainder had received more conventional radical therapy, either with radical cystectomy (15 patients), radical radiation therapy to a dose of 65–75 Gy (16 patients) or preoperative irradiation, 40–60 Gy, followed by radical cystectomy. Their 5-year survival rates of 63% and 38% for B1 (T2) and B2 (T3a) disease, respectively, compared favourably with the survival rates for the contemporary groups of patients treated more conventionally. Indeed, the survival rates for patients treated with either conventional treatment modality alone were somewhat less good. Acknowledging the possible influence of patient selection in a retrospective series they nevertheless were unable to find any difference in known prognostic factors, such as tumour size, between the groups. Those patients treated by TUR tended to be older and of poorer performance status.

Koloszy[28] performed a systematic histopathological examination of material removed by extensive TUR. Bladder tumours were resected to a depth at which the tissue appeared normal endoscopically. Further resection of the base, sides and adjacent margins of the resection 'crater' was then performed and as many as six separate specimens of this material reviewed. Residual tumour beyond the initial resection was seen in 35% of 662 tumours, 195 of which were invasive. Solid tumours were more frequently associated with residual tumour (76%) than papillary (21.5%). The incidence of residual tumour also correlated with T category and was 36% in T1 lesions, 56% in T2 and 83% in T3. Alken and Köhrmann[2] reported a similar incidence of residual tumour following initial TUR and recommend secondary resection, some 6–8 weeks after initial TUR, as routine.

Both Herr[19] and Kolosky[28] have demonstrated that wider re-excision following initial TUR may demonstrate residual disease but also that if the margins of an extensive TUR are clear of tumour 'cure of the particular bladder tumour resected may be obtained'. Nevertheless, others feel that TUR alone, however radical, may never be regarded as curative and advocate the adjuvant use of radiotherapy or chemotherapy.

Kondás and Szentogyörgyi[29] reported a retrospective series of 761 patients in whom 1250 bladder tumours were resected endoscopically between 1973 and 1990. The majority of these tumours were G1 and/or T1, but their series did include 58 T3a tumours (eight G2 and 50 G3) and 130 T2 tumours, of which 73 were G3. Following 'removal of tumours which presented muscle invasion' 17 patients were subjected to intra-arterial chemotherapy and a further 40 to radiotherapy. In their discussion, while acknowledging the role of TUR as an organ-preserving surgical method, they advocate cystectomy for those tumours infiltrating beyond the halfway point of the bladder wall muscle layer. However, they acknowledge the difficulty in identifying this cut-off point by pointing out that their 5-year survival figures for T3a tumours (47%) appear better than for T2 (31%). Despite the obviously 'radical' nature of their approach to TUR, 'The removal is performed each time by aimed perforation which we carry on, in primary or secondary form, in the adipose tissue, too' and thorough histological examination of specimens from the resection margins, they caution that TUR alone may not be held to be curative for T3a tumours and advocate the additional administration of radiotherapy or intra-arterial chemotherapy in such cases. In a later series from the same institution[30] 103 patients with muscle invasive bladder cancer are described. Twenty-two with T2 tumours and five with T3a tumours were treated by TUR only. Of the T2 tumours, only seven were relapse-free at 5 years, while only one of the T3 tumours had not relapsed. In this paper they conclude that cystectomy remains the treatment of choice for muscle invasive bladder cancer, that TUR should be performed only if cystectomy cannot be carried out and, again, that TUR should be combined with chemotherapy or radiotherapy when muscle invasion is demonstrated.

They suggest that selected patients, with primary, solitary G1–G3 tumours with a maximum diameter of 2–3 cm (at the base) located in the fixed portion of the bladder, with no evidence of extravesical extension, negative 'marginal' biopsies, negative random biopsies of the remaining bladder mucosa and negative urine cytology at 10 days be considered as candidates for curative TUR.

PARTIAL OR SEGMENTAL CYSTECTOMY

The role of partial cystectomy in the management of invasive bladder cancer is controversial since the multicentric or field-change nature of transitional cell carcinoma and the high propensity of transitional cells to implant on any raw surface cause a significant incidence of recurrence both in the bladder and outside. Conventionally, partial cystectomy for invasive bladder cancer is an acceptable operation for 'a patient with a solitary tumour in the base or posterior wall of the bladder who is elderly or who poses a significant surgical risk or a patient who refuses total cystectomy'[12].

The role of partial cystectomy in the management of bladder cancer is, however, incompletely evaluated[59]. There have been no randomized trials comparing partial cystectomy, stage for stage, with other treatment modalities. Five-year survival rates quoted in series of patients with bladder tumours treated by partial cystectomy do not compare unfavourably with contemporary series treated by radical cystectomy[59], implying that for carefully selected patients with invasive bladder cancer, partial cystectomy may be an appropriate alternative to a more radical operation (Table 10.3).

Preoperative radiotherapy has been advocated as a means of reducing the incidence of intraoperative tumour-cell implantation and subsequent local recurrence, which tends to occur at the margins of resection rather than elsewhere within the bladder[41,52,65].

Cancers arising in bladder diverticula account for between 1.5% and 10% of all bladder cancers[36] and tend to be of higher grade. Because the wall of a diverticulum is thin and lacks the usual muscle layer, tumours penetrate earlier and this, together with a propensity for early dissemination, leads to low rates of disease-free and 5-year survival (both less than 10%) following radical cystectomy[10,36]. This prompts

Table 10.3 Partial cystectomy: percentage survival rates by T category (modified from Sweeney *et al.*[59]); see Sweeney *et al.* for details of references

Author	T0	T1	T2	T3	T4	Overall
Long *et al.* (1962)	80	67	43	9	0	–
Magri (1962)	–	80	38	26	0	42
Utz *et al.* (1973)	–	68	47	29	0	39
Evans and Tester (1975)	–	69	43	14	0	–
Novick and Stewart (1976)	–	67	53	20	–	46
Brannan *et al.* (1978)	100	69	54	33	0	57
Cummings *et al.* (1978)	–	79	80	6	–	60
Resnick and O'Connor (1978)	75	71	77	13	20	35
Faisal and Freiha (1979)	75	58	29	7	0	40
Merrell *et al.* (1979)	100	100	67	25	0	48
Schoberg *et al.* (1979)	69	69	29	12	100	43
Lindahl *et al.* (1984)	–	59	38	–	–	42
Kaneti *et al.* (1984)	–	68	40	33	0	48
Kondás *et al.* (1993)	–	–	34	41	–	36

the suggestion that tumours arising in diverticula might be managed adequately by diverticulectomy or partial cystectomy.

Urachal adenocarcinoma of the bladder accounts for 20–40% of primary bladder adenocarcinomas, which themselves account for less than 2% of bladder cancers. These tumours usually present late and consequently are associated with a poor (6–15%) 5-year survival. Although *en bloc* radical cystectomy is advocated by some as the treatment of choice, this aggressive approach has not been shown to influence survival favourably, leading authors to advocate *en bloc* partial cystectomy[43]. Herr[22] has suggested that, because these tumours often push into the bladder lumen without invading the bladder wall and display a sharp demarcation between the tumour and adjacent bladder epithelium, a wide safe excision to include an ample margin of normal bladder is made easier. As discussed below, the advent of cisplatin-based combination chemotherapy regimes with high response rates, used in a neoadjuvant setting, may encourage the further exploration of this uncommonly performed surgical procedure.

SYSTEMIC CHEMOTHERAPY IN REGIONALLY ADVANCED BLADDER CANCER

Combination chemotherapy regimes based on cisplatin and methotrexate, with the addition of vinblastine alone (CMV) or with Adriamycin® (MVAC), produce overall response rates, in transitional cell carcinoma, of the order of 50%, with complete remissions being achieved in approximately 20% of patients[46]. Several authors have observed an apparently higher response rate in the primary bladder tumour than in metastatic lesions[25,35]. In series where cisplatin-based combination chemotherapy has been given as a prelude to total or radical cystectomy, investigators have observed a proportion of apparently tumour-free bladders[1,2,47,48,49], either free of all tumour (T0) or free of invasive tumour (pTcis, pTa), with from 12% to 43% being pT0. In the MRC/EORTC multinational trial of neoadjuvant CMV in muscle invasive bladder cancer, 33% of patients who had received neoadjuvant chemotherapy were pT0 in the cystectomy specimen compared to only 13% in patents who had not received neoadjuvant treatment (R. R. Hall, personal communication).

Data from other series in which patients have undergone cystectomy following neoadjuvant chemotherapy allow estimates of the proportion of patients that can be rendered tumour-free within the bladder[49] (Table 10.4). The occurrence of complete responses of bladder tumours to chemotherapy and the subsequent reluctance of some patients to undergo cystectomy or of surgeons to remove an apparently tumour-free bladder has allowed the observation, in a number of centres, of occasional patients who have remained tumour-free for extended periods of time, with some apparently cured. The potential proportion of patients cured by chemotherapy alone cannot be greater than that rendered pT0 in the bladder and it

Table 10.4 'Complete' pathological response in the bladder using combination chemotherapy (reproduced, with permission, from Seidman and Scher 1991[49])

Agents	No. of trials	Patients evaluable	CR at cystectomy	95% Confidence interval (%)
DDP/5-FU	1	16	7 (43%)	19–66
DDP/MTX	4	110	28 (25%)	17–34
CMV	5	107	32 (30%)	21–39
CAP	4	83	18 (22%)	13–31
MVAC	11	228	64 (28%)	22–34

DDP, cisplatin; 5-FU, 5-fluorouracil; MTX, methotrexate; CAP, cyclophosphamide, Adriamycin®, cisplatin; CMV, cisplatin, methotrexate, vinblastine; MVAC, methotrexate, vinblastine, Adriamycin®, cisplatin.

follows from this that, while a proportion of muscle invasive bladder cancers may be curable by combination chemotherapy, the majority are not.

In a series of 101 selected patients with muscle invasive bladder cancer (four T2, 67 T3, 12 T4a, 18 T4b) who were participating in phase II studies of cisplatin-based chemotherapy (cisplatin, methotrexate; CMV; or epirubicin, cisplatin, methotrexate (EpiC-M)), where ultimate treatment of the primary bladder tumour was at the discretion of the responsible physician, we have observed sustained complete remissions lasting for up to 10 years following systemic chemotherapy alone, with no attempt at conventional definitive treatment of the bladder tumour.

As these patients were the subjects of phase II trials, no attempt was made to resect the primary bladder cancer, which was biopsied to establish the diagnosis and to document muscle invasion but then left in situ, to act as a marker lesion. After two cycles of chemotherapy a further endoscopic assessment of the bladder tumour was performed. Patients whose tumours had not shown at least a partial response went off study and underwent radiotherapy or total cystectomy. Those patients whose tumours had responded received two more cycles of chemotherapy and were again assessed endoscopically. If tumour remained in the bladder, again patients went off study and either received radiotherapy or underwent cystectomy. Those patients with no visible tumour and with negative deep resection biopsies at the site of former tumour were deemed complete responders and went on to receive one or (more usually) two further cycles of chemotherapy to consolidate their response. Thereafter, they were followed closely and those subsequently relapsing within the bladder went on to receive radiotherapy or a cystectomy. Using this approach 17% of 103 patients have remained tumour-free without conventional definitive treatment of their bladder cancers. This is intriguing, as it confirms that transitional cell carcinoma of the bladder is not merely chemosensitive but may, in selected patients, prove chemo-curable.

In a series of 47 patients with muscle invasive bladder cancer treated with neoadjuvant MVAC, Sternberg and her colleagues[57] had 18 patients who 'did not accept or were not adequately encouraged to undergo pathologic staging'. In this context, 'pathologic staging' means complete or partial cystectomy. Seven of the 18 patients had T2 disease; six, T3a; and five, T3b. Thirteen of these 18 patients were T0 at postchemotherapy TUR. With a median follow-up of more than 33 months (range 14–51+ months) 10 of these 13 (77%) remain tumour-free and have had no recurrences. In four patients, recurrence has been with superficial disease only; treated successfully in three of four cases with BCG. Of the 18 patients not undergoing partial or complete cystectomy following chemotherapy, 15 (83%) remained alive. At the time of their report the median survival of this group of patients had not been reached with a median follow-up of 36+ months (range 11–55+ months).

Srougi and Simon[56] treated 36 patients with T2 to T4a NX M0 TCC bladder with three or four cycles of MVAC, following which they were reassessed either with TUR or open excisional biopsies of the bladder with iliac lymphadenectomy. Fourteen of 30 patients (47%) who completed the treatment schedule achieved a complete remission. The authors note that this was more likely with T2 than with T3 or T4a tumours. Ten of the 14 patients (71%) in whom the tumour-free bladder was preserved had one or more local recurrences; eight underwent salvage cystectomy and two (with superficial disease) were treated conservatively. Interestingly, all recurrences in this group were in the bladder and no patient had disease elsewhere in the urinary tract or outside the bladder. At 5 years six patients with a complete response (43%) still had an intact bladder.

In reporting the Memorial Sloan-Kettering Cancer Centre (MSKCC) series of patients treated with neoadjuvant MVAC, Scher *et al.*[47] give details of 71 patients treated with between one and six (median of three) cycles of MVAC for muscle invasive bladder cancer. Of 23 patients who refused surgical exploration within 3 months of completing MVAC, one had had disease resected prior to chemotherapy, was therefore not considered evaluable for response but remained free of disease at 26 months. Eleven of the remaining 22 (50%) were T0, seven (32%) had CIS, while four (18%) had persistent muscle invasive disease. At the time of their report (with follow-up on the whole group of 71 of between 2 and 42+ months (median 24 months), six of 11 T0 cases (55%) remained disease-free. Hatcher *et al.*[17] observed seven of 39 patients with muscle invasive bladder cancer treated with cisplatin-based chemotherapy who were disease-free in the bladder immediately after chemotherapy. A further six patients had their response consolidated by bladder-conserving surgery (see below). At the time of their report 46% (7/13) of these patients remained disease-free. It is not clear how many of the persistently disease-free patients had received only chemotherapy. Thus, in anecdotal cases within large series of patients treated with chemotherapy without bladder preservation as the main aim, and in large series where chemotherapy was given with the aim of achieving cure and bladder preservation without conventional definitive treatment of the primary bladder tumour, between 43% and 77% remain disease-free, but there is a tendency for the more mature series to contain smaller numbers of durable complete remissions.

Of further concern is the observation that clinical staging with CT scanning, bimanual examination and TUR understages bladder disease following chemotherapy when compared with pathological staging at definitive surgery. In the MSKCC series, six of 20 patients (30%) who were free of invasive disease immediately prior to definitive surgery had pathological confirmation of residual muscle invasive disease in the surgical specimen.

The relatively small proportion of patients remaining disease-free is less than would be observed with radical radiotherapy alone, or in combination with chemotherapy and, consequently, the sole use of systemic chemotherapy for the routine management of muscle invasive TCC cannot be recommended. Nevertheless, we are not aware of any other adult carcinoma where chemotherapy 'cures' are observed in such a substantial minority of patients, and this observation, combined with the high response rates observed in metastatic sites, encourages the exploration of the use of chemotherapy as part of other treatment strategies aimed at curing bladder cancer while allowing preservation of the physiologically functional urinary bladder.

The combination of bladder-conserving surgery and systemic chemotherapy

Socquet[54] used high-dose methotrexate with folinic acid rescue (HDMTX-CF) after partial cystectomy or partial cystectomy combined with regional lymphadenectomy for 20 patients with T3a N0 and five with T3a N+ TCC bladder. The N0 patients received six and the N+ patients nine courses of HDMTX-CF. Of these 25 patients, only two (both with N+ disease) had relapsed or died at the time of his report, with a follow-up between 12 and 36 months. It is difficult teasing out the exact outcome for these 25 patients from this paper. Nevertheless, of the whole group of 33 patients with T3a, T3b, T4a N0 or N+ disease treated with surgery and HDMTX-CF, 90% were alive at 30 months. Eight patients with T3b or T4a disease underwent total cystectomy. The author compares his results favourably with those from series treated more conventionally.

A collaborative North of England group[14] developed Socquet's work, treating a series of patients with T3 bladder cancer with the same cytotoxic drug regime but gave eight cycles of HDMTX-CF after an endoscopic resection removed 'all visible and palpable tumour' in 54 patients. In a minority of patients, further resection for residual tumour was performed 2–3 weeks later. A further three patients, whose tumours lay at the dome of the bladder, underwent partial cystectomy rather than endoscopic removal of tumour. Four cycles of chemotherapy were given before a repeat cystoscopy/TUR. In those patients in whom invasive tumour persisted, conventional treatment of the clinician's choice was initiated. The survival and disease-free survival in this series also compared favourably with historical control groups, conventionally treated.

With the realization that combinations of cisplatin and methotrexate were capable of producing superior response rates in metastatic disease, it was decided to assess the effect of using cisplatin and methotrexate following radical transurethral resection[15,16]. Cisplatin and methotrexate have been used as a two-drug combination, or with vinblastine (CMV), or epirubicin (EpiC-M) in a further 55 patients.

We have now updated our experience in these groups of patients treated by 'complete' TUR and subsequent systemic chemotherapy. A total of 116 patients, treated with either HDMTX-CF (61 (52.6%) – 54 of whom have been previously reported by Hall *et al.*[14]) or cisplatin combination chemotherapy (55 (47.4%)) after 'complete' TUR have now been followed for a median of 11.6 years (range 3.9–15.2). The median follow-up on the HDMTX-CF is now 13 years and for the cisplatin combination treated group is 8 years. The median age of these patients was 67 years (range 37–88) and the male:female ratio was 96 (82.8%):20 (17.2%). Details of the tumours treated are given by T category (Table 10.5), histological grade (Table 10.6) and tumour size (Table 10.7). The majority of tumours were less than 5 cm in diameter but only 12.9% were T2, the remainder T3 or T4a. Two tumours were not histologically graded (both in the older HDMTX-CF series), but of the remainder, the majority (78.4% of the total) were G3. In all but one case, the

Table 10.5 Tumour category of patients treated by TUR and chemotherapy

Tumour category	Overall (%)	Methotrexate only (%)	Cisplatin combinations (%)
T2	15 (12.9)	5 (8.2)	10 (18.2)
T3	79 (68.1)	47 (77.1)	32 (58.2)
T3b	15 (12.9)	8 (13.1)	7 (12.7)
T4a	6 (5.2)	1 (1.6)	5 (9.1)
T4b	1 (0.9)	0	1 (1.8)

Table 10.6 Histological grade of tumours treated by TUR and chemotherapy

Tumour grade	Overall (%)	Methotrexate only (%)	Cisplatin combinations
G2	23 (19.8)	10 (16.4)	13 (23.6)
G3	91 (78.4)	49 (80.3)	42 (76.2)
Gx	2 (1.7)	2 (3.3)	0

Table 10.7 Size of tumours treated by complete TUR and chemotherapy

Tumour size (cm)	Overall (%)	Methotrexate only (%)	Cisplatin combinations (%)
≤ 2.0	18 (15.5)	2 (3.3)	16 (29.1)
2.1–3.0	32 (27.6)	17 (27.9)	15 (27.3)
3.1–4.0	21 (18.1)	13 (21.3)	8 (14.5)
4.1–5.0	17 (14.7)	9 (14.7)	8 (14.5)
5.1–6.0	6 (5.2)	5 (8.2)	1 (1.8)
6.1–7.0	1 (0.9)	0	1 (1.8)
Unknown	21 (18.1)	15 (24.6)	6 (10.9)

nodal status of these patients is unknown (NX). However, in one case fat resected from the depths of the TUR contained a lymph node which was histologically positive for TCC (N1). In the case of patients treated with cisplatin combinations, the tumour was resected completely to normal detrusor muscle and/or fat and three cycles of chemotherapy given. A further cystoscopic assessment was then carried out and biopsies taken. Patients free of tumour in the bladder at this time received a further two or three cycles of chemotherapy to 'consolidate' their response; those in whom tumour was present at this second assessment went off study and were treated by cystectomy or radiotherapy.

The median disease-specific survival for the whole series of 116 patients is in excess of 6 years, while for patients remaining disease-free in the bladder the median survival has not been reached at 13 years. Figure 10.1 shows actuarial disease-specific survival for patients treated with HDMTX-CF and with cisplatin combination therapy. The 2-, 5- and 10-year actuarial disease-specific survival for the former group is 69%, 39% and 33%, respectively. For patients receiving cisplatin combination chemotherapy the corresponding figures are 82%, 70% and 61%. Only 28% of patients treated with combination chemotherapy required salvage therapy, with radiotherapy or cystectomy, for invasive locally recurrent disease. These results compare favourably with those from other, conventionally treated, series.

Martinez-Piñeiro *et al.*[32], in a smaller group of patients followed 'radical TUR of the tumour and bladder wall into the perivesical fat' by between three and six cycles of MVAC, CMV or other combination chemotherapy. With a median follow-up of 36.5 months (range 12–60), nine patients (69%) were alive with no evidence of disease, seven (53.8%) without any other form of surgery; in eight patients (61.5%) the bladder has been preserved. The ability of the combination of radical TUR and cisplatin-based chemotherapy to produce a satisfactory clinical outcome in suitably selected patients, achieving a significant proportion free of invasive tumour in the bladder has been supported by Noguiera-March *et al.*[39] (who used MVAC) and

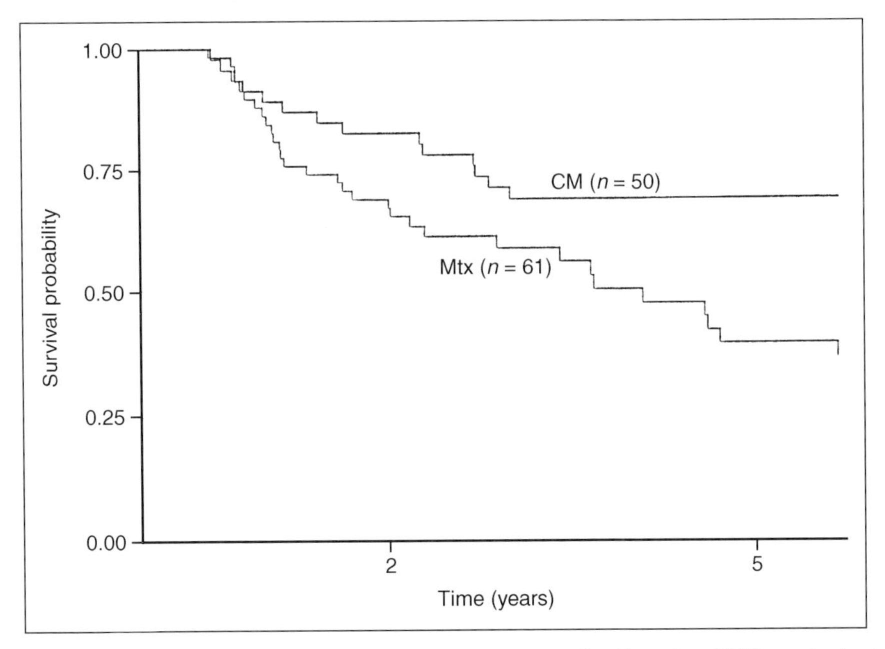

Figure 10.1 Disease-specific survival of patients treated either by TUR and cisplatin/methotrexate (CM) or TUR plus methotrexate and folinic acid (MTX). These were sequential cohorts of patients, non-randomized.

Amiel *et al.*[3], who used a combination of cisplatin and 5-fluorouracil (5-FU) following TUR.

Others have preferred to give systemic chemotherapy first and then to consolidate a complete response, or good partial response, by using TUR or partial cystectomy to clear the site of previous tumour. In a preliminary report of the MSKCC experience with initial systemic chemotherapy followed by partial cystectomy[23] details are given of 32 patients who underwent partial cystectomy following a median of four cycles of MVAC (range 2–5). With a median follow-up period of 19.5 months from surgery, 24 (75%) of their patients remained free of disease with functioning bladders. The authors make the point that only four patients had tumours thought suitable for partial cystectomy before MVAC, the others being sufficiently downstaged, by chemotherapy, that segmental resection became a viable option. Prior multifocal disease was, interestingly, not considered a contra-indication to this approach provided 'complete responses were documented in all or enough sites'. The presence, or otherwise, of concomitant CIS is not commented upon, but in their discussion the authors speculate that such patients may be at higher risk of developing new invasive disease. Certainly such an increased risk has been noted following the use of radiotherapy to treat muscle invasive bladder cancer[63].

Hatcher *et al.*[17] performed partial cystectomy on four, and TUR on two, of 39 patients who received presurgical combination chemotherapy. A further seven patients, clinically T0 after chemotherapy, were merely observed, while the remaining 26 underwent radical cystectomy. Overall they demonstrated no impact of surgical intervention on subsequent survival. The results of our own and other series of patients treated with a combination of cisplatin-containing chemotherapy and conservative surgery suggest that the approach should be tested in a randomized comparison with more conventional approaches.

DISCUSSION

In recent years the use of conservative surgery in combination with radiotherapy and/or chemotherapy with the aim of organ conservation has become standard management for many malignancies. Mastectomy is now performed on only a minority of women with breast cancer and even for less common tumours, such as limb sarcomas, a combination of chemotherapy and limb-sparing surgery has become standard. Surgical oncologists have become convinced that for these and other sites, the enhanced quality of life that accompanies the lack of mutilation and preservation of function afforded by organ sparing is accomplished without the cost of reduced survival rates. The picture with regard to bladder cancer is rather different. Organ preservation has been accepted as a curative alternative to cystectomy in only a minority of countries, and for many cystectomy remains the 'gold standard'[32,33] which should not be denied even the elderly patient thought unfit for surgery[31] who is thus 'not only . . . deprived of his right to definitive curative treatment but also is exposed to higher morbidity and mortality and worse quality of life than are those who undergo operations'.

Urologists world-wide have been less ready to embrace the concept of organ conservation than surgeons dealing with cancer in other primary sites. Zietman and Shipley[64] speculate that this may, in part, be due to the relative rarity of bladder cancer compared with breast cancer and the consequent delay in accumulating convincing data to support bladder conservation. In many countries radiotherapy has assumed a secondary role to surgery and, as definitive treatment, has largely been reserved for palliative treatment of cases considered too advanced for radical surgery or too frail. No satisfactory randomized comparison of the two conventional treatment modalities exists and improvements in surgical and anaesthetic techniques, patient selection, staging and postoperative support of the, often elderly, patient combine to produce surgical outcomes superior to those of 15 or 20 years ago. In addition, the recently acquired ability to construct neobladders adds a technically exciting and challenging dimension to surgery. This, combined with the cosmetic advantages of a reconstructed bladder and the perception of function approaching that of a conserved bladder must combine to make bladder conservation an option which, for some, is not worthy of consideration. After all, surgeons exist to perform surgical procedures. Another inadequately explored influence on therapeutic choice may be the system of remuneration for individual clinicians or the units within which they work.

In countries in which there has been, historically, a prejudice in favour of cystectomy, there is a reawakening of interest in radiotherapy as a bladder-conserving radical treatment for muscle invasive bladder cancer[7,9,24,27,60] when given in combination with cisplatin-based chemotherapy. Only one of these studies[7] was a prospective randomized comparison of radiation alone with a combination of chemotherapy and radiotherapy. In this study no survival benefit was demonstrated for the addition of single-agent cisplatin to radiotherapy. However, the numbers in the study were small and the trial lacked the power to detect a true survival benefit of 10–15%. There was, however, an apparent reduction in the pelvic recurrence rate of some 25%. The remaining studies failed to randomize combination therapy against radiation alone, and to conclude that the apparent advantages of chemoradiotherapy are due to the addition of cytotoxics may be premature. At least one modern radiotherapy series[13] has reported response rates in the bladder which approximate to those quoted in chemoradiotherapy series and which are clearly superior to the figure 'of around 40%' quoted from older series in support of chemoradiotherapy[64]. This emphasizes the importance of patient selection and the role of chance in determining treatment outcome and highlights the need for comparisons to be made within the context of randomized trials of sufficient size to detect real and clinically meaningful differences.

The interest that the results of integrated chemoradiotherapy has provoked in bladder conservation comes at an interesting time. Not only does cystectomy remain an option, with the possibility of increasingly sophisticated neobladder construction allowing retention of an outwardly normal anatomy and more nearly normal urinary function than with a urinary conduit, but various non-radiotherapeutic strategies for conserving the bladder are emerging as worthy of consideration, at least for selected patients. Nevertheless, recent reviews from both sides of the Atlantic have failed to acknowledge the potential value of non-radiotherapeutic approaches to bladder conservation[8,64], concentrating instead on the traditional rivalries of the surgeon and the radiotherapist. There is, perhaps, a sufficient body of literature available on the non-radiotherapeutic approach to enable us to place the various available strategies in context alongside the more traditional approaches.

With currently available agents, systemic chemotherapy does not appear to be a serious alternative to conventional treatments for the cure of muscle invasive bladder cancer. The initial high response rates observed when cisplatin-containing chemotherapy combinations were given to patients with metastatic transitional cell carcinoma fostered the view that TCC might, like germ-cell cancers, prove a reliably curable adult solid tumour. For a small minority of patients with metastatic disease, aggressive cisplatin-based chemotherapy produces cures[58]; usually patients with lymph-node metastases whose response to chemotherapy is consolidated by resection of sites of residual or previous bulk disease. Response rates are high and complete response is more likely to occur in the primary tumour site than in metastases, following chemotherapy. Nevertheless two factors weigh against cytotoxic chemotherapy finding acceptance as the sole curative therapy for bladder cancer, at least with currently available drug combinations. Although there are reports of patients with muscle invasive bladder cancer remaining disease-free in the bladder for long intervals after sole treatment with chemotherapy, the proportion in whom this is attainable is smaller than with the conventional bladder-sparing technique of radical radiotherapy. Therefore, while the fact that transitional cell carcinoma of the bladder is chemocurable in a sizeable minority of cases is undoubted, the relatively low rate of complete response prevents its consideration as sole therapy. There are reservations about the lack of reliability of endoscopic staging in detecting the presence of persistent cancer in the bladder, as judged by the persistence of tumour in cystectomy specimens after negative staging[47].

Local removal of the primary bladder tumour by TUR or partial cystectomy is capable of curing a proportion of patients with muscle invasive bladder cancer. However, authors who have adopted this approach counsel either extreme selectivity[19,20] or the routine adjuvant use of radiation or cytotoxic chemotherapy[29,62]. It would seem prudent not to recommend TUR or partial cystectomy as sole treatment, other than for those few patients too frail for any other therapeutic approach to be viable or for a highly selected minority with particular tumour characteristics which define a group at low risk of local relapse. For those rare patients with urachal carcinomas or cancer arising in a diverticulum which is diagnosed early enough to be locally resectable, formal, open local excision would appear to be a viable alternative to cystectomy[10,22,36,43].

Few series have reported the results of local resection combined with systemic chemotherapy. However, those that do exist foster the perception that this combined approach offers the potential, in appropriate patients, for cure rates similar to those observed with conventional radical treatments with a greater likelihood of physiologically 'normal' bladder function. It is inconceivable that the quality of bladder function after a combination of bladder-sparing surgery and chemotherapy will be worse than after surgery, chemotherapy **and** radiotherapy. Indeed, the likelihood is that the omission of radiotherapy from the combination will improve the functional end result.

In the United States the concept that cure of muscle invasive bladder cancer with conservation of a functioning bladder may be achieved safely with multimodality treatment is gaining, at best, grudging acceptance. In Britain we approach the disease with a different prejudice, that radical radiotherapy alone is a safe and effective alternative to cystectomy for the majority of patients with bladder cancer, and a scepticism about the necessity for the addition of chemotherapy.

In paediatric malignancies, previously treated by a combination of surgery, radiotherapy and chemotherapy, concern about the long-term effects of radiotherapy on growth and development has led clinicians to explore treatment strategies in which radiation has been reduced or omitted. As a result, the observation has been made, for a number of childhood tumours, that radiation may safely be omitted from the treatment regime with no loss of therapeutic efficacy and appreciable gains in the reduction of long-term morbidity. Such an approach is worthy of consideration as an alternative to the conventional radical options of radiotherapy and chemotherapy in muscle invasive bladder cancer. As collaborative groups contemplate clinical trials to compare cystectomy with bladder-conserving radiotherapy, with or without chemotherapy[8], consideration should be given to using this opportunity to examine, in the setting of a randomized comparison, the role of non-radiotherapeutic bladder-conserving regimes, probably combinations of conservative surgery and cisplatin-containing chemotherapy.

There will, of course, be patients for whom bladder preservation represents a less good option for treatment of muscle invasive bladder cancer. Wolf and co-workers[63] identified patients with concomitant urothelial dysplasia and muscle invasive bladder cancer as being at high risk of developing new invasive tumour following complete response of the original primary tumour to full-dose radiotherapy. Of 114 patients with invasive bladder tumours treated by radiotherapy alone, 32 patients had complete primary tumour response and mucosal biopsies taken at preselected sites during initial cystoscopy. Concomitant carcinoma in situ was present in 10, seven of whom developed new primary tumours 9–24 months after completion of radiotherapy. New invasive tumours also developed in four of nine patients with concomitant dysplasia Grade 2. The remaining 13 patients, who achieved complete response following radiotherapy and who did not have pretreatment concomitant CIS or dysplasia, remained free of new tumours between 9 and 75 months after treatment. The authors recommend that patients with concomitant CIS ± dysplasia should not be considered for full-dose radiotherapy as definitive treatment of their muscle invasive bladder cancer. In our institution, treating patients with muscle invasive bladder cancer and concomitant CIS with systemic chemotherapy, we have never observed the *in situ* disease to respond to systemic therapy and have on several occasions observed the apparent *de novo* development of CIS, during chemotherapy, in patients whose invasive bladder cancer was responding. The concern about CIS predicting for failure after radiotherapy is not shared by all authors (Chapter 8), and none of the reports of durable remission of primary bladder tumours following chemotherapy, with or without endoscopic or segmental resection, seems to mirror our anecdotal experiences. Nevertheless, in defining groups of patients suitable for non-radiotherapeutic bladder-conserving strategies, the early recognition of those who may be at greater risk of new tumour development within the preserved bladder is important. Local resection, partial cystectomy and chemotherapy do not, perhaps, carry the same risk of increasing the difficulty of salvage cystectomy, following attempted bladder preservation, as full-dose radiotherapy[63]. Nevertheless, in developing alternative organ-conserving strategies cognisance must be taken of the likelihood of groups being at higher risk of relapse, intensive observation policies pursued in such groups, and early attempts made to define prognostic indices which allow the identification both of those likely to benefit from a bladder-conserving approach and those for whom such an approach is dangerous. In addition to the currently available indicators of tumour behaviour, such as histological type and grade, multifocality and the presence of concomitant *in situ* disease, the growing array of molecular and genetic markers of tumour behaviour should be documented and used prospectively to develop indices that enable us to select treatment appropriate to the particular tumour.

The impact of our initial therapeutic decision upon further management of the patient must also be considered. Some patients in whom bladder preservation is attempted will relapse locally and be offered salvage cystectomy. Salvage cystectomy may, increasingly often, be followed by construction of a neobladder. Among the other questions to be answered in defining the role of non-radiotherapeutic, bladder-conserving strategies in the management of bladder cancer is that of whether the prior use of aggressive chemotherapy will jeopardize surgical reconstruction.

There will be patients whose death from muscle invasive bladder cancer can only be prevented by cystectomy. However, those destined to die whatever the therapeutic intervention (of bladder cancer or intercurrent disease before recurrence of their bladder cancer) should be selected for bladder preservation, as should those whose death can equally well be prevented by a bladder-conserving approach. The means of identifying such groups must be developed. We must not allow prejudice in favour of a particular therapeutic intervention, perhaps the only one we are licensed to perform, to blind us to the possibility that a more unconventional approach may be as safe and offer better preservation of function. The relative contribution of the components of the combination of aggressive local surgical resection, chemotherapy and radiotherapy to the success of this stratagem should be examined systematically. If the removal of tumour is a less important contribution of TUR to overall success than the induction of trauma-related cytokines, are there other ways that we can achieve or enhance this effect? It is possible that amplification of the anti-tumour effect could be achieved by, for example, dose intensification of a chemotherapy regime or alterations in radiotherapy fractionation with acceptable or no increase in morbidity. It is also possible that reducing or omitting one of the components of the triad could be achieved with no loss of effectiveness but a reduction in morbidity.

The realization that bladder preservation is a feasible option for patients with muscle invasive bladder cancer opens exciting possibilities that should be approached without bias. Such approaches have proved successful in other tumour sites and the acceptance that organ preservation is a possibility for many patients with muscle invasive bladder cancer should stimulate us to emulate those caring for patients with cancer in other sites and strive to optimize the means of achieving it.

REFERENCES

1. Abi-Aad AS, Stenzl A, Figlin R and de Kernion JB (1993) Local response and long term results of pre-operative M-VAC regimen in regionally advanced transitional cell carcinoma of the bladder. *Eur J Can*cer **29A**: 1223–1224.

2. Alken P and Köhrmann KU (1994) The essentials of transurethral resection of bladder tumours. *Arch Ital Urol Androl* **65:** 629–632.
3. Amiel J, Quintens H, Thyss A *et al.* (1988) Association resection transurethrale (RTU)-chimiotherapie systematique comme traitement initiale des tumeurs infiltrantes de vessie (pT2-pT3,NX,M0). *J Urol (Paris)* **94**: 333–336.
4. Barnes RW, Bergman RT, Hadley HL *et al.* (1967) Control of bladder tumours by endoscopic surgery. *J Urol* **97:** 864–868.
5. Barnes RW, Dick AL, Hadley HL *et al.* (1977) Survival following transurethral resection of bladder carcinoma. *Cancer Res* **37:** 2895–2897.
6. Brendler CB, Steinberg GD, Marshall FF *et al.* (1990) Local recurrence and survival following nerve-sparing radical cystoprostatectomy. *J Urol* **144**: 1137–1140.
7. Coppin CM, Gospodarowicz MK, James K *et al.* (1996) Improved local control of invasive bladder cancer by concurrent cisplatin and pre-operative or definitive radiation. The National Cancer Insititute of Canada Clinical Trials Group. *J Clin Oncol* **14**: 2901–2907.
8. Duchesne GM (1994) Radical treatment for primary bladder cancer – where are we and where do we go from here? A review. *Clin Oncol* **6:** 121–126.
9. Dunst J, Sauer R, Schrott KM *et al.* (1994) Organ-sparing treatment of advanced bladder cancer: a 10-year experience. *Int J Radiat Oncol, Biol Phys* **30:** 261–266.
10. Faisal MH and Freiha FS (1981) Primary bladder neoplasm in bladder diverticula: a report of 12 cases. *Br J Urol* **53**: 141–143.
11. Flocks RH (1951) Treatment of patients with carcinoma of the bladder. *J Am Med Ass* **145:** 295–301.
12. Freiha F (1992) Open bladder surgery, in *Cambell's Urology*, (eds PC Wash, AB Retik, TA Stamey and ED Vaughan), W. B. Saunders Co., Philadelphia.
13. Gospodarowicz MK, Rider WD, Keen CW *et al.* (1991) Bladder cancer: long-term follow-up results of patients treated with radical radiation. *Clin Oncol* **3:** 155–161.
14. Hall RR, Newling DW, Ramsden PD *et al.* (1984) Treatment of invasive bladder cancer by local resection and high dose methotrexate. *Br J Urol* **56:** 668–672.
15. Hall RR and Roberts JT (1989) Chemotherapy in advanced bladder cancer, in *Therapeutic Progress in Urological Cancers*, Alan R Liss, New York.
16. Hall RR, Roberts JT and Marsh MM (1993) Radical TUR and chemotherapy aiming at bladder preservation. *Prog Clin Biol Res* **353:** 163–168.
17. Hatcher PA, Hahn RG, Richardson RL and Zincke H (1994) Neoadjuvant chemotherapy for invasive bladder carcinoma: disease outcome and bladder preservation and relationship to local tumour response. *Eur Urol* **5:** 209–215.
18. Henry K, MIller J, Mori M *et al.* (1988) Comparison of transurethral resection to radical therapies for stage B bladder tumours. *J Urol* **140:** 964–967.

19. Herr HW (1987) Conservative management of muscle-infiltrating bladder tumours: Prospective experience. *J Urol* **138:** 1162–1163.
20. Herr HW (1992) Transurethral resection in regionally advanced bladder cancer. *Urol Clin N Am* **19:** 695–700.
21. Herr HW (1992) Editorial Comment on Solsona *et al. J Urol* **147:** 1515.
22. Herr HW (1994) Urachal carcinoma: the case for extended partial cystectomy. *J Urol* **151:** 365–366.
23. Herr HW and Scher HI (1992) Neoadjuvant chemotherapy and partial cystectomy for invasive bladder cancer. *Cancer Treat Res* **59:** 99–103.
24. Housset M, Maulard C, Chretien Y *et al.* (1993) Combined radiation and chemotherapy for invasive transitional cell carcinoma of the bladder: a prospective study. *J Clin Oncol* **11:** 2150–2157.
25. Igawa H, Ohkuchi T, Ueki T *et al.* (1990) Usefulness and limitations of methotrexate, vinblastine, doxorubicin and cisplatin for the treatment of advanced urothelial cancer. *J Urol* **144:** 662–665.
26. Karim ABMF (1994) Organ preservation for patients with cancer. *Eur Cancer Centre Newsletter* **3:** 11–14.
27. Kaufman DS, Shipley WU, Griffin PP *et al.* (1993) Selective bladder preservaton by combination treatment of invasive bladder cancer. *N Engl J Med* **329:** 1377–1382.
28. Koloszy Z (1991) Histological 'self control' in transurethral resection of bladder tumours. *Br J Urol* **67:** 162–164.
29. Kondás J and Szentogyörgyi E (1992) Transurethral resection of 1250 bladder tumours. *Int Urol Nephrol* **24(i)**: 35–42.
30. Kondás J, Váczil L, Szecsó L and Kondér G (1993) Transurethral resection for muscle-invasive bladder cancer. *Int Urol Nephrol* **25:** 557–563.
31. Leibovitch I, Avigad I, Ben-Chaim J *et al.* (1993) Is it justified to avoid cystoprostatectomy in elderly patients with invasive transitional cell carcinoma of the bladder? *Cancer* **71:** 3098–3101.
32. Martinez-Piñeiro JA, Jiminez L and Martinez-Piñeiro L (1991) Aggressive TURB combined with systemic chemotherapy for locally invasive TCC of the urinary bladder *Eur J Cancer* **27** (Suppl. 2): 5104, Abstract 605.
33. Martinez-Piñeiro JA and Martinez-Piñeiro L (1994) Is bladder preservation possible in infiltrating cancer. *Arch Esp Urol* **47:** 337–342.
34. Mathur VK, Krahn HP, Ramsey EW (1991) Total cystectomy for bladder cancer. *J Urol* **125**: 784–786.
35. Meyers FJ, Palmer JM, Freiha FS *et al.* (1985) The fate of the bladder in patients with metastatic bladder cancer treated with cisplatin, methotrexate and vinblastine: a Northern California Oncology Group Study. *J Urol* **134:** 1118–1121.
36. Micic S and Ilic V (1983) The incidence of neoplasm in vesical diverticulae. *J Urol* **129:** 734–735.
37. Milner WA (1954) The role of conservative surgery in the management of bladder tumours. *Br J Urol* **26:** 375–386.

38. Montie JE, Straffon RA and Stewart RH (1984) Radical cystectomy without radiation therapy for carcinoma of the bladder. *J Urol* **131**: 477–482.
39. Nogueira-March JL *et al.* (1991) Radical TUR and M-VAC in the treatment of infiltrating bladder tumours. *Proc. 22nd Congress, Societé Internationale d'Urologie, Seville*, Abstract 215.
40. O'Flynn JD, Smith JD and Hanson JG (1975) Transurethral resection for the assessment and treatment of vesical neoplasms. A review of 800 consecutive cases. *Eur Urol* **1:** 38–41.
41. Ojeda L and Johnson DE (1983) Partial cystectomy; can it be incorporated into an integrated therapy program? *Urology* **22:** 115–117.
42. Pagano F, Bassi P, Galetti TP *et al.* (1991) Results of contemporary radical cystectomy for invasive bladder cancer: a clinico-pathological study with an emphasis on the inadequacy of the tumor, nodes and metastasis classification. *J Urol* **145**: 45–50.
43. Pode D and Fair WR (1991) Urachal tumours. *AUA Update Ser* **10:** 33–38.
44. Regaud C, Coutard H and Hautant A (1922) Contribution au traitement des cancers endolarynges par les rayons. *XX Internat Congr D'Otol,* pp. 19–22.
45. Scher HI (1990) Chemotherapy for invasive bladder cancer: neoadjuvant versus adjuvant. *Sem Oncol* **17:** 555–565.
46. Scher HI (1992) Systemic chemotherapy in regionally advanced bladder cancer: theoretical considerations and results. *Urol Clin N Am* **19:** 747–759.
47. Scher HI, Herr HW, Sternberg C *et al.* (1989) Neoadjuvant chemotherapy for invasive bladder cancer. Experience with M-VAC regime. *Br J Urol* **64:** 250–255.
48. Scher HI, Yagoda A, Herr HW *et al.* (1988) Neoadjuvant M-VAC (methotrexate, vinblastine, doxorubicin and cisplatin); effect on the primary bladder lesion. *J Urol* **139:** 470–474.
49. Seidman AD and Scher HI (1991) The evolving role of chemotherapy for muscle-infiltrating bladder cancer. *Sem Oncol* **18:** 585–595.
50. Shipley WU, Prout GR, Kaufman SD and Perrone TL (1987) The importance of initial transurethral resection for improved survival with full-dose irradiation. *Cancer* **60:** 514–520.
51. Skinner DG, Daniels JR, Russell CA *et al.* (1990) The role of adjuvant chemotherapy following cystectomy for invasive bladder cancer: a prospective comparative trial. *J Urol* **145**: 459–464.
52. Skinner DG and Lieskovsky G (1988) Management of invasive and high-grade bladder cancer, in *Diagnosis and Management of Genitourinary Cancer*, (eds DG Skinner and G Lieskovsky), WB Saunders, Philadelphia, pp. 295–312.
53. Slack NH, Bross ID and Prout GR (1977) Five year follow up results of a collaborative study of therapies for carcinoma of the bladder. *J Surg Oncol* **9**: 393–405.

54. Socquet Y (1981) Combined surgery and adjuvant chemotherapy with high dose methotrexate and folinic acid rescue (HDMTX-CF) for infiltrating tumours of the bladder. *Br J Urol* **53:** 439–443.
55. Solsona E, Iborra I, Ricós JV *et al.* (1992) Feasibility of transurethral resection for muscle infiltrating carcinoma of the bladder: prospective study. *J Urol* **147**:1513–1517.
56. Srougi M and Simon SD (1994) Primary methotrexate, vinblastine, doxorubicin and cisplatin chemotherapy and bladder preservation in locally invasive bladder cancer: a five year follow up. *J Urol* **151:** 593–597.
57. Sternberg CN, Arena MG, Calabresi F *et al.* (1993) Neoadjuvant M-VAC (methotrexate, vinblastine, doxorubicin and cisplatin) for infiltrating transitional cell carcinoma of the bladder. *Cancer* **72:** 1975–1982.
58. Sternberg CN, Yagoda A, Scher HI *et al.* (1989) M-VAC for advanced transitional cell carcinoma of the urothelium: efficacy and patterns of response and relapse. *Cancer* **64:** 2448–2458.
59. Sweeney P, Kursh ED and Resnick MI (1992) Partial cystectomy. *Urol Clin N Am* **19:** 701–711.
60. Tester W, Porter A, Absell S *et al.* (1993) Combined modality programme with possible organ preservation for invasive bladder carcinoma: results of RTOG protocol 85–12. *Int J Radiat Oncol, Biol Phys* **25:** 783–790.
61. Varsons G and Yahalom J (1991) Lactation following conservation surgery and radiotherapy for breast cancer. *J Surg Oncol* **46**: 141–144.
62. Veronesi A, Lo Re G, Carbone A *et al.* (1994) Multimodal treatment of locally advanced transitional cell bladder carcinoma in elderly patients. *Eur J Cancer* **30**: 918–920.
63. Wolf H, Olsen PR and Højgaard K (1985) Urothelial dysplasia concomitant with bladder tumours: a determinant for future new occurrences in patients treated by full-course radiotherapy. *Lancet* **1:** 1005–1007.
64. Zietman AL and Shipley WU (1994) Organ-sparing treatment for bladder cancer: time to beat the drum. *Int J Radiat Oncol Biol Phys* **30:** 741–742.
65. Zingg EJ (1985) Treatment of muscle-invasive bladder cancer, in *Bladder Cancer*, (eds EJ Zingg and DMA Wallace), Springer, Berlin, pp. 189–234.

Commentary

Saad Khoury

I have no major disagreement with this excellent chapter by Dr Roberts. Over the past decade developments in the systemic chemotherapy and radiotherapy of infiltrating bladder cancers have had an impact on the routine care of patients with this stage of disease, and could possibly change what is considered to be standard treatment over the next few years. Preliminary data seem to suggest that bladder preservation may be a possibility in some selected patients who may be managed successfully by the combination of chemotherapy and TUR, with or without radiotherapy.

However, whether or not it is possible to preserve bladder function without compromising cure is still a matter of debate. Until this question has been resolved by ongoing and future randomized studies, cystectomy remains the standard treatment for non-metastatic infiltrating bladder cancer. Nevertheless bladder preservation could be attempted:

1. In some patients when the tumour is at the dome of the bladder and can be removed with a safe margin by partial cystectomy.
2. If the patient refuses cystectomy for any reason.
3. In poor-risk patients, due to other disease or age.
4. In the framework of randomized trials designed to evaluate whether the recent developments can improve the standard treatment and justify bladder preservation. These studies must be encouraged as this is the only way to validate new standards of care for the future.

11

Systemic chemotherapy for metastatic cancer of the uro-epithelial tract

Derek Raghavan

Although more than 80% of incident cases of bladder cancer are clinically non-metastatic at presentation, many patients with superficial or invasive disease subsequently develop metastases[69]. Furthermore, there is a marked similarity in the natural history and response to therapy of the gamut of other metastatic uro-epithelial cancers, including those arising from the urethra, ureters and renal pelvis, to that of metastatic bladder cancer, and in particular a similar response to cytotoxic chemotherapy[50]. Accordingly, it is appropriate to group tumours arising from the uro-epithelial tract together in any discussion of management of metastatic disease. Thus, the management of metastatic uro-epithelial cancer is a much more common problem than would be inferred from published incidence figures for metastatic bladder cancer[84]. The median survival of patients presenting with metastatic uro-epithelial cancer, who do not receive cytotoxic chemotherapy is only 3–6 months[42,64,102]. Since the introduction of cytotoxic chemotherapy into the management of metastatic uro-epithelial cancer, the median survival figures have more than doubled[64,106], but more than 80% of patients with this problem are still destined ultimately to die of cancer, and more effective strategies of treatment are required.

BIOLOGY OF METASTATIC UROTHELIAL CANCER: IMPLICATIONS FOR THE RESULTS OF TREATMENT

The metastatic potential of primary bladder cancer is relatively predictable, with clear correlations with depth of invasion (stage), extent of histological differentiation (grade), and a range of less dominant biological predictors, including ploidy, expression of the receptor for the epidermal growth factor and for transferrin, presence of blood group antigens, microvessel density, and expression of certain oncogenes or mutations of suppressor genes[35,53,69,73]. In addition, other clinical predictors, such as the presence of hydronephrosis, anaemia, the presence of carcinoma in situ, and perhaps the size of the primary tumour have been reported to correlate with subsequent metastasis, at least in univariate analysis of large series[69,71]. The biology of upper-tract malignancies is similar, but with a less predictable correlation between T stage and metastasis, and the added variable of antecedent analgesic nephropathy as an aetiological association[50], with the concomitant problems of renal dysfunction for patients for whom chemotherapy is planned.

The specific biology of **metastatic** bladder cancer and the other uro-epithelial malignancies has not been reported in detail, although there appear to be significant similarities to the features of high-grade, invasive primary bladder cancer, with apparent overlap of histology, ploidy, tumour antigen expression, expression of oncogenes and suppressor genes, and many other characteristics.

The common sites of metastasis of bladder cancer and the other uro-epithelial malignancies include regional lymph nodes, bone, lung, skin and liver, and less frequently brain, meninges and the organs within the peritoneal cavity[4,13,39,66]. The distribution of metastases is of particular importance when considering treatment as the site(s) of involvement correlate with prognosis, with an improved median survival and survival tail in patients with metastases limited to the lymph nodes and skin, and a substantially worse prognosis in patients with liver and bone metastases[25,43].

Another important biological factor in assessing the outcome of management of metastatic uro-epithelial cancer is the histological type. There are clear differences in the outcome of treatment of transitional cell carcinoma and the non-transitional cell types (adenocarcinoma and squamous carcinoma) of metastatic uro-epithelial cancer[43]. In addition, although the dominant histology of most metastatic deposits is transitional cell carcinoma, there is considerable occult heterogeneity within deposits of transitional cell cancer, with respect to grade, histology, growth parameters, ploidy, karyotype, gene expression and tumour markers[10,53,68,73], providing an added source of variation in the outcomes of treatment.

The traditional view has been that there are separate cell lineages of the different histological types of uro-epithelial cancer[39], although studies in our laboratory have suggested the existence of a common stem cell of origin of transitional cell carcinoma, squamous carcinoma and adenocarcinoma of the bladder[74,75]. This observation could explain the occasional discrepancies noted between the histology of primary tumours and subsequent metastatic disease, and emphasizes the importance of reconfirming the histology of apparent metastatic deposits. In this context, it should not be forgotten that second malignancies occur in up to 10% of patients with bladder cancer, and thus confirmation of histology of metastases is even more important in determining that the appropriate type of chemotherapy or other treatment is selected.

SINGLE-AGENT CHEMOTHERAPY

The history of systemic chemotherapy for metastatic bladder cancer extends back to the 1950s, and has been reviewed in detail elsewhere[12,42,64,106]. Anecdotal responses were often documented after treatment with the early cytotoxic agents, although the quality of such responses is difficult to assess as the criteria of assessment were inadequate, and the duration of response was rarely reported. Nevertheless, it should not be forgotten that occasional patients have been cured of metastatic disease, or at least have sustained remissions of 5 years or longer, with single-agent chemotherapy.

More carefully defined criteria of response have now been developed to allow more accurate and reproducible assessment of the efficacy of systemic chemotherapy for metastatic tumours of all types[52], and these have been further adapted by a series of consensus conferences for application to metastatic uro-epithelial cancer[100]. Using these more rigorous standards, the reported objective response rates are lower than in past times (Table 11.1) with most single agents yielding objective responses in about 15–20% of cases, including complete responses in only 5–10%[2,7,12,24,28,31,42,51,64,70,72,81,86,98,99,101,104,105,106]. Although the duration of response often is not recorded in the non-comparative (phase I–II trial) literature, most randomized trials of single-agent chemotherapy have documented response durations of only 4–6 months[31,38,43,86].

One of the complicating features in assessing the published literature regarding single-agent chemotherapy for uro-epithelial cancer is the lack of constancy of dosing. For example, trials of cisplatin have used doses ranging from 50 to 120 mg/m^2, but there are no published data that define the optimal level for the management of uro-epithelial cancer. Similarly, optimal dose levels for application to bladder cancer have not been defined for mitomycin C, doxorubicin

Table 11.1 Responses to single agents in bladder cancer

Agent	Number of patients	Average CR+PR (%)
Amsacrine	59	10
Cisplatin	320	30
Cyclophosphamide	98	31
Doxorubicin	235	23
5-Fluorouracil	75	35
Methotrexate	236	29
Mitomycin C	48	21
PALA	12	0
Vinblastine	38	16

CR, complete response; PR, partial response.

and methotrexate. Given the discrepancies reported in optimal dosing for other tumour types, there is no clear extrapolation that can be applied to uro-epithelial cancer.

Also of importance, the most stringent assessments of outcome of single-agent chemotherapy appear to have been reported from studies employing randomized comparisons, rather than in single-arm, non-comparative phase I or phase II trials (Table 11.2)[31,38,43,86,98]. Whether this is due to a greater precision of assessment in multicentre studies, larger patient numbers, or assessment by a broader range of investigators who are less 'invested' in producing high response rates is unclear. Conversely, many of these randomized trials have been executed in more recent times, and the improved outcome could be a function of the phenomen of 'stage migration'[19], i.e. the situation in which improved staging (for example, with the introduction of computed axial tomography) has revealed hitherto occult metastases. In this situation, smaller-volume disease is incorporated within the metastatic category, and could contribute to improved outcome in the more recent trials. One of the highest response rates reported for the single-agent use of cisplatin (33%) was reported by the Australian Bladder Cancer Study Group[31], a multicentre group; this result was a surprise to the authors, and was thought to represent case selection, with a higher proportion than expected of patients with lymph node-dominant disease in both the single-agent arm and in the patients treated with the combination of cisplatin and methotrexate (response rate 45%).

Although there are no clear data to define the 'best' single agent, there is a general consensus that the list of most active conventional cytotoxic drugs against uro-epithelial cancer includes cisplatin, methotrexate, mitomycin C and doxorubicin. Whether the newer agents, such as the taxanes, gemcitabine, new-generation platinum co-ordination complexes and camptothecin derivatives will be added to this list remains to be seen (as discussed below).

Table 11.2 Response to single-agent cisplatin: results from randomized Co-operative Group trials

Research group	Cisplatin dosage (mg/m^2)	No. of patients	No. of responses	%
National Bladder Cancer Co-Operative Group[86]	70	50	10	20
SECSG[98]	60	48	8	17
ECOG[38]	60	45	7	16
Australian Bladder Cancer Study Group[31]	80	55	17	31
Intergroup[43]	70	120	14	12

SECSG, Southeastern Cancer Study Group; ECOG, Eastern Co-operative Oncology Group.

COMBINATION REGIMENS

Building upon the demonstrable activity of the single agents discussed above, a wide range of two-, three- and four-drug combination regimens were developed in an attempt to increase response rates and duration of survival[1,2,24,31,38,42,64,86,98,106]. For more than a decade, the use of combination regimens appeared to improve response rates, but without a corresponding increase in survival (Table 11.3)[1,11,24,31,37,38,85,86,89,94,98].

It was not until the development of the MVAC and CMV regimens, which combine methotrexate, vinblastine and cisplatin (with or without doxorubicin), that highly effective treatment for some patients with hepatic and bone metastases became available[30,92,93]. At the Memorial Sloan-Kettering Cancer Center in particular, the MVAC regimen was shown to yield biopsy-proven complete remissions in patients with liver and bone metastases[92,93].

Table 11.3 Response of advanced bladder cancer to treatment with combination chemotherapy

Regimen	% Response rate	Median survival (months)	% Long-term survival	Reference
Randomized trials				
Cisplatin	20	<12	<20	
Cyclophosphamide/cisplatin	12	<12	<20	Soloway *et al.*[86]
Cisplatin	17	6	10	
CAP	33	7	10	Khandekar *et al.*[38]
Cisplatin	33	7	<20	
Methotrexate/cisplatin	45	10	<20	Hillcoat *et al.*[31]
Cisplatin	16	6		
CAP	21	6		Troner *et al.*[98]
Cisplatin	9	9	<15	
MVAC	33	13	<30	Loehrer *et al.*[43]*
CAP	46	9	?–12	
MVAC	65	11	?–30	Logothetis *et al.*[45]
Non-randomized trials				
CAP	13–84	–	–	Kedia *et al.*[37] Sternberg *et al.*[89] Young and Garnick[106]
CMV	56	–	–	Harker *et al.*[30]
MVAC	70	13	20	Sternberg *et al.*[93]

CAP, cisplatin/doxorubicin/cyclophosphamide; CMV, cisplatin/methotrexate/vinblastine; MVAC, methotrexate/vinblastine/Adriamycin®/cisplatin.
*Only randomized trial to show statistically significant benefit for combination chemotherapy in advanced bladder cancer both in median survival and the tails of the survival curves.

Table 11.4 Cumulative results of treatment with MVAC regimen

Series	Year	No. of cases	CR (%)	RR (%)	Median survival
Sternberg *et al.*[93]	1989	121	26	72	13.4
Tannock *et al.*[96]	1989	30	13	43	10.0
Igawa *et al.*[34]	1990	58	17	57	8.0
Logothetis *et al.*[45]	1990	55	35	65	11.0
Boutan-Laroze *et al.*[8]	1991	67	19	57	13.0
Loehrer *et al.*[43]	1992	120	13	38	12.5

Adapted from Levine and Raghavan[40].

Unlike the trials of the previous older combination regimens, a multicentre, international randomized trial, conducted in the United States, Canada and Australia, which effected a comparison of single-agent cisplatin with the MVAC regimen, demonstrated for the first time a survival benefit from the combination regimen[43]. Whereas cisplatin alone yielded a median survival of 8 months, the MVAC regimen gave a median survival of 12 months, analogous to the late experience from the Memorial Sloan-Kettering Cancer Center and other non-randomized trials (Table 11.4)[8,34,40,43,93,96]. The tail of the survival curve at 2 years has also confirmed the superiority of this combination[42,43]. In another randomized trial, Logothetis *et al.*[45] demonstrated similar superiority of MVAC when compared to the combination of cisplatin, doxorubicin and cyclophosphamide (CAP), not a surprising result in view of the lack of difference in outcome between single-agent cisplatin and the CAP combination in previous trials. However, a late analysis of the Intergroup trial has shown that less than 10% of treated patients were alive at 5 years, and these included predominantly patients treated with the MVAC regimen[78].

Despite the improved outcomes from the MVAC and CMV regimens in patients with metastatic transitional cell carcinoma, these advances have not extended to patients with non-transitional cell histology[43]. The problems of metastatic adeno-carcinoma and squamous carcinoma of the bladder remain unsolved, although some improvement in objective response rate has been seen after treatment with the combination of cisplatin and infusional 5-fluorouracil (sometimes modulated with calcium leucovorin or interferon)[48,91].

DOSE INTENSITY

As noted above, there is remarkably little published information regarding dose–response relationships or the optimal dosing of cytotoxics in the management of bladder cancer or other uro-epithelial malignancies. Nevertheless, extrapolating from the data interpreted to support increased dose intensity in the treatment of other cancers[33], several clinical trials have addressed the impact of moderate increases in dosing, under cover of colony stimulating factors[23,44,46,47,56,79,80,90].

Gabrilove *et al.*[23] first assessed the possible impact of granulocyte colony-stimulating factor (G-CSF) on the use of the MVAC regimen and demonstrated a dose-dependent reduction of number of days with neutropenia, antibiotic requirement and fever, as well as an unexpected reduction in the severity of mucositis. In addition, under the influence of G-CSF, more patients were eligible to receive day 14 booster doses of chemotherapy. Subsequently, Seidman *et al.*[80] attempted to use G-CSF to allow delivery of MVAC on a 14- or 21-day schedule. In this study, an increment in relative dose intensity of 33% was achieved, but without any obvious improvement in response rate compared to conventional use of MVAC.

When using granulocyte–macrophage colony-stimulating factor (GM-CSF) in an attempt to achieve normal tissue cytoprotection from MVAC chemotherapy, Moore *et al.*[56] reported that significant mucositis occurred in eight of 52 treatment cycles, and that the GM-CSF was associated with significant toxicity, including hypotension, syncope and local tissue reaction. Although responses were recorded in eight of 11 cases, there was marked variability in the extent of disease, with the programme being applied to neoadjuvant or classical adjuvant therapy in 50% of the treated cases. Although GM-CSF reduced haematological toxicity somewhat, the effect decreased with subsequent courses of treatment. Similarly, Sternberg *et al.*[90] studied the impact of G-CSF on the ability to increase dose intensity and were able to escalate MVAC to a relative dose intensity of 1.98, but without a major increment in objective response rate.

Logothetis *et al.*[46] treated 32 patients with bladder cancer that was refractory to combination chemotherapy (including MVAC or combination regimens) in a phase I dose escalation trial of high-dose MVAC with GM-CSF. Although 40% of patients had objective responses (including 23% complete remissions), it should be noted that most of the treated patients had previously experienced a tumour response, implying the possibility of a significant selection bias. In a subsequent preliminary report of an incomplete randomized trial, Logothetis *et al.*[47] reported a response rate of 83% to dose-escalated MVAC, noting that GM-CSF did not appear to contribute to increased dose intensity because of the range of non-haematological toxicity. From this preliminary report, it was not possible to assess the true impact of dose intensity, nor the possibility of case selection bias.

Another important set of data addressing this issue from the viewpoint of delivered dose intensity has recently been reported from the Memorial Sloan-Kettering Cancer Center. Scher and colleagues assessed the effect of relative cumulative dose intensity on survival of 132 patients treated with MVAC chemotherapy and followed for a median of 6 years[79]. This analysis, although addressing only intensity predicated on adherence to the definitive treatment schedule (with day 14 and 21 booster doses), did not reveal any major impact of dose intensity on patient survival.

Thus, to date, this approach has been somewhat disappointing in the context of bladder cancer, although in most of the published trials, the dose escalation has been modest, and a formal phase III comparison has not been reported. However, the most significant criticism of this approach has been reported by the Eastern Co-operative Oncology Group[44]. In a multicentre phase I–II trial of modest dose escalation of MVAC under cover of G-CSF, eight toxic deaths were recorded among 35 patients. Although the relatively small number of cases treated prior to the premature closure of this trial did not allow great statistical power, this study did report an objective response rate of only 60% (including 17% complete response), well within the range of the previous cumulative data for conventional MVAC[40]. Given the advanced age of most patients with uro-epithelial cancer, and the smoking-related disorders characteristic of this group, it seems unlikely that increased dose intensity will play a major role in future treatment strategies for this disease.

NEW AGENTS AND APPROACHES

A detailed discussion of all promising new agents and combination regimens is beyond the scope of this chapter. Instead, some of the more promising innovations are reviewed. Analogous to other tumour types, much of the development of cytotoxic therapy for uro-epithelial cancer has occurred in the usual sequential fashion that progresses from phase I to phase III trials. Thus, the application of the taxane compounds, **paclitaxel** and **docetaxel**, to the treatment of patients with advanced uro-epithelial cancer has occurred as part of broadly based phase II screening programmes. For example, Roth *et al.* have reported a response rate of 42% in 26 previously untreated patients treated with 250 mg/m^2 given by 24-hour continuous infusion[72]. This series included five complete responses (20%), and several of the complete and partial responses lasted more than 1 year. Of note, the sites of involvement correlated with response rate, with an improved outcome associated with nodal disease.

Similarly, **ifosfamide** has been shown to have activity in a broad range of solid tumours. In a formal phase II trial in 60 patients with uro-epithelial tumours who had previously received one type of chemotherapy, 47 underwent complete assessment, of whom 10 had objective responses (21%)[104].

Based on the substantial success of the metallic salt, cisplatin, in the management of bladder cancer, other similar compounds have been assessed, including second- to fourth-generation platinum complexes and gallium. The substituted platinum complex, **carboplatin**, has been shown to have only modest activity against uro-epithelial cancer[51], but is still occasionally used in combination regimens in an attempt to reduce side-effects because of its acceptable profile of toxicity[103]. Seligman and Crawford[81] demonstrated the activity of **gallium nitrate**, administered by continuous infusion, against bladder cancer previously treated by MVAC or single-agent cisplatin. In this study, four of eight patients had objective tumour responses, especially in soft tissue masses, although significant toxicity was documented, including hypocalcaemia, hypomagnesaemia, microcytic anaemia, renal dysfunction and optic neuritis. Of interest, it has been postulated that gallium interacts with tumour cells through transferrin receptors; these receptors have been shown to be expressed heavily in transitional cell cancer.

In an attempt to capitalize on the single-agent activities of some of the above drugs, the Eastern Co-operative Oncology Group has assessed the combination of **vinblastine, ifosfamide and gallium** in a series of 25 previously untreated patients with uro-epithelial cancers[16]. Five patients (20%) achieved complete responses with chemotherapy alone, although early relapses were also reported. Although the activity of this regimen is interesting, its utility will be limited by significant toxicity, including myelosuppression and renal dysfunction as well as significant neurological toxicity and one case of temporary blindness (a recognized effect of gallium).

Perhaps one of the most significant recent advances in the field has been the introduction of **gemcitabine**, an analogue of cytosine arabinoside, into treatment programmes for advanced bladder cancer. Its activity was demonstrated initially in early phase trials in Italy[63] in previously treated patients, and has been confirmed in North American series which have demonstrated an objective single-agent response rate of around 30%[55,87]. Recently, combination regimens have been assessed, incorporating gemcitabine with cisplatin, revealing an objective response rate of approximately 70%[88], including responses in bone and liver.

PALLIATIVE CHEMOTHERAPY: AN OXYMORON?

The impact of cytotoxic chemotherapy on reducing the symptoms of metastatic uro-epithelial cancer is a function of the efficacy of the treatment in tumour cytoreduction, balanced by the symptoms created by the treatment itself. Although the toxicity of single-agent and combination chemotherapy regimens has been meticulously characterized, the impact of these regimens on symptom reduction and quality of life *per se* has been less clearly defined in the literature. Most structured trials have shown clearly that single-agent chemotherapy is much less toxic than combination regimens, but that there is a concomitant reduction in objective response with single-agent therapy[31,38,43].

Although there are no specifically defined guidelines regarding the use of cytotoxics in **palliation** of metastatic disease, the following general principles should apply:

1. The patient should have symptoms that require palliation and that are not being relieved adequately by alternative methods.
2. The patient should be sufficiently robust to tolerate the treatment and its potential toxicity.
3. The planned treatment should have a track record of anticancer efficacy that is adequate to yield a reasonable chance of reduction of tumour-related symptoms;

known prognostic factors should be taken into consideration when assessing the likelihood of patient benefit.

4. The toxicity of planned treatment should not exceed the tumour-related symptoms.

For example, randomized trials have shown clearly that the combination of cisplatin, doxorubicin and cyclophosphamide yields only a modest increase in response rate and no survival benefit compared to single-agent cisplatin or cyclophosphamide, but also that the combination is substantially more toxic[38,86]. It is thus illogical to use this combination in an attempt to palliate the symptoms of metastatic uro-epithelial cancer. By contrast, the MVAC regimen has been shown to yield statistically significant increments in response rate and survival, compared to cisplatin alone, but at the expense of greater toxicity[43]. When considering the application of this toxic regimen, it is clear that poor performance status, significant weight loss and a large tumour burden (especially with bone or liver metastases) do not usually correlate with successful MVAC chemotherapy, and that only occasional patients will derive true benefit from this approach[25,43].

Thus, careful and thoughtful consideration should be given to the use of this regimen for palliation in the debilitated patient. On the rare occasions on which I use the MVAC or CMV regimens in the context of a highly symptomatic but debilitated patient, I carefully apply early stopping rules, assessing the balance of response, symptomatic benefit and toxicity after only one-to-two doses before deciding on whether to continue and the length of the planned course of therapy. Of particular importance, the patient must have adequate cardiac, renal and hepatic function before embarking on such an exercise.

As an alternative, Waxman *et al.*[103] have reported a small series in which the combination of less toxic analogues of the components of the MVAC regimen (i.e. methotrexate, vinblastine, mitoxantrone and carboplatin) were used in this context. While the toxicity was apparently less than the MVAC regimen or full-dose CMV, the true efficacy of this approach has not been validated in any large, comparative clinical trials. With the small number of patients and the wide confidence intervals, it is possible that the response rate reported for this combination is not significantly different from the response rate achieved from methotrexate and vinblastine alone.

It appears that the regimens predicated on some of the newer available agents with lesser profiles of toxicity, such as the taxanes and gemcitabine, allow effective palliation of patients with advanced disease with less expenditure in terms of iatrogenic toxicity. Thus in non-randomized trials, the extent of nausea, vomiting, malaise and febrile neutropenia appears to be less for the combination of gemcitabine and cisplatin[88] or paclitaxel and carboplatin than for the MVAC and CMV regimens. This observation is now being tested in a structured randomized comparison of the MVAC regimen against gemcitabine–cisplatin.

The difficulty of the decision-making process is illustrated by a recent study that assessed how physicians themselves would prefer to be treated if they had symptomatic bladder cancer with lung and bone metastases[54]. Notwithstanding my views on the limitations of the physician-surrogate model[65], it is relevant that there was little consensus among the surveyed urological oncologists. Respondents chose MVAC chemotherapy, other chemotherapy and palliative treatment without chemotherapy in approximately equal proportions. Factors influencing the decision included country of origin (the British strongly favoured palliative measures, compared with a more active stance by North Americans) and medical specialty (medical oncologists opting for chemotherapy and radiation oncologists choosing palliation in a greater proportion of instances)[54]. Interestingly, only 31% of the medical oncologists favoured MVAC, compared with 59% who advocated alternative chemotherapy regimens.

TREATMENT OF THE ASYMPTOMATIC PATIENT: A DILEMMA

An even more difficult management issue is the approach to the asymptomatic patient with emerging metastatic disease, a problem that is not unique to the patient with uro-epithelial cancer. There is a real lack of objective information to facilitate a rational decision process in this setting as, to date, no trials have addressed this important question.

It is clear from published single-agent and combination chemotherapy trials that the bulk and sites of disease correlate with outcomes of chemotherapy[25,43]. Thus patients with good performance status, normal biochemical function (in particular liver function tests and bone enzymes), and lymph node and soft tissue metastases have a higher response rate and improved median survival, compared to those with more advanced disease (liver, bone and lung metastases) and worse performance status. While this could be interpreted as evidence supporting early intervention for the patient with metastatic disease, irrespective of the symptomatic status, the benefit of this approach has not been validated in any structured trial. However, in the management of colorectal cancer, a small, but statistically significant, survival benefit for early chemotherapy (compared to watchful expectancy) has been demonstrated[58].

In the context of advanced bladder cancer, one randomized trial that could be interpreted to support the use of early intervention with chemotherapy was reported by Freiha *et al.*[20]. This trial assessed the impact of adjuvant chemotherapy (methotrexate, vinblastine and cisplatin) on the survival of patients with pelvic lymph node involvement at cystectomy. This study was marred by premature closure in association with a disease-free survival benefit. Of importance, patients who did not receive adjuvant chemotherapy were permitted to undergo salvage chemotherapy at the time of relapse. The group who received adjuvant chemotherapy had a statistically significant improvement in disease-free survival, but also a trend towards improved overall survival, suggesting a possible benefit for early intervention. However, as the total number of patients in the trial was very small, the power of the study was limited and a well-designed formal trial that tests this hypothesis is still required.

An alternative stance is to watch such patients closely, defining whether they have rapidly progressive disease or an indolent variant. In the latter instance, patients can often be followed expectantly for several months before the need to initiate chemotherapy – given that chemotherapy only yields clinically sustained responses in up to 50% of cases, but is associated with signficant toxicity in a greater proportion, the period of time free of the symptoms of tumour and treatment can be prolonged by this approach. As the absolute difference in median survival was only 5 months in the above-mentioned Nordic Gastrointestinal Tumor Adjuvant Therapy Group trial[58], with the survival curves crossing at only 2 years, it could easily be inferred that this study in colorectal cancer suggests watchful expectancy as a more rationale option.

In my own practice, I approach this problem through a frank and detailed discussion with the patients, taking into consideration their own views and preferences, as well as prognostic factors such as age, intercurrent disease, and performance status, as well as geography (their access to medical support for problems that may arise from aggressive chemotherapy). Such a discussion must perforce review the available knowledge of response rates, duration of response, and the risks of toxicity from the planned treatment. Thus, in a septuagenarian with the range of smoking-related disorders that are characteristic of patients with uro-epithelial cancer, who also has asymptomatic metastases, I tend to favour a more conservative approach, reserving chemotherapy for the patient with rapid tumour progression or the development of major symptoms. It should, however, be emphasized that this is in stark contrast to a much more aggressive approach adopted for the fit, elderly patient with locally invasive, non-metastatic disease[67]; furthermore, it should not be forgotten than age *per se* is not an adverse prognosticator for the patient with metastatic disease[25,43]. Finally, the algorithm appears to be changing with the introduction of more effective, but less toxic, chemotherapy regimens, and my own clinical approach is now evolving to a more aggressive stance.

In the present era, characterized by a predominance of patients with strong views on management, it seems unlikely that a well-structured, randomized trial (that allows for the different strata of disease extent) will ever answer this important question, and thus simple common sense will be required in response to this dilemma.

RESISTANCE TO CHEMOTHERAPY: PREDICTION AND INTERVENTION

A considerable amount of information has been amassed regarding the factors that govern the resistance of tumour cells to cytotoxic chemotherapy, as well as potential predictors of such resistance. Although a detailed discussion is beyond the scope of this chapter, it is pertinent to review briefly three principal factors known to be involved in the resistance of uro-epithelial cancer to the effects of MVAC and equivalent chemotherapy regimens.

Multidrug resistance

It has been known for several years that some cytotoxic agents can be exported from tumour cells through a mechanism based on the cellular surface, the so-called multidrug efflux pump, which is characterized by the expression of a specific 170 kDa protein complex, p-glycoprotein[36]. It was initially demonstrated that resistance to the cellular effects of such diverse agents as actinomycin D, colchicine, the vinca alkaloids and doxorubicin is reduced in cells (both normal and malignant) that express a protein complex on the cell surface, coded by a series of multidrug resistance (*mdr*) genes, and that this occurs as a result of reduced intracellular concentrations of the agents due to increased cellular efflux[36]. Several workers have attempted to assess the presence of the multidrug phenotype in uro-epithelial cells, but for reasons that are unclear, this has proved to be a particularly difficult problem. Although many other tumours have been studied in detail, revealing clear correlations between the expression of the mdr phenotype and drug resistance, in preclinical and clinical studies, the expression of p-glyoprotein in bladder cancer has been highly variable and inconstant[28,83,95], and thus the interpretation of data has been made difficult. Nevertheless, it has been shown that expression of p-glycoprotein may be upregulated in resistant populations of bladder cancer cells after treatment with the MVAC regimen[62].

Furthermore, it is clear from studies of other tumour types, that multidrug resistance can occur in the absence of expression of the 170 kDa p-glycoprotein, and that other proteins may be associated with very similar patterns of resistance[6], which may explain this phenomenon in the absence of expression of p-glycoprotein.

Ultimately, in addition to the predictive function, the true relevance of this work is likely to be realized if the multidrug resistance phenotype can be functionally overcome. For example, the calcium-channel blockers, such as verapamil, have been shown to reverse multidrug resistance[15], although the toxic side-effects of this approach have precluded routine use. Although clinical trials have not yet been published in bladder cancer, work initiated in our laboratories suggests that verapamil can overcome the impact of the *mdr* phenotype[60]. Surprisingly, it appears that, at least in the cell lines studied *in vitro*, the effect is largely mediated through enhancement of cellular uptake, rather than by inhibition of efflux[60]. Much more work will be required before the true clinical relevance of these observations is understood.

Glutathione and metallothioneins

The mechanisms of resistance to cisplatin and the other platinum co-ordination complexes have been studied in detail, particularly in relation to ovarian cancer and malignant melanoma. Although several mechanisms have been identified, including factors that influence cellular accumulation, signal transduction, ionic fluxes and intracellular enzyme function[97], the function of the intracellular scavenger, glutathione, has recently been the focus of particular attention in the context of the resistance of bladder cancer to the effects of cisplatin.

Glutathione (GSH) is a tripeptide (γ-glutamyl cysteinyl glycine) found in most mammalian cells, which has many functions, including regulation of protein and DNA synthesis and detoxification. Inhibitors of GSH synthesis, such as buthionine sulphoxamine, have been shown to cause a decrease in intracellular levels of GSH with a concomitant increase in the cytotoxicity of some anticancer agents, such as cisplatin[32], melphalan[59] and paclitaxel[41].

Although much of the experimental data regarding the significance of glutathione in cisplatin resistance has been derived from models of ovarian cancer, we have demonstrated recently that very high levels of glutathione are present in cell lines derived from bladder cancer, and correlate with cisplatin resistance, and further that human bladder cancer biopsy specimens also express substantial levels of glutathione[27].

Another potential mechanism for the detoxification of cisplatin in bladder cancer cells is the induction of expression of metallothionein, a cysteine-rich protein of low molecular weight which shows high affinity for metals such as zinc, copper and platinum[76]. Metallothionein has previously been implicated in cisplatin resistance in human ovarian cancer[3]. More recently, Satoh *et al.* (1993) have demonstrated that the induction of metallothionein synthesis, by the administration of zinc sulphate to mice bearing the MBT-2 bladder tumour, is associated with protection against the tumoricidal effect of cisplatin[76].

Expression of EGFR and P53

Recently, a complex body of work has emerged, relating the expression of several oncogene products to the phenomenon of resistance to cytotoxic agents. The exact nature of this interaction is not yet clear, and is particularly difficult to define as several of these products code for specific aspects of cellular growth control irrespective of exposure to cytotoxic agents. For example, it has been shown that the interaction of epidermal growth factor (EGF) and its specific receptor (EGFR) are involved in the regulation of growth of bladder cancer[9]. It has also been demonstrated *in vitro* that treatment with EGF can increase cellular sensitivity of epithelial tumours to cisplatin, presumably by an effect on a signal transduction pathway[14]. It has also been reported that specific monoclonal antibodies can block the function of the epidermal growth factor receptor[18], and further that treatment with these anti-EGFR monoclonal antibodies plus cisplatin can cause a synergistic anti-tumour effect[17]. These data are particularly difficult to interpret in view of the previously documented impact of expression of EGFR on the natural history of bladder cancer[57], and the demonstration that *erb*B-2 gene amplification and overexpression is an adverse prognostic determinant in bladder cancer[77].

Equally as complex is the relationship between the expression of p53, a suppressor gene product, and the growth and cytotoxicity of bladder cancer. Although the expression of p53 appears ubiquitous, some of the carcinogenic stimuli for uro-epithelial cancer, such as cigarette smoking and schistosomiasis, appear to influence the nature of mutation patterns of p53[29]. Alterations of the p53 gene are among the most frequent genetic abnormalities found in human cancer, and appear to have a broad range of postulated roles in the regulation of cellular growth, including involvement in cellular repair and apoptosis. Apoptosis, or programmed cell death, is regarded as one of the forms of physiological cell death as it represents a genetically determined cellular sequence that is part of the normal tissue homeostatic mechanism. It has been shown that p53 can be induced to accumulate within the cell by exposure to cytotoxic agents, such as cisplatin and mitomycin C[21], and conversely that p53-dependent apoptosis modulates the cytotoxicity of radiotherapy, 5-fluorouracil and doxorubicin[49]. However, in our laboratory, we have demonstrated that cisplatin cytotoxicity appears not to be mediated via apoptosis in bladder cancer cell lines[26].

These issues may be of particular importance as it has already postulated that p53 expression may constitute an independent prognosticator of response to the MVAC regimen[5]. As p53 may be induced by cytotoxic exposure, it is possible that the timing of tissue sampling may be critical in determining the expression of this potential prognostic factor, especially if intravesical or systemic chemotherapy has been used previously. Furthermore, it has been suggested that infection with high-risk human papillomavirus (HPV) types 16, 18 and 33 may synergize with expression of p53 as prognostic determinants in bladder cancer[22], further emphasizing the need for caution at the present time before molecular determinants are interpreted as predictors of outcome of treatment.

However, despite the need for continued meticulous investigation, it does appear that we are beginning to understand the mechanisms of resistance to the therapeutic options, and perhaps developing rational strategies for predicting such resistance (allowing the use of alternative agents) or overcoming the mechanisms to facilitate effective anti-cancer therapy.

SUMMARY AND FUTURE PROSPECTS

In a relatively short period of time, important progress has been made in the management of metastatic uro-epithelial cancer, with an increase in median survival from 4 to 5 months to more than 12 months, and, of more importance, an increase of 3-year survival figures from less than 5% to 15–20%. Combination chemotherapy has now been shown to be superior to single-agent chemotherapy in randomized clinical trials, and more precise management approaches have been targetted to the different histological patterns of disease. There is now an improved understanding of the biology of superficial bladder cancer, and in particular the immunological factors that contribute to its successful treatment. It may prove to be possible to apply immunotherapy to the maintenance of complete remission induced by cytotoxics, or even to use the emerging techniques of gene transfer in this context. In addition, several promising new cytotoxic agents have been shown to have activity against transitional cell carcinoma and preclinical tests have identified other promising options[27,82], and there appears to be an emerging understanding of the heterogeneity of tumour function and of the mechansms of resistance to chemotherapy. As the techniques of molecular biology unravel these mysteries, it seems likely that cure will ultimately be achieved for most patients with this hitherto resistant management problem.

REFERENCES

1. Ahmed T, Yagoda A, Needles B *et al.* (1985) Vinblastine and methotrexate for advanced bladder cancer. *J Urol* **133**: 602–604.
2. Al-Sarraf M, Frank J, Smith JAA *et al.* (1985) Phase II Trial of cyclophosphamide, doxorubicin and cisplatin (CAP) versus amsacrine in patients with transitional cell carcinoma of the urinary bladder: a Southwest Oncology Study Group Study. *Cancer Treat Rep* **62**: 189–194.
3. Andrews PA, Murphy MP and Howell SB (1987) Metallothionein-mediated cisplatin resistance in human ovarian carcinoma cells. *Cancer Chemother Pharmacol* **19**: 149–154.
4. Babaian RJ, Johnson DE, Llamas L *et al.* (1980) Metastases from transitional cell carcinoma of the urinary bladder. *Urology* **16**: 142–144.
5. Bajorin D, Sarkis A, Reuter V *et al.* (1994) Invasive bladder cancer treated with neoadjuvant MVAC: the relationship of p53 nuclear overexpression with survival. *Proc Am Soc Clin Oncol* **13**: 232.
6. Barrand MA, Heppell-Parton AC, Wright KA *et al.* (1994) A 190-kilodalton protein overexpressed in non-p-glycoprotein-containing multidrug-resistant cells and its relationship to the MRP gene. *J Natl Cancer Inst* **86**: 110–117.
7. Blumenreich MS, Yagoda A, Natale RG *et al.* (1982) Phase II trial of vinblastine sulfate for metastatic urothelial tract tumors. *Cancer* **50**: 435–438.
8. Boutan-Laroze A, Mahjoubi M, Droz JP *et al.* (1991) M-VAC (methotrexate, vinblastine, doxorubicin and cisplatin) for advanced carcinoma of the bladder. The French Federation of Cancer Centers Experience. *Eur J Cancer* **27**: 1690–1694.
9. Brown JL (1990) The clonal analysis of a human bladder cancer cell line. Thesis submitted for Doctorate of Philosophy, Faculty of Medicine, University of Sydney.

10. Brown JL, Russell PJ, Phillips J *et al.* (1990) Clonal analysis of a bladder cancer cell line: an experimental model of tumour heterogenicity. *Br J Cancer* **61**: 369–376.
11. Carmichael J, Cornbleet MA, MacDougall RH *et al.* (1985) Cisplatin and methotrexate in the treatment of transitional cell carcinoma of the urinary tract. *Br J Urol* **57**: 299–302.
12. Carter SK and Wasserman TH (1975) The chemotherapy of urological cancer. *Cancer* **36**: 729.
13. Cooling CI (1959) Review of 150 post-mortems of carcinoma of the urinary bladder, in *Tumours of the Bladder* (ed. DM Wallace), E & S Livingstone, Edinburgh, pp. 171–186.
14. Christen RD, Hom DK, Porter DC *et al.* (1990) Epidermal growth factor regulates the *in vitro* sensitivity of human ovarian carcinoma cells to cisplatin. *J Clin Invest* **86**: 1632–1640.
15. Dalton WS, Grogan TM, Meltzer PS *et al.* (1989) Drug resistance in multiple myeloma and non-Hodgkin's lymphoma: detection of P-glycoprotein and potential circumvention by addition of verapamil to chemotherapy. *J Clin Oncol* **7**: 415–424.
16. Einhorn LH, Roth BJ, Ansari R *et al.* (1994) Vinblastine, ifosfamide and gallium (VIG) combination chemotherapy in urothelial carcinoma. *Proc Am Soc Clin Oncol* **13:** 229A.
17. Fan Z, Baselga J, Masui H *et al.* (1993) Antitumor effect of anti-epidermal growth factor receptor monoclonal antibodies plus cis-diamminedichloroplatinum on well established A-431 cell xenografts. *Cancer Res* **53**: 4637–4642.
18. Fan Z, Masui H, Atlas I *et al.* (1993) Blockade of epidermal growth factor receptor function by bivalent and monovalent fragments of 225 anti-epidermal growth factor receptor monoclonal antibodies. *Cancer Res* **53**: 4322–4328.
19. Feinstein AR, Sosin DM and Wells CK (1985) The Will Rogers phenomenon. Stage migration and new diagnostic techniques as a source of misleading statistics for survival in cancer. *N Engl J Med* **312**: 1604–1608.
20. Freiha F, Reese J and Tirti FM (1996) A randomized trial of radical cystectomy plus cisplatin, vinblastine and methotrexate chemotherapy for muscle invasive bladder cancer. *J Urol* **155**: 495–500.
21. Fritsche M, Haessler C and Brandner G (1993) Induction of nuclear accumulation of the tumor-suppressor protein p53 by DNA-damaging agents. *Oncogene* **8**: 307–318.
22. Furihata M, Inoue K, Ohtsuki Y *et al.* (1993) High-risk human papillomavirus infections and overexpression of p53 protein as prognostic indicators in transitional cell carcinoma of the urinary bladder. *Cancer Res* **53**: 4823–4827.
23. Gabrilove JL, Jakubowski A, Scher H *et al.* (1988) Effect of granulocyte colony-stimulating factor on neutropenia and associated morbidity due to

chemotherapy for transitional-cell carcinoma of the urothelium. *N Engl J Med* **318**: 1414–1422.

24. Gagliano R, Levin H, El-Bolkainy MN *et al.* (1983) Adriamycin versus adriamycin plus *cis*-diamminedichloroplatinum in advanced transitional cell bladder carcinoma. *Am J Clin Oncol* **6**: 215–219.
25. Geller NL, Sternberg CN, Penenberg D *et al.* (1991) Prognostic factors for survival of patients with advanced urothelial tumors treated with methotrexate, vinblastine, doxorubicin and cisplatin chemotherapy. *Cancer* **67**: 1525–1531.
26. Glaves D, Hitt S and Raghavan D (1998). Apoptosis is not the mechanism of cisplatin-induced cell death in bladder cancer cell lines. *J Urol,* submitted for publication.
27. Glaves D, Murray M and Raghavan D (1996) Novel bi-functional anthracycline/nitrosurea chemotherapy for human bladder cancer: an analysis in a pre-clinical survival model. *Clin Cancer Res* **2**: 1315–1319.
28. Goldstein LJ, Galski H, Fojo A *et al.* (1989) Expression of a multidrug resistance gene in human cancers. *J Natl Cancer Inst* **81**: 116–124.
29. Habuchi T, Takahashi R, Yamada H *et al.* (1993) Influence of cigarette smoking and schistosomiasis on p53 gene mutation in urothelial cancer. *Cancer Res* **53**; 3795–3799.
30. Harker WG, Meyers FJ, Freiha FS *et al.* (1985) Cisplatin, methotrexate and vinblastine (CMV): an effective chemotherapy regimen for metastatic transitional cell carcinoma of the urinary tract. A Northern California Oncology Group study. *J Clin Oncol* **3**: 1463–1470.
31. Hillcoat BL, Raghavan D, Matthew J *et al.* (1989) A randomized trial of cisplatin versus cisplatin plus methotrexate in advanced cancer of the urothelial tract. *J Clin Oncol* **7**: 706–709.
32. Hospers GAP, Mulder NH, De Jong B *et al.* (1988) Characterization of a human small cell lung carcinoma cell line with acquired resistance to *cis*-diamminedichloroplatinum (II) *in vitro*. *Cancer Res* **48**: 6803–6807.
33. Hryniuk WM, Figueredo A and Goodyear M (1987) Applications of dose intensity to problems in chemotherapy. *Semin Oncol* **14**: 3–11.
34. Igawa M, Ohkuchi T, Ueki T *et al.* (1990) Usefulness and limitations of methotrexate, vinblastine, doxorubicin and cisplatin for the treatment of advanced urothelial cancer. *J Urol* **144**: 662–665.
35. Johansson SL and Cohen SM (1997) Pathology of bladder cancer, in *Principles and Practice of Genitourinary Oncology*, (eds D Raghavan, HI Scher, S Leibel and PH Lange), Lippincott-Raven, Philadelphia, pp. 207–213.
36. Juranka PF, Zastawny RL and Ving V (1989) P-glycoprotein: multidrug resistance and a superfamily of membrane-associated transport proteins. *FASEB J* **3**: 2583–2592.
37. Kedia KR, Gibbons C and Persky L (1981) The management of advanced bladder carcinoma. *J Urol* **125**: 655–658.

38. Khandekar JD, Elson PJ, DeWys WD *et al.* (1985) Comparative activity and toxicity of cis-diamminedichloroplatinum (DDP) and a combination of doxorubicin, cyclophosphamide and DDP in disseminated transitional cell carcinomas of the urinary tract. *J Clin Oncol* **3**: 539–545.
39. Koss LG (1975) Tumors of the urinary bladder, in *Atlas of Tumor Pathology*, (2nd series, Fascicle 11), Armed Forces Institute of Pathology, Washington.
40. Levine E and Raghavan D (1993) MVAC for bladder cancer: time to move forward again. *J Clin Oncol* **11**: 387–389.
41. Liebmann JE, Hahn SM, Cook JA *et al.* (1993) Glutathione depletion by 1-buthionine sulfoximine antagonizes taxol cytotoxicity. *Cancer Res* **53**: 2066–2070.
42. Loehrer PJ and De Mulder PHM (1997) Management of metastatic bladder cancer, in *Principles and Practice of Genitourinary Oncology*, (eds D Raghavan, HI Scher, S Leibel and PH Lange), Lippincott-Raven, Philadelphia, pp. 299–305.
43. Loehrer PJ, Einhorn LH, Elson PJ *et al.* (1992) A randomized comparison of cisplatin alone or in combination with methotrexate, vinblastine and doxorubicin in patients with metastatic urothelial carcinoma a cooperative group study. *J Clin Oncol* **10**: 1066–1072.
44. Loehrer PJ, Elson P, Dreicer R *et al.* (1994) Escalated dosages of methotrexate, vinblastine, doxorubicin and cisplatin plus recombinant human granulocyte colony-stimulating factor in advanced urothelial carcinoma: an Eastern cooperative Oncology Group trial. *J Clin Oncol* **12**: 483–488.
45. Logothetis CJ, Dexeus FH, Finn L *et al.* (1990) A prospective randomized trial comparing M-VAC and CISCA chemotherapy for patients with metastatic urothelial tumors. *J Clin Oncol* **8**: 1050–1055.
46. Logothetis CJ, Dexeus F, Sella S *et al.* (1990) Escalated therapy for refractory urothelial tumors: methotrexate, vinblastine, doxorubicin, cisplatin plus unglycosylated recombinant tumor granulocyte–macrophage colony stimulating factor. *J Natl Cancer Inst* **82**: 667–671.
47. Logothetis CJ, Finn L, Amato R *et al.* (1992) Escalated (ESC) MVAC +/- rhGM-CSF (Schering Plough) in metastatic transitional cell carcinoma (TCC): preliminary results of a randomized trial. *Proc Am Soc Clin Oncol* **11**: 202A.
48. Logothetis CJ, Hossan E, Sella A *et al.* (1991) 5-Fluorouracil and recombinant human interferon alpha 2a in the treatment of metastatic chemotherapy refractory urothelial tumors. *J Natl Cancer Inst* **83**: 285–288.
49. Lowe SW, Ruley HE, Jacks T *et al.* (1993) p53-Dependent apoptosis modulates the cytotoxicity of anticancer agents. *Cell* **74**: 957–967.
50. Malden LT, Raghavan D, Eisinger D *et al.* (1988) In *The Management of Bladder Cancer*, (ed. D Raghavan), Edward Arnold, London, pp. 299–316.

51. Medical Research Council (1987) A phase II study of carboplatin in metastatic transitional cell carcinoma of the bladder. *Eur J Cancer* **23**: 375–377.
52. Miller AB, Hoogstraten B, Staquet M *et al.* (1981) Reporting results of cancer treatment. *Cancer* **47**: 207–214.
53. Miyao N, Tsai YC, Kerner SP *et al.* (1993) Role of chromosome 9 in human bladder cancer. *Cancer Res* **53**: 4066–4070.
54. Moore MJ, O'Sullivan B and Tannock IF (1988) How expert physicians would wish to be treated if they had genitourinary cancer. *J Clin Oncol* **6**: 1736–1745.
55. Moore MJ, Tannock IF, Ernst DS *et al.* (1997) Gemcitabine: a promising new agent in the treatment of advanced urothelial cancer. *J Clin Oncol* **15**: 3441–3445.
56. Moore MJ, Tannock IF, Iscoe N *et al.* (1992) A phase II study of methotrexate, vinblastine, doxorubicin and cisplatin (MVAC) + GM-CSF in patients (pts) with advanced transitional cell carcinoma. *Proc Am Soc Clin Oncol* **11**: 199.
57. Neal DE, Marsh C, Bennett MK *et al.* (1985) Epidermal-growth-factor receptors in human bladder cancer: comparison of invasive and superficial tumours. *Lancet* **1**: 366–368.
58. Nordic Gastrointestinal Tumor Adjuvant Therapy Group (1992) Expectancy or primary chemotherapy in patients with advanced asymptomatic colorectal cancer. A randomized trial. *J Clin Oncol* **10**: 904–911.
59. Ozols RF, Louie KG, Plowman J *et al.* (1987) Enhanced melphalan cytotoxicity in human ovarian cancer *in vitro* and in tumour-bearing nude mice by buthionine sulphoximine depletion of glutathione. *Biochem Pharmacol* **36**: 147–153.
60. Palavidis Z (1993) Mechanisms of drug resistance in bladder cancer; the role of p-glycoprotein. Thesis for Master of Science in Medicine, University of New South Wales, Sydney.
61. Pendyala L, Raghavan D, Velagapudi S *et al.* (1997) Translational studies of glutathionine in bladder cancer cell lines and human specimens. *Clin Cancer Res* **3**: 793–798.
62. Petrylak DP, Scher HI, Reuter V *et al.* (1992) P-glycoprotein (PGP) expression in invasive and metastatic urothelial tract cancer (UTC). *Proc Am Soc Clin Oncol* **11**: 200.
63. Pollera CF, Ceribelli A, Crecco M *et al.* (1994) Weekly gemcitabine in advanced bladder cancer: a preliminary report. *Ann Oncol* **5**: 132–134.
64. Raghavan D (1990) Chemotherapy for advanced bladder cancer: 'Midsummer Night's dream' or 'Much Ado about Nothing'? *Br J Cancer* **62**: 337–342.
65. Raghavan D (1991) Clinician surrogates and equipoise: an analogy to lawyers who represent themselves? *Eur J Cancer* **27**: 1072–1074.

66. Raghavan D and Chye RWM (1991) Treatment of carcinomatous meningitis from transitional cell carcinoma of the bladder. *Br J Urol* **67**: 438–440.
67. Raghavan D, Grundy R, Greenaway TM *et al.* (1988) Pre-emptive (neoadjuvant) chemotherapy prior to radical radiotherapy for fit septuagenarians with bladder cancer: age itself is not a contra-indication. *Br J Urol* **62**: 154–159.
68. Raghavan D, Russell PJ and Brown JL (1992) Experimental models of histogenesis and tumor cell heterogenicity. *Br J Cancer* **61**: 369–376.
69. Raghavan D, Shipley WU, Garnick MB *et al.* (1990) The biology and management of bladder cancer. *N Engl J Med* **322**: 1129–1138.
70. Richards B, Newling D, Fossa S *et al.* (1983) Vincristine in advanced bladder cancer: a European Organisation for Research and Treatment of Cancer (EORTC) phase II study. *Cancer Treat Rep* **67**: 575–577.
71. Rose MA and Shipley WU (1988) Radiation therapy in invasive bladder cancer: principles, results, patient selection and innovations. In *The Management of Bladder Cancer,* (ed. D Raghavan), Arnold, London, pp. 154–173.
72. Roth BJ, Dreicer R, Einhorn LH *et al.* (1994) Significant activity of paclitaxel in advanced transitional cell carcinoma of the urothelium: a phase II trial of the Eastern Cooperative Oncology Group. *J Clin Oncol* **12**: 2264–2270.
73. Russell PJ, Brown JL, Grimmond SM *et al.* (1990) The molecular biology of urological tumours. *Br J Urol* **65**: 121–130.
74. Russell PJ, Jelbart M, Wills E *et al.* (1988) Establishment and chacterization of a new human bladder cancer cell line showing features of squamous and glandular differentiation. *Int J Cancer* **41**: 74–82.
75. Russell PJ, Wills EJ, Phillips J *et al.* (1988) Features of squamous and adenocarcinoma in the same cell in a xenografted human transitional cell carcinoma: evidence of a common histogenesis? *Urol Res* **16**: 79–84.
76. Satoh M, Kloth DM, Kadhim SA *et al.* (1993) Modulation of both cisplatin nephrotoxicity and drug resistance in murine bladder tumor by controlling metallothionein synthesis. *Cancer Res* **53**: 1829–1832.
77. Sauter G, Moch H, Moore D *et al.* (1993) Heterogenicity of *erb*B-2 gene amplification in bladder cancer. *Cancer Res* **53**: 2199–2203.
78. Saxman SB, Propert K, Einhorn LH *et al.* (1997) Long term follow up of phase III intergroup study of cisplatin alone or in combination with methotrexate, vinblastine and doxorubicin in patients with metastatic urothelial carcinoma. *J Clin Oncol* **15**: 2564–2569.
79. Scher HI, Geller NL, Curley T *et al.* (1993) Effect of relative cumulative dose-intensity on survival of patients with urothelial cancer treated with M-VAC. *J Clin Oncol* **11:** 400–407.
80. Seidman AFD, Scher HI, Gabrilove JL *et al.* (1993) Dose intensification of MVAC with recombinant granulocyte colony-stimulating factor as initial therapy in advanced urothelial cancer. *J Clin Oncol* **11**: 408–414.

81. Seligman PA and Crawford ED (1991) Treatment of advanced transitional cell carcinoma of the bladder with continuous infusion gallium nitrate. *J Natl Cancer Inst* **83**: 1582–1584.
82. Sharma A, Glaves D, Porter CW *et al.* (1997) Antitumor efficacy of N^1N^{11}-diethyl norspermine on a human bladder tumor xenograft in nude athymic mice. *Clin Cancer Res* **3**: 1239–1244.
83. Shinohara N, Liebert M, Wedemeyer G *et al.* (1993) Evaluation of multiple drug resistance in human bladder cancer cell lines. *J Urol* **150**: 505–509.
84. Silverberg E, Boring CC and Squires TS (1990) Cancer statistics, 1990. *Ca-A Cancer J Clinic* **40**: 9–26.
85. Smalley RV, Bartolucci AA, Hemstreet G *et al.* (1981) A phase II evaluation of a 3-drug combination of cyclophosphamide, doxorubicin and 5-fluorouracil in patients with advanced bladder carcinoma or stage D prostatic carcinoma. *J Urol* **125**: 191–195.
86. Soloway MS, Einstein A, Corder MP *et al.* (1983) A comparison of cisplatin and the combination of cisplatin and cyclophosphamide in advanced urothelial cancer. A National Bladder Cancer Collaborative Group A Study. *Cancer* **52**: 767–772.
87. Stadler WM, Kuzel TM, Roth BJ *et al.* (1997) Phase II study of single-agent gemcitabine in previously untreated patients with metastatic urothelial cancer. *J Clin Oncol* **15**: 3394–3398.
88. Stadler WM, Murphy D, Kaufman D *et al.* (1997) Phase II trial of gemcitabine (BEM) plus cisplatin (CDDP) in metastatic urothelial cancer (UC). *Proc Am Soc Clin Oncol* **16**: 323A.
89. Sternberg JJ, Bracken RB, Handel PB *et al.* (1977) Combination chemotherapy (CISCA) for advanced urinary tract carcinoma: a preliminary report. *J Am Med Ass* **238**: 2282–2287.
90. Sternberg C, De Mulder P, van Oosterom AT *et al.* (1992) Intensified M-VAC chemotherapy and recombinant human granulocyte–macrophage colony stimulating factor (GM-CSF) in patients with advanced urothelial tract tumors. *Proc Am Soc Clin Oncol* **11**: 210.
91. Sternberg SN and Swanson DA (1997) Non-transitional cell bladder cancer, in *Principles and Practice of Genitourinary Oncology*, (eds D Raghavan, HI Scher, S Leibel and PH Lange), Lippincott-Raven, Philadelphia, pp. 315–330.
92. Sternberg SN, Yagoda A, Scher HI *et al.* (1988) M-VAC (methotrexate, vinblastine, doxorubicin and cisplatin) for advanced transitional cell carcinoma of the urothelium. *J Urol* **139**; 461–469.
93. Sternberg SN, Yagoda A, Scher HI *et al.* (1989) Methotrexate, vinblastine, doxorubicin and cisplatin for advanced transitional cell carcinoma of the urothelium. Efficacy and patterns of response and relapse. *Cancer* **64**: 2448–2458.

94. Stoter G, Splinter TAW, Child JA *et al.* (1987) Combination chemotherapy with cisplatin and methotrexate in advanced transitional cell cancer of the bladder. *J Urol* **137**: 663–667.
95. Suguwara I (1990) Expression and functions of P-glycoprotein (mdr-1) in normal and malignant tissues. *Acta Path Japon* **40**: 545–553.
96. Tannock I, Gospodarowicz M, Connolly J *et al.* (1989) M-VAC (methotrexate, vinblastine, doxorubicin and cisplatin) chemotherapy for transitional cell carcinoma: the Princess Margaret Hospital experience. *J Urol* **142**: 289–292.
97. Timmer-Bosscha H, Mulder NH and de Vries EGE (1992) Modulation of *cis*-diamminedichloroplatinum (II) resistance: a review. *Br J Cancer* **66**: 227–238.
98. Troner M, Birch R, Omura GA *et al.* (1987) Phase III comparison of cisplatin alone versus cisplatin, doxorubicin and cyclophosphamide in the treatment of bladder (urothelial) cancer: a South Eastern Cancer Study Group trial. *J Urol* **137**: 660–662.
99. Turner AG, Hendry WF, Williams GB *et al.* (1977) The treatment of advanced bladder cancer with methotrexate. *Br J Urol* **49**: 673–678.
100. Van Oosterom AT, Akaza H, Hall RR *et al.* (1986) Response criteria, phase II/ phase III trials in invasive bladder cancer, in *Developments in Bladder Cancer*, (eds L Denis, T Niijima, GR Prout Jr and FH Schroeder), Alan R Liss, New York, pp. 301–310.
101. Van Oosterom AT, Fossa SD, Mulder JH *et al.* (1985) Mitozantrone in advanced bladder carcinoma. A phase II study of the EORTC Genitourinary Tract Cancer Cooperative Group. *Eur J Cancer Clin Oncol* **21**: 1013–1014.
102. Wallace DM (1959) Clinico-pathological behaviour of bladder tumours, in *Tumours of the Bladder*, (ed. DM Wallace), E & S Livingstone, Edinburgh, pp. 157–170.
103. Waxman J, Abel P, James N *et al.* (1989) New combination chemotherapy programme for bladder cancer. *Br J Urol* **63**: 68–71.
104. Witte R, Loehrer P, Dreicer R *et al.* (1993) Ifosfamide (IFX) in advanced urothelial carcinoma: an ECOG trial. *Proc Am Soc Clin Oncol* **12**: 230.
105. Yagoda A, Watson RC, Gonzalez-Vitale JC *et al.* (1976) *Cis*-dichlorodiammineplatinum (II) in advanced bladder cancer. *Cancer Treat Rep* **60**: 917–923.
106. Young DC and Garnick MB (1988) Chemotherapy in bladder cancer: the North American experience, in *The Management of Bladder Cancer,* (ed. D Raghavan), Edward Arnold, London, pp. 245–263.

12

Clinical trials and statistics in bladder cancer

Richard J. Sylvester

A clinical trial may be defined as a carefully designed, prospective, medical study which attempts to answer a precisely defined set of questions with respect to a particular treatment or treatments[26].The first step in planning a new trial is to define its objectives precisely and to determine the type of study to be carried out. Clinical trials in bladder cancer can generally be divided into one of three types:

- Phase I trials: toxicity screening studies where, after testing in animals, a new drug is administered for the first time in man in order to determine the maximum tolerated dose.
- Phase II trials: new drugs are screened for potential anti-tumour activity in a limited number of patients.
- Phase III trials: different treatment regimens are compared with respect to their relative efficacy and/or toxicity.

Phase IV trials are also carried out within the pharmaceutical industry. These are essentially just post-marketing phase III trials in which a very specific hypothesis about an established treatment is tested.

The most important document in any trial is the protocol, a self-contained description of the rationale, objectives and logistics of the study. It provides the scientific basis for the trial and describes in detail how it is to bc carried out[30]. The topics to be included in a typical protocol are given in Table 12.1.

PHASE I TRIALS

After testing in animals, the initial use of a new drug in humans is done in a phase I clinical trial. A phase I trial is essentially a human toxicology study and anti-tumour activity is not usually an end point. They seldom involve patients with bladder

Table 12.1 Recommended contents for a clinical trial protocol

0.	Title page
1.	Background and introduction
2.	Objectives of the trial
3.	Patient selection criteria (inclusion/exclusion criteria)
4.	Trial design and schema
5.	Therapeutic regimens, toxicity, dose modifications
6.	Required clinical evaluations, laboratory tests and follow-up
7.	Criteria of evaluation, end points
8.	Patient registration and randomization procedure
9.	Forms and procedures for collecting data
10.	Reporting adverse events
11.	Statistical considerations
12.	Quality of life assessment
13.	Cost evaluation assessment
14.	Data monitoring committee
15.	Quality assurance
16.	Ethical considerations
17.	Investigator commitment statement
18.	Administrative responsibilities
19.	Trial sponsorship/financing
20.	Trial insurance
21.	Publication policy
22.	Administrative signatures
23.	List of participants with expected yearly accrual
24.	References
25.	Appendices (as appropriate)
	TNM classification
	Performance status scale (Karnofsky, WHO, ECOG)
	Trial schema, follow-up
	Body surface area (m^2)
	Surgical details
	Radiotherapy details
	Toxicity grading scales: common toxicity criteria, SOMA scale
	Adverse drug reactions
	Flowsheet (checklist) of required investigations
	Pathology review
	Drug storage/supply
	Case report forms
	Investigator commitment statement
	Informed consent statement
	Patient information sheet
	Declaration of Helsinki
	Conflict of interest form

cancer but are an essential step in developing new drugs that may be used against bladder cancer at a later date. The goals of a phase I study are:

1. To determine the optimal dose that will be used at the next stage in a phase II trial when screening for drug activity. In practice the maximum tolerated dose (MTD) is sought. This is generally defined as the dose that provides the first evidence of treatment-limiting toxicity.
2. To determine both the type and degree of toxicity associated with the drug[33].

Patients entered in phase I trials should have disseminated disease which is no longer amenable to treatment with standard forms of therapy. They should also have given their informed consent, have a life expectancy of at least 12 weeks, should not have received treatment within the previous 4–6 weeks to ensure that the toxic effects of the previous treatment have subsided, and should have normal bone marrow, hepatic, renal and cardiac function.

As the toxicity of anti-cancer agents usually depends on the schedule employed, phase I trials may be performed using both daily and intermittent schedules where the treatment is repeated every 3–4 weeks. The starting dose to be used in phase I trials is determined by the toxicity observed in animal studies: for example one-third of the toxic dose low (TDL) in the dog or one-tenth of the LD_{10} in mice[33]. Generally three patients are treated at each non-toxic dose level. When toxicity is reached, four to six patients are usually treated at that dose and at all subsequent dose levels until the MTD is reached. A number of dose escalation techniques have been used. One of the oldest is based on a modified Fibonacci search scheme. With this method dose escalations are initially rapid with smaller increments being used at higher, more toxic levels. Using this scheme the MTD is generally reached in a maximum of about nine escalations. In order to avoid problems of interpretation related to cumulative toxicity, patients should only be treated at a single dose level. Thus the dose should not be escalated in a given patient. For a more comprehensive review of the goals and methodology of phase I trials, see[7,8,23,33].

New drugs for intravesical use have usually been previously evaluated systemically in phase I and II trials in metastatic disease (not necessarily transitional cell carcinoma (TCC)) and have shown some anti-cancer activity. They are then tested for intravesical use in phase I trials (to determine dose and toxicity) and in phase II trials (to identify activity) in patients who fail to respond to established intravesical agents and are unsuitable for cystectomy.

PHASE II TRIALS

For bladder cancer a distinction must be made between phase II trials in patients with metastatic disease, phase II marker lesion studies in patients with superficial (Ta T1) bladder cancer, and phase II trials in patients with carcinoma in situ (CIS).

Metastatic disease

After the completion of the phase I study and the determination of the MTD, the drug is then submitted to its initial screen for possible anti-tumour activity in man in a phase II trial. The goal of a phase II study is to identify in a small number of patients those drugs of potential promise that should be passed on for more intensive testing in phase III trials. The end points of such a trial are the overall response rate (complete plus partial response) and toxicity[32]. Thus patients must have measurable disease in order to be able to make an objective assessment of the drug's activity. Patients entered in phase II trials will generally be heavily pretreated with chemotherapy and will have progressed on standard treatment.

It is important to decide with as few patients as possible whether [28]:

1. a new drug is possibly active in a specified proportion of patients and should thus undergo additional study to assess its effectiveness further; or
2. a new drug is probably inactive and should be dropped from further study.

If the drug is found to be active, then either:

1. the phase II trial should be continued in a small number of patients in order to estimate the response rate or assess the toxicity more precisely;
2. the drug should be assessed in combination with other drugs in a pilot or feasibility study; or
3. the drug should be tested in a randomized phase III trial in patients with metastatic disease in order to compare the drug, either alone or in combination, to the classical or best available treatment.

Anywhere from 14 to 40 patients are generally treated in a phase II trial of a new agent in patients with metastatic disease. The actual number of patients required for such a trial depends on the underlying response rate of interest and the size of the type I and type II errors, α and β. Most phase II designs minimize the size of the type II error since in a phase II trial it is a more serious error to reject an effective drug from further study. Generally β is set equal to 0.05 and α to 0.10.

For ethical reasons some type of sequential approach is desirable so that ineffective drugs may be rejected from further study as quickly as possible if the preliminary results suggest that they are ineffective and will ultimately be rejected. Thus two-stage designs are often employed whereby the drug is rejected from further study if no responses or only a 'very few' responses are observed in the first patients treated.

The use of Simon's optimal two-stage designs for phase II clinical trials is recommended[24]. Initially n_1 patients are entered. If there are r_1 or fewer responses, the drug is rejected from further study. If there are more than r_1 responses, then the trial is continued to a total of n patients. If there are a total of r or fewer responses in the n patients, then the drug is rejected from further study. Otherwise the drug is accepted.

Two different designs were proposed by Simon: the **optimal** design and the **minimax** design. The optimal design minimizes the sample size if the treatment regimen has low activity, whereas the minimax design minimizes the maximum number of patients that are treated. The choice between these two designs depends on practical considerations such as the drug's toxicity and the expected rate of patient accrual.

The main end point in phase II trials in patients with metastatic bladder cancer is the overall response rate (complete plus partial response). Suppose, for example, it is desired to have a high probability ($\beta = 0.05$, or power $= 1 - \beta = 0.95$) of accepting a new drug for further study if it has a 30% response rate (P_1), and a low probability ($\alpha = 0.10$) of accepting the drug for further study if it has a response rate of only 10% (P_0). In this case the Simon **optimal** two-stage design (which minimizes the expected sample size if the drug has a response rate of only 10%) would require a minimum of 20 patients and a maximum of 40. If only two responses occur in the first 20 patients, the trial is stopped. If six or fewer responses are seen in 40 patients, the drug is deemed inactive. If seven or more responses are seen, the drug is considered worthy of further study. The Simon **minimax** design (which minimizes the maximum number of patients treated) would require a minimum of 4/26 responses to continue and 6/33 responding patients to be considered active enough for further study (Table 12.2).

If the percentage of patients with a given degree of toxicity was the main end point, then the Simon design could still be applied where the response rate is replaced by the percentage of patients not experiencing the given degree of toxicity.

More recently, optimal three-stage designs have been proposed. While they have the possibility of stopping earlier when no successes are observed, they are not as practical and are administratively more complicated. The advantages and

Table 12.2 Simon optimal two-stage designs for $P_0 = 0.10$, $P_1 = 0.30$, $\alpha = 0.10$, $\beta = 0.05$

	Optimal design	Minimax design
r_1	2	3
n_1	20	26
r	6	5
n	40	33

P_0 = response rate for which one wishes to have a high probability of rejecting the drug from further study.
P_1 = response rate for which one wishes to have a high probability of accepting the drug for further study.

Initially n_1 patients are entered. If there are $\leq r_1$ responses, the drug is rejected from further study. If there are $> r_1$ responses, the trial is continued to a total of n patients. If there are $\leq r$ responses out of n patients, the drug is rejected from further study.

disadvantages of this and other alternative designs for phase II trials are discussed by Schröder *et al.*[20].

Phase II marker lesion studies in TA T1 bladder cancer

Prior to using a new intravesical treatment in a prophylactic manner to prevent the recurrence of tumour after complete resection by TUR, its ablative effect should be tested in phase II trials. These give information about the appropriate dose of drugs and treatment schemes without exposing hundreds of patients to potentially ineffective treatment in phase III trials. In such a trial, all Ta T1 lesions except for a single marker tumour are resected at entry on study[15,31]. The main end point in such a trial is the complete response rate (CR) of the marker tumour based on negative cystoscopy, cytology and biopsy after a given period of treatment.

Contrary to some clinicians' anxieties, marker tumour trials do not expose patients to any risk[31] and have been recommended as the most appropriate trial design in this situation by the Report of the Fourth International Consensus Conference on Bladder Cancer[15]. The clinical issues related to marker tumour studies have been discussed in Chapter 4.

Table 12.3 Simon optimal two-stage designs for $P_1 = 0.60$, $\alpha = 0.10$, $\beta = 0.05$

P_0		Optimal design	Minimax design
0.25	r_1	2	2
	n_1	9	9
	r	6	6
	n	17	17
0.30	r_1	4	4
	n_1	13	14
	r	11	10
	n	28	25
0.35	r_1	6	6
	n_1	17	19
	r	18	15
	n	42	34
0.40	r_1	11	17
	n_1	27	39
	r	29	26
	n	62	55

Suppose, for example, it is desired to have a high probability (β = 0.05, or power = 1 – β = 0.95) of accepting the drug for further study if it has a 60% CR rate (P_1), and a low probability (α = 0.10) of accepting the drug for further study if it has a response rate of only 30% (P_0). In this case the Simon optimal design (which minimizes the sample size if the drug has a response rate of only 30%) and the Simon minimax design (which minimizes the maximum number of patients treated) would require a minimum of 13 patients and a maximum of 25 patients, respectively (see Table 12.3, which also shows sample sizes for other response rates, P_0). As can be seen from this table, the sample sizes n_1 and n increase as P_0 increases.

Phase II studies in carcinoma in situ

The statistical design of phase II trials in patients with carcinoma in situ is very similar to that described above for marker lesion studies. Based on the results of cystoscopy, biopsy and cytology after a given period of intravesical or oral treatment, the complete response (CR) rate is assessed[21,22]. If a CR rate of 60% is sought, the sample size calculation is the same as that given above for marker lesion studies. For a CR rate of 70%, see Table 12.4.

Table 12.4 Simon optimal two-stage designs for P_1 = 0.70, α = 0.10, β = 0.05

P_0		Optimal design	Minimax design
0.35	r_1	3	4
	n_1	9	11
	r	9	8
	n	19	17
0.40	r_1	6	9
	n_1	14	19
	r	13	12
	n	26	24
0.45	r_1	8	9
	n_1	18	20
	r	19	18
	n	35	33
0.50	r_1	13	15
	n_1	26	31
	r	31	29
	n	54	50

PHASE III TRIALS

The effectiveness of a new treatment relative to other treatments cannot be determined from a phase II trial. In order to assess a treatment's efficacy properly, large-scale randomized phase III trials are required, the goal of which generally falls into one of the following categories:

1. **To determine the effectiveness of a treatment relative to the natural history of the disease**. For example, to compare a new systemic chemotherapy to no treatment in an adjuvant setting after cystectomy for muscle invasive bladder cancer, or to compare TUR alone with TUR plus intravesical chemotherapy for Ta T1 bladder tumours.
2. **To determine if a new treatment is more effective than the best current standard treatment** (at the risk of increasing toxicity). For example, when testing new combinations in the treatment of metastatic disease, or a new intravesical agent versus mitomycin C (MMC) for Ta T1 tumours.
3. **To determine if a new treatment is as effective as the best current standard therapy but is associated with less severe toxicity or a better quality of life.** For example, comparing a less toxic regimen to MVAC, or comparing epirubicin to BCG for CIS.

A simple two-arm trial with the treatments being as different as possible gives the greatest chance of detecting significant treatment differences. Three- or four-arm trials are generally inefficient and are not to be recommended, except for factorial designs[25] which may allow one to answer two questions for the price of one.

RANDOMIZATION

In most cancer clinical trials that are carried out, the size of the treatment effect is generally smaller than the difference in prognosis according to the possible levels of the patient's prognostic factors. In the measurement of the size of the treatment effect, both

1. systematic errors (bias) and
2. random errors (random variation)

must be small in comparison to the actual size of the treatment effect. In order to minimize the possibility of both bias and random variation, randomized phase III trials need to be large in order to measure accurately the size of the treatment effect and to detect reliably small to moderate, but medically important, treatment differences.

All phase III trials must contain a control group against which a new treatment is to be compared. Unfortunately, as most phase III clinical trials yield negative results (that is, no significant treatment differences are observed), it is important that the treatment in the control group be chosen so that negative results can be properly interpreted. As a general rule, patients in the control group should receive the standard treatment.

If the control group is not chosen by randomization, one must assume that the treatment groups to be compared are identical with respect to the prognosis of the patients entered, or that one can correct for all factors, both known and unknown, that can affect a patient's prognosis. Since most of these factors are ill understood for bladder cancer and may not be assessed at entry on study, randomization is the only method of treatment assignment that allows one to draw conclusions without making any special assumptions[25].

In trials studying the same question, one cannot base comparisons between trials on historical controls since there is no guarantee that patients will have similar characteristics and prognoses in different trials. Patient entry criteria may change from one trial to the next, and diagnostic techniques and methods for the evaluation of treatment efficacy may vary with time. Even trials carried out during the same time period may not be comparable. For example, two **simultaneous** EORTC trials studying the prophylactic treatment of Ta T1 bladder cancer in supposedly identical patient populations showed remarkable differences in patient characteristics. Patients in one trial had a worse prognosis than those entered in the other: 72% versus 37% of patients had recurrent tumours and 31% versus 16% of patients had more than three tumours at entry. Differences in patient characteristics may have also been present for other important factors that were not measured at entry on study and thus for which no adjustment can be made. It is thus very dangerous to base comparisons on historical controls. The main advantages of randomization are:

1. It minimizes bias in the assignment of treatments.
2. It tends to balance treatment groups with respect to the distribution of the prognostic factors.
3. It guarantees the validity of the statistical tests of significance.
4. It eliminates the effect of time trends due to changes in diagnostic techniques and methods for the evaluation of treatment response.

STRATIFICATION

Since randomization only ensures that the treatment groups are comparable **on the average** with respect to both the number of patients entered and the distribution of a given factor (whether measured or unmeasured), stratification at entry, either by the static technique of randomized blocks or by the dynamic technique of minimization[14,25], ensures that the treatment groups are balanced both with respect to the number of patients entered and the distribution of any important prognostic factor.

In multicentre clinical trials it is generally recommended to stratify at entry by institution and by the one or two most important 'objective' prognostic factors.

Overstratification, especially using the method of randomized blocks, is inefficient and may by chance be worse than no stratification at all. Use of the minimization technique is generally to be preferred. In Ta T1 bladder tumour trials it is usual to stratify for the number of tumours (single, multiple) and the tumour status (primary, recurrent), in CIS for its type (primary, secondary or concurrent) and in muscle invasive bladder cancer for the T category and, if appropriate, by the type of definitive treatment.

SAMPLE SIZE CALCULATION

The greatest defect in most clinical trials is that too few patients are entered in order to have a high probability (power) of detecting a clinically meaningful difference should it exist[10]. In calculating the sample size for phase III trials, it is necessary to make a distinction between trials where the goal is to detect a treatment difference and those where the goal is to show treatment equivalence.

Hypothesis testing to show a treatment difference

The hypotheses to be tested are:

null hypothesis (H_0: no treatment difference); and
alternative hypothesis (H_A: specified treatment difference).

The statistical terminology is as follows:

α is the risk of a false-positive result = Prob (Reject $H_0 | H_0$); and
β is the risk of a false-negative result = Prob (Accept $H_0 | H_A$).

Both should be as small as possible.

The **power** of a clinical trial is the probability that a true therapeutic benefit will be detected at a given level of significance and is defined as:

$$\text{power} = 1 - \beta = \text{Prob (Reject } H_0 | H_A).$$

In planning trials, the risk of obtaining a false-positive result (α) is usually taken to be ≤ 0.05. While in theory β (the risk of a false-negative) is designed to be ≤ 0.20, in practice it may be much higher due to an insufficient number of patients having been randomized or treated, or too few events having been observed during the follow-up before the time of the analysis. Some determinants of sample size are:

1. the size of the type I error α (≤ 0.05);
2. the size of the type II error β (≤ 0.20);
3. a realistic estimate of the event rate in the control arm for the primary end point of interest;

4. an estimate of the expected treatment effect based on either the **medically worthwhile** treatment effect or the **medically plausible** treatment effect; and
5. patient availability and accrual rate.

Unfortunately the plausible treatment effect is generally **less than** the medically worthwhile treatment effect. As a consequence not enough patients are generally entered in phase III clinical trials to have a high power of detecting medically plausible treatment effects. Thus it is not surprising that most phase III trials have a negative conclusion, that is that there is no significant difference in efficacy between the treatments. For this reason it is crucial that clinicians and statisticians agree during the design stage of a new trial on the size of benefit that would lead to the clinical application of the trial's result. There is no point in embarking on a trial that can only demonstrate a difference that is not scientifically or clinically plausible, or that will require an unrealistically large number of patients.

When assessing the outcome, a one-sided test of significance assumes that the treatment will have a beneficial effect or no effect. In practice it is possible that the experimental treatment could have an unfavourable or negative effect (e.g. reducing overall survival because of unexpected treatment-related deaths). This possibility requires the application of two-sided tests of significance.

Although two-sided significance tests require more patients than one-sided tests, in practice it is generally recommended to calculate sample sizes based on a two-sided test because one cannot always a priori rule out an unfavourable treatment effect. One exception, as discussed below, where one-sided tests are appropriate, is in the case of testing for equivalence.

When interpreting the results of a trial attempting to show a difference in efficacy between two treatments, a non-significant result does not imply that the two treatments are equivalent. P by itself does not give any information about the possible size of the treatment effect. For this, confidence intervals are required. Thus it is important to avoid interpreting a simple lack of statistical significance as evidence of equivalence unless a sufficient number of patients have been studied. To test for equivalence, special techniques are required, as described below.

Hypothesis testing to show equivalence

Discoveries of new treatments that have made a major impact on survival or tumour recurrence rates are few and far between, but patients may benefit by the choice of the least toxic option among treatments that seem equivalent in other respects. For this reason it is desirable to show that a new, more conservative or less toxic treatment is 'equivalent' in efficacy to a standard, more intensive therapy. One wishes to select the more conservative treatment as being equivalent when it is not worse than the standard treatment by more than some small amount (Δ) judged to be acceptable by the investigators[3,6]. This implies the use of **one-sided** statistical tests of significance. In the case of testing for equivalence the null hypothesis and the alternative hypothesis are now reversed[3]:

null hypothesis (H_0: specified treatment difference); and
alternative hypothesis (H_A: no treatment difference),

where

α = Prob (Reject $H_0|H_0$) = Prob(concluding no difference when there is a difference)
β = Prob (Accept $H_0|H_A$) = Prob(concluding treatment difference when no difference)
Power = $1 - \beta$ = Prob (Reject $H_0|H_A$) = Prob(correctly concluding no difference).

The following examples illustrate the basic principles of the design and analysis of phase III bladder cancer trials.

DESIGN AND ANALYSIS OF PROPHYLACTIC SUPERFICIAL (TA T1) BLADDER CANCER TRIALS

After removal of all tumours by TUR, the short-term effect of prophylactic treatment is assessed either by the disease-free interval or the recurrence rate[5,27,29]. The long-term effect is measured by the time to progression to muscle invasive disease or by the duration of survival.

Disease-free interval (time to first recurrence)

This is calculated as the time from randomization to the date of the first (histologically confirmed) bladder recurrence. Patients without recurrence are censored at the date of the last follow-up cystoscopy. The time to first recurrence in the treatment groups is estimated using the Kaplan–Meier technique[18] and compared using the two-sided logrank test[4].

Approximately 50% of the patients entered in a trial can be expected to recur within 5 years[12,13,17], the exact figure depending on the prognosis of the particular patients entered[13]. Based on the combined analysis of EORTC and MRC superficial bladder cancer trials[17], it is not realistic to expect an absolute treatment difference of more than 10% in the percentage of patients with recurrence at any given time point. As in trials comparing the duration of survival, the power to detect a difference depends not on the number of patients entered but rather on the number of patients having observed the event of interest, in this case recurrence. Thus the number of patients planned for the trial should be such that the required number of recurrences will have been observed after patients have been followed for a given length of time. Table 12.5 gives the number of recurrences and the number of patients per treatment group required to detect an increase from 50% to 60% of patients disease-free at 5 years.

As can be seen from this table, for a power of 80% a total of 2 × 173 = 346 recurrences must be observed. If the total duration of the study is fixed at 5 years (entry plus follow-up), then fewer patients are required as the patient entry rate increases, going from 2 × 411 = 822 patients if all patients are entered in 1 year to 2 × 690 = 1380 patients if 5 years are required to enter all patients. If the patient entry period is fixed at 3 years, then the required number of patients goes from 2 × 173 = 346 if all patients are followed until recurrence, to 2 × 1066 = 2132 patients if there is no further follow-up after the 3-year patient entry period. A reasonable compromise is to follow-up all patients for 2 years after the 3-year patient entry period, thus requiring a total of 2 × 504 = 1008 patients.

Table 12.5 Number of recurrences and patients per treatment group required to detect an increase from 50% to 60% of patients disease-free at 5 years at error rates $\alpha = 0.05$ and $\beta = 0.10, 0.20$ based on a two-sided logrank test and different periods of patient entry and duration of additional follow-up after the last patient has been entered

Power		80%	90%
Recurrences		173	232
Entry (years)	Follow up (years)	Patients	
5	0	690	923
4	1	579	775
3	2	504	675
2	3	451	604
1	4	411	550
3	0	1066	1428
3	1	670	897
3	2	504	675
3	3	414	554
3	4	357	478
3	5	318	426
3	10	230	308
3	20	187	250
3	∞	173	232

Recurrence rate per year

This is estimated as the total number of follow-up cystoscopies at which a recurrence is detected divided by the total years of follow-up. It has the advantage of taking into account all (successive) recurrences observed during the follow-up period. The recurrence rate in the different treatment groups is compared using a non-parametric permutation test[27]. Sample size calculations are not generally based on a comparison of the recurrence rate since the sample sizes would generally be smaller than those obtained for comparison of the disease-free interval.

It is important to recognize that the percentage of patients with recurrence at a given point in time is not a good end point since the choice of the time point is arbitrary and the percentage recurring depends on the maturity of the follow-up data[27].

Time to muscle invasive disease

This is calculated as the time from randomization to progression to stage T2 or higher within the bladder. Patients without muscle invasion at the time of analysis are censored at the date of the last available follow-up cystoscopy. Time to muscle invasion is estimated using the Kaplan–Meier technique and compared using the two-sided logrank test.

About 10–15% of patients treated in intravesical chemotherapy trials can be expected to develop muscle invasive disease within about 10 years[13,17] and the medically plausible difference between treatments that could be expected is about 5% or less[17]. In this case approximately 1000 patients must be followed for an average of 10 years in order to have a power of about 85% to detect an absolute difference of 5% (from 90% to 95%) in the percentage of patients who are invasion-free at 10 years, based on a two-sided logrank test.

Hardly surprisingly, time to muscle invasion is seldom chosen as the primary end point in superficial bladder cancer trials. However, it could be argued that future trials should address this end point because prevention of muscle invasion is the eventual goal of treatment for Ta T1 TCC.

Duration of survival

This is calculated as the time from randomization to the date of death. Patients still alive or lost to follow-up are censored at the last date they were known to be alive. Alternatively, for calculating the time to death from malignant disease, patients dying due to another cause are censored at their date of death. The duration of survival in different treatment groups is estimated using the Kaplan–Meier technique and compared using the two-sided logrank test.

Pawinski *et al.*[17] found that only about 15% of all patients with Ta T1 bladder cancer entered in trials died of malignant disease within 10 years. The overall 10-year survival was 50–60% and only 40% of deaths were due to cancer. As any plausible treatment benefit in reducing cancer deaths is of the order of 5% or less at 10 years, 'time to death due to malignant disease' is an unrealistic end point for most superficial bladder trials. While the short-term effects based on the time to first recurrence and the recurrence rate may be assessed in individual trials of prophylactic treatment, assessment of the long-term relative treatment effects with respect to muscle invasion and cancer death almost certainly require the pooling of trial data in individual patient data based meta-analyses[17].

For a general discussion of the problems related to the design and analysis of superficial bladder cancer trials, see Byar *et al.*[5] and Sylvester[27]. For more information on prognostic factors for superficial bladder cancer, see Aso *et al.*[1,2] and Kurth *et al.*[12].

DESIGN AND ANALYSIS OF BLADDER CARCINOMA IN SITU TRIALS

The main end points in such trials are the complete response rate and the duration of response in the complete responders.

Considering BCG to be the standard treatment with a complete response rate of 80%, to test the efficacy of a less toxic dose of BCG or a new alternative agent, the trial is designed to test for 'equivalence'. In this case the new regimen will be considered as being 'equivalent' to BCG if it has a response rate not less than 80%–Δ%. In calculating the number of patients required for this trial, it will be assumed that the new regimen has a response rate which is worse than BCG by at least Δ%. The data will then be used in an effort to reject this hypothesis in the direction of no difference.

Table 12.6 gives the number of patients required in each arm of the study to test for equivalence of a new regimen assuming a complete response rate of 80% on

Table 12.6 Number of patients required on each treatment to test for equivalence of a new less toxic regimen assuming a complete response rate of 80% on BCG for various values of Δ based on a one-sided test at $\alpha = 0.05$ and $\beta = 0.10, 0.20$

Difference Δ	$\beta = 0.10$	$\beta = 0.20$
0.05	1232	900
0.08	511	375
0.10	339	250
0.12	243	180
0.15	163	121

BCG for various values of Δ based on a one-sided test at $\alpha = 0.05$ and $\beta = 0.10, 0.20$. As can be seen from this table the choice for Δ is crucial. As an example, in a trial comparing epirubicin to BCG for CIS on the basis that epirubicin will be considered as equivalent (i.e. a practical treatment option) if it is less toxic and the CR rate is not more than 10% worse than BCG, a total of at least 500 patients will be required.

DESIGN AND ANALYSIS OF MUSCLE INVASIVE BLADDER CANCER TRIALS

Whether a bladder-preserving technique is to be compared with radical cystectomy or if cystectomy plus adjuvant chemotherapy is compared with cystectomy alone, the statistical design is essentially the same.

For T2–T4a TCC, the median duration of survival after definitive local treatment is approximately 2 years[11]. The ultimate goal of new modalities of neoadjuvant and adjuvant treatment is to improve the overall duration of survival. A secondary goal is to reduce the local recurrence rate, mainly in patients receiving bladder-preserving treatment but also after cystectomy. Assuming a 2-year survival rate of 50% in the control arm, then 510 patients followed for an average of 2 years and 230 deaths would be required in each treatment arm in order to detect an absolute difference in survival of 10% (from 50% to 60%) in the 2-year survival rate at error rates $\alpha = 0.05$ and $\beta = 0.10$.

The duration of survival in the different treatment groups is estimated using the Kaplan–Meier technique and compared using the two-sided logrank test. In the recent international MRC/EORTC trial of neoadjuvant cisplatin, methotrexate and vinblastine (CMV), clinicians agreed before the trial that they would use neoadjuvant CMV in routine practice if it could be shown to increase survival by 10%. The trial was therefore designed on this basis using 3-year rather than 2-year survival as the end point. Nine hundred and seventy-six patients were recruited over a period of 5.5 years. Prognostic factors in muscle invasive bladder cancer have been considered by a number of authors[9,19,29].

DESIGN AND ANALYSIS OF METASTATIC BLADDER CANCER TRIALS

The main end points in phase III trials of metastatic disease are the duration of survival and the overall response rate. The response rate is often taken as the primary end point in such trials since one is first of all interested in showing the anti-tumor effect of new regimens. For practical clinical purposes improved survival is the desired aim. However, in metastatic patients it is much more difficult to show an improvement in the overall duration of survival using chemotherapy regimens currently available. With combination chemotherapy such as MVAC, the overall response rate is approximately 50%. Using this as the 'standard', Table 12.7 gives the number of patients required per treatment group to detect an increase in response rate with an alternative regimen based on a two-sided test at error rates $\alpha = 0.05$ and $\beta = 0.10, 0.20$.

Metastatic patients treated with MVAC have a median survival of approximately 1 year. Table 12.8 gives the number of **deaths** required per treatment group to detect an increase in the median survival with a new experimental regimen based on a two-sided logrank test at error rates $\alpha = 0.05$ and $\beta = 0.10, 0.20$. If only 80% of the patients are followed until death, then the figures in the table must be increased by 25% to get the number of patients required on each treatment. For further explanation concerning the calculation of sample size, see Machin and Campbell[16].

Table 12.7 Number of patients per treatment group required to detect an increase in response rate with a new experimental regimen from a baseline of 50% on MVAC, based on a two-sided test at error rates $\alpha = 0.05$ and $\beta = 0.10, 0.20$.

Response rate	$\beta = 0.10$	$\beta = 0.20$
55%	2095	1565
60%	519	388
65%	227	170
70%	124	93
75%	77	58

Table 12.8 Number of **deaths** per treatment group required to detect an increase in the median duration of survival with a new experimental regimen from a baseline median of 12-months on MVAC, based on a two-sided logrank test at error rates $\alpha = 0.05$ and $\beta = 0.10, 0.20$.

Median survival	$\beta = 0.10$	$\beta = 0.20$
14 months	885	661
15 months	423	316
16 months	254	190
17 months	174	130
18 months	128	96

CONCLUSION

The biggest obstacle to the proper assessment of new treatments has been the underestimation of the number of patients required to detect plausible, but medically important treatment differences. This has, in large part, been due to an overly optimistic estimate of the size of the treatment benefit to be expected.

Meta-analyses of small, undersized clinical trials cannot replace well-conducted, large-scale clinical trials. In order to achieve the sample sizes that are now recognized as being required to conduct meaningful studies with sufficient power to detect realistic differences, the need for urologists and clinicians to participate in collaborative, multicentre clinical research has become even more compelling. Only through a true collaborative effort can real progress be made, with the conduct of true international intergroup studies, in order to reach the large sample sizes that are required in the shortest period of time.

REFERENCES

1. Aso Y, Anderson L, Soloway M *et al.* (1986) Prognostic factors in superficial bladder cancer, in *Developments in Bladder Cancer*, (eds L Denis, T Niijima, G Prout and FH Schröder), Proceedings of the First International Consensus Development Conference on Guidelines for Clinical Research in Bladder Cancer, Antwerp, Belgium, Alan R Liss, New York, pp. 257–269.
2. Aso Y, Bouffioux C, Flanigan R *et al.* (1994) Prognostic factors in Ta, T1 (superficial) bladder cancer, in *Proceedings of the Second and Third International Consensus Development Symposia*, (eds T Niijima, Y Aso, W Koontz, G Prout and L Denis), SCI, France, pp. 176–186.
3. Blackwelder WC (1982) Proving the null hypothesis in clinical trials. *Control Clin Trials* **3**: 345–353.
4. Breslow N (1984) Comparison of survival curves. *Cancer Clinical Trials: Methods and Practice*, Oxford University Press, Oxford, pp. 381–406.
5. Byar D, Kaihara S, Sylvester R *et al.* (1986) Statistical analysis techniques and sample size determination for clinical trials of treatments for bladder cancer, in *Developments in Bladder Cancer*, (eds L Denis, T Niijima, G Prout and FH Schröder), Proceedings of the First International Consensus Development Conference on Guidelines for Clinical Research in Bladder Cancer, Antwerp, Belgium, Alan R Liss, New York, pp. 49–64.
6. Dunnett CW and Gent M (1977) Significance testing to establish equivalence between treatments, with special reference to data in the form of 2×2 tables. *Biometrics* **33**: 593–602.
7. EORTC new drug development committee (1985) EORTC guidelines for phase I trials. *Eur J Cancer Clin Oncol* **21**: 1005–1007.

8. EORTC (1994) *Trial Methodology: A Practical Guide to EORTC Studies*. European Organisation for Research and Treatment of Cancer, Brussels, pp. 56–78.
9. Fossa SD, Koontz W, Matsumoto K *et al.* (1994) Prognostic factors in muscle-invasive bladder cancer, in *Proceedings of the Second and Third International Consensus Development Symposia,* (eds T Niijima, Y Aso, W Koontz, G Prout and L Denis), SCI, France, pp. 187–198.
10. George SL (1984) The required size and length of a phase III clinical trial, in *Cancer Clinical Trials: Methods and Practice*, (eds ME Buyse, MJ Staquet and RJ Sylvester), Oxford University Press, Oxford, pp.287–310.
11. Ghersi D, Stewart LA, Parmar MKB *et al.* (1995) Does neoadjuvant cisplatin-based chemotherapy improve the survival of patients with locally advanced bladder cancer: a meta-analysis of individual patient data from randomized clinical trials. *Br J Urol* **75**: 206–213.
12. Kurth KH, Denis L, Bouffioux C *et al.* (1995) Factors affecting recurrence and progression in superficial bladder tumours. *Eur J Cancer* **31A:** 1840–1846.
13. Kurth KH, Denis L, Ten Kate FJW *et al.* (1992) Prognostic factors in superficial bladder tumours. *Probl Urol* **6:** 471–483.
14. Lagakos SW and Pocock SJ (1984) Randomization and stratification in cancer clinical trials: an international survey, in *Cancer Clinical Trials: Methods and Practice*, (eds ME Buyse, MJ Staquet and RJ Sylvester), Oxford University Press, Oxford, pp. 276–286.
15. Lamm DL, van der Meijden APM, Akaza H *et al.* (1995) Intravesical chemotherapy and immunotherapy; how do we assess their effectiveness and what are their limitations and uses? Report of the fourth International Consensus Conference on Bladder Cancer, Antwerp, Belgium, *Int J Urol* **2** (Suppl. 2): 22–35.
16. Machin D, Campbell MJ, Fayers PM and Pinol APY (1997) *Sample Size Tables for Clinical Studies,* 2nd edn, Blackwell Scientific, Oxford.
17. Pawinski A, Sylvester RJ, Bouffioux C *et al.* (1996) A combined analysis of EORTC/MRC randomized clinical trials for the prophylactic treatment of TaT1 bladder cancer, *J Urol* **156**: 1934–1941.
18. Peto J (1984) The calculation and interpretation of survival curves. *Cancer Clinical Trials: Methods and Practice*, Oxford University Press, Oxford, pp. 361–380.
19. Richards B, Aso, Y, Bollack C *et al.* (1986) Prognostic factors in iniltrating bladder cancer, in *Developments in Bladder Cancer*, (eds L Denis, T Niijima, G Prout and FH Schröder), Proceedings of the First International Consensus Development Conference on Guidelines for Clinical Research in Bladder Cancer, Antwerp, Belgium, Alan R Liss, New York, pp. 271–286.
20. Schröder FH, Norming U, Blumenstein BA *et al.* (1997) Phase II Studies on prostate cancer. *Urology* **49**: (Suppl. 4A), 3–14.

21. Schröder FH, Sylvester R, Barton B *et al.* (1994) Response criteria and endpoints employed in phase III studies of superficial bladder cancer, in *Proceedings of the Second and Third International Consensus Development Symposia*, (eds T Niijima, Y Aso, W Koontz, G Prout and L Denis), SCI, France, pp. 199–207.
22. Schröder FH, Sylvester RJ, Gustafson H *et al.* (1986) Response criteria for phase III studies of superficial bladder cancer, in *Developments in Bladder Cancer*, (eds L Denis, T Niijima, G Prout and FH Schröder), Proceedings of the First International Consensus Development Conference on Guidelines for Clinical Research in Bladder Cancer, Antwerp, Belgium, Alan R Liss, New York, pp. 311–321.
23. Schwartsmann G, Wanders J, Koier IJ *et al.* (1991) EORTC new drug development office coordinating and monitoring programme for Phase I and II trials with new anticancer agents. *Eur J Cancer* **27**: 1162–1168.
24. Simon R (1989) Optimal two-stage designs for phase II clinical trials. *Control Clin Trials* **10**: 1–10.
25. Staquet M and Dalesio O (1984) Designs for phase III trials, in *Cancer Clinical Trials: Methods and Practice*, (eds ME Buyse, MJ Staquet and RJ Sylvester), Oxford University Press, Oxford, pp. 261–275.
26. Sylvester R (1984) Planning cancer clinical trials, in *Cancer Clinical Trials: Methods and Practice*, (eds ME Buyse, MJ Staquet and RJ Sylvester), Oxford University Press, Oxford, pp. 47–63.
27. Sylvester R (1984) The analysis of results in prophylactic superficial bladder cancer studies, in *Superficial Bladder Tumors*, (eds FH Schröder and B Richards), Alan R Liss, New York, pp. 3–11.
28. Sylvester R (1988) General overview of phase II designs in cancer, in *Biometry: Clinical Trials and Related Topics*, (ed. T Okuno), Proceedings of the ISI Satellite Meeting on Biometry, Osaka, Japan, Excerpta Medica, Amsterdam, pp. 17–28.
29. Sylvester R, Barton B, Hisazumi H *et al.* (1994) Prognostic factors for randomization, stratification and endpoints for the evaluation of bladder trials, in *Proceedings of the Second and Third International Consensus Development Symposia*, (eds T Niijima, Y Aso, W Koontz, G Prout and L Denis), SCI, France, pp. 17–26.
30. Sylvester R, Griffin P, Koiso K *et al.* (1986) Standardization of protocol format, in *Developments in Bladder Cancer*, (eds L Denis, T Niijima, G Prout and FH Schröder), Proceedings of the First International Consensus Development Conference on Guidelines for Clinical Research in Bladder Cancer, Antwerp, Belgium, Alan R Liss, New York, pp. 3–13.
31. van der Meijden APM, Hall RR, Kurth KH *et al.* (1996) Phase II trials in superficial bladder cancer. The marker tumour concept. *Br J Urol,* **77**: 634–637.

32. van Oosterom A, Tannock I, Matsumura Y *et al.* (1994) Response criteria in phase II/phase III studies of invasive bladder cancer, in *Proceedings of the Second and Third International Consensus Development Symposia*, (eds T Niijima, Y Aso, W Koontz, G Prout and L Denis), SCI, France, pp. 208–218.
33. von Hoff DD, Kuhn J and Clark GM (1984) Design and conduct of phase I trials, in *Cancer Clinical Trials: Methods and Practice*, (eds ME Buyse, MJ Staquet and RJ Sylvester), Oxford University Press, Oxford, pp. 210–220.

13

Smoking and bladder cancer

Suzanne Cholerton and Reginald R. Hall

Statements such as 'smoking causes bladder cancer', 'smoking is the single most important cause of urothelial cancer'[37,116], 'its role is undeniable'[91], and 'bladder cancer is a largely preventable disease'[77] are not uncommon in the medical literature. Morrison *et al.*[92] estimated that cigarette smoking accounted for 25–60% of all bladder cancers in industrialized, developed countries and d'Avenzo[40] suggested that 46% of bladder cancer deaths in Italy in 1990 could have been prevented by the elimination of cigarette smoking. Can this extreme view of a causal relationship between cigarette smoking and bladder cancer be justified?

In the early reports of their original study of 1951 Doll and Hill[47,48] did not find an association between smoking and bladder cancer, but in 1958 Hammond and Horn[60] reported a significant excess of histologically proven bladder cancers in smokers, with a ratio[46] of 2.17 of observed versus expected cases. Several other retrospective and prospective studies found a similar association but Doll and Hill[49] advised caution in interpreting results obtained from heterogeneous populations and concluded 'it does not appear that there can be a strong relationship'.

By 1971 a leading article in the *British Medical Journal* stated that 'smoking may increase the risk of bladder cancer, though by how much is not known'[15]. A review by Clayson 4 years later[35] concluded that most published surveys showed that the incidence of bladder cancer was raised two- or threefold in male smokers, and patients who continued to smoke after the diagnosis of bladder cancer had a poorer prognosis[35]. The failure of British surveys to show this association was thought possibly due to the high lung cancer rate in smokers, these men dying of their first smoking-related cancer before they could develop bladder cancer. The general absence of a correlation between smoking and bladder cancer in women was considered to be due to their low frequency of cigarette smoking. Clayson also noted the lack of experimental evidence to support an association between smoking and bladder cancer but pointed out that tobacco smoke contained traces of 2-naphthylamine which was known to be carcinogenic[66].

Between 1980 and 1995, 16 case-controlled studies have compared the smoking status of more than 10 000 men and women known to have bladder cancer with a variety of age- and sex-matched controls. The results are summarized in Table 13.1 and reveal a strong apparent association between bladder cancer and smoking.

Smoking is only one of many possible causes of bladder cancer, the majority of which are unknown or cannot be quantified. However, the published studies show a consistent association which is significant when allowance has been made for other known or possible contributory factors such as age, sex, race, alcohol, coffee, artificial sweeteners, urinary infection, diet, education and family income level and occupation. Siemiatyki *et al.*[115] considered the possibility that the association between smoking and bladder cancer may have reflected inadequately controlled confounding factors due to occupational carcinogen exposure. However, their studies of over 300 potential covariates in 857 lung cancers, 484 bladder cancers and 2240 controls did not support this thesis. Specific smoking-related factors that appear to influence the size of the increased risk of bladder cancer are the quantity of tobacco smoked, the duration of smoking, the type of tobacco and inhalation on the part of the smoker.

Table 13.1 Increased risk of bladder cancer with smoking: relative risk/odds ratio for smokers versus non-smokers in bladder cancer patients compared with control populations

Author	Year[Ref]	Country	Cases	Controls	Relative risk/odds ratio*
Howe *et al.*	1980[68]	Canada	632	632	Males 3.0, females 2.4
Morrison *et al.*	1984[92]	Japan/UK/USA	1435	1825	Males 1.9, females 2.4
Claude *et al.*	1986[32]	Germany	431	431	Males 2.3, females 2.9
Hartge *et al.*	1987[62]	USA	1151	1416	2.9 (2.6–3.3)
Iscovich *et al.*	1987[71]	Argentina	117	234	7.2 (3.0–20.1)
Møller-Jensen *et al.*	1987[89]	Denmark	388	787	2.9 (1.8–4.8)
Nomura *et al.*	1989[96]	Hawaii	261	261	Males 6.2, females 2.8
Clavel *et al.*	1989[33]	France	477	477	3.95
Burch *et al.*	1989[20]	Canada	826	792	Two-fold
Akdas *et al.*	1990[1]	Turkey	194	194	–
D'Avanzo *et al.*	1990[40]	Italy	337	392	3.3 (2.2–5.0)
Lopez-Abente *et al.*	1991[85]	Spain	430	791	3.79 (2.41–5.97)
Momas *et al.*	1994[91]	France	219	794	3.64 (2.21–5.98)
Sorahan *et al.*	1994[116]	UK	989	3658	3.12 (2.38–4.09)
Siemiatycki *et al.*	1994[115]	Canada	484	533	Two- to three-fold
McCarthy *et al.*	1995[86]	USA	1507	–	1.68 (1.22–2.32)

*The figures in parentheses ( ) = 95% confidence intervals.

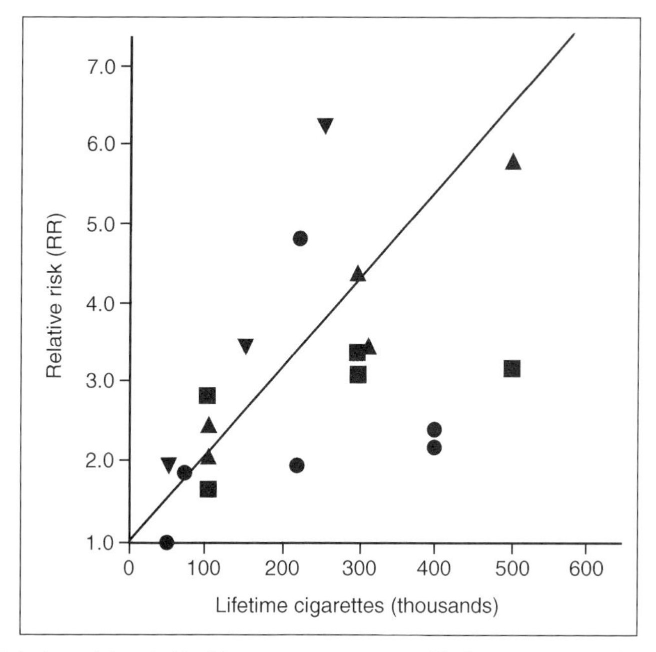

Figure 13.1 Relative risk of bladder cancer versus lifetime consumption of cigarettes. ●, Williams and Horm[134]; ▲, Howe *et al.*[68]; ■, Claude *et al.*[32]; ▼, Mommsen and Aagaard[90].

QUANTITY OF TOBACCO

Smoking has been quantified in three different ways. **Current smoking habit** is the most reliable to ascertain (smoker versus non-smoker) but ignores the quantity and the long-term effects of smoking, and the possibility that stopping smoking may affect the outcome. **Smoking history** (never smoked, ex-smoker, ever smoked, duration of smoking) and **estimation of quantity** address these issues but may be less reliable or accurate.

A simple comparison between current smokers and non-smokers shows a highly significant increased risk of bladder cancer, with odds ratios ranging from 1.68 to 6.2[32,33,85,86,96]. The degree of risk relates to the number of cigarettes (or packets) smoked per day[68,92,134]. The age of starting to smoke is generally not important but duration of smoking (years) and the lifetime total consumption are significant. The longer the duration and the greater the total consumption the higher the risk[32,68,90,91,92,116,134]. Momsen and Aagaard[90] derived the following equation for risk versus tobacco consumption with reference to all tobacco use (cigarettes, cigars and pipes) but it is similar to that of cigarette smoking only, combining data for men and women:

$$\text{Relative risk} = \frac{[0.945 + 1.103 \times (\text{lifetime cigarette equivalent consumption})]}{100}$$

This is illustrated in Figure 13.1 where the line representing the above equation has been superimposed on the data available from the published reports of four of the other studies[104]. Based on the pooled data from these studies, Robinson[104] has summarized the risk in practical terms, showing the numbers of years of smoking at specified levels which are likely to lead to negligibly increased risk, slightly increased risk and a relative risk of about 2 (Table 13.2). It should be noted that levels of total life-consumption much higher than those in this table have been reported (for example in the South of France[91]) for which the relative risk of bladder cancer was very much greater.

Table 13.2 Risk of bladder cancer as a function of smoking years (P. Robinson, personal communication)

	Total cigarettes	Years to give stated risk when smoking				
		20/week	5/day	10/day	20/day	40/day
Negligible excess risk	20 000	20	12	6	3	1.5
Possible slight excess risk	40 000	40	22	11	6	3
Significant excess risk	80 000	75	45	22	11	6

SMOKING CESSATION

Stopping smoking appears to reduce the risk of bladder cancer significantly. Ex-smokers have a reduced risk compared with current smokers. Compared with never-smokers the risk for ex-smokers was 1.9 compared with 3.3 for current smokers[40]. Those who stopped smoking more than 15 years previously have shown a significant reduction of risk from 1.13 to 0.38[32], and those who quit 20 or more years previously

Table 13.3 Reduction in risk of bladder cancer with time after stopping smoking (from Sorahan *et al.*[116])

Smoking status	Relative risk	95% CI for RR
Non-smoker	1.0	–
Current smoker	3.12	2.38–4.09
Quit 1–9 years ago	1.94	1.43–2.63
Quit 10–19 years ago	1.54	1.12–2.11
Quit ≥20 years ago	1.24	0.90–1.70

CI, confidence intervals; RR, relative risk.

had only about 10% of the excess risk experienced by current smokers, with a regular trend for risk reduction with time since quitting, which remained after adjustment for lifetime consumption of cigarettes (Table 13.3)[116].

TYPE OF TOBACCO

The odds ratios for bladder cancer and smoking reported from southern France, Italy, Spain and Greece are higher than those observed in the USA and UK[91]. It has been suggested that this difference could be due to the type of tobacco used, black (air-cured) tobacco being more common in these countries than blond (flue-cured)[82]. Lopez-Abente[85] noted a diminution in risk for smokers of low-tar and low-nicotine ('light') cigarettes. Clavel *et al.*[34] observed that black tobacco and inhaling both doubled the bladder cancer risk and that for ex-smokers the residual risk remained highest for former smokers of black tobacco. *In vitro* tests of urinary mutagenicity and bladder carcinogen levels have been found to be higher in smokers of black tobacco than blond[18]. The use of filter-tipped cigarettes was associated with a lower risk compared with non-filter-tipped[85,116]. Smokers of cigars and pipes generally have only slightly elevated risk of bladder cancer compared to non-tobacco users; however, inhalation of pipe smoke appears to increase the risk[20,134]. Overall, inhalation with smoking increases the risk[19,33]. Men who smoked two or more packs per day and inhaled deeply had nearly seven times the risk of non-smokers[92].

SMOKING AND HISTOLOGICAL GRADE, STAGE AND RECURRENCE OF BLADDER CANCER

Smoking has been shown to correlate with histological grade and stage of bladder cancer at the time of initial diagnosis, and with the likelihood of recurrence. In a prospective study of 126 bladder cancer patients Chinegwundoh and Kaisary[31] reported that heavy smokers accounted for 41.8% (33/79) of G1 and G2 tumours compared with 81% (17/21) of G3 cancers. The difference in total lifetime cigarette consumption between G1, G2 and G3 was significant. Thompson *et al.*[125] analysed thc influence of smoking on tumour characteristics in 386 patients. Most tumours were superficial (Ta, T1) but a third of the patients with a smoking history had muscle invasive bladder cancer compared with only 14% of non-smoking patients. The grade of tumour (degree of histological differentiation) was relatively evenly distributed within smoking patients, whereas non-smokers tended to have lower-grade tumours, this difference being significant. During follow-up those patients who had never smoked had a significantly lower number of recurrences. Smokers had a greater tendency to multiple tumour recurrence. In a study of 100 patients Carpenter[22] also observed that smokers had a significantly higher pattern of tumour recurrence. Of 79 smokers, 81% had recurrent tumours compared with 56% of 27 non-smokers. In smokers, 93% of recurrences were multiple, compared with 53% in non-smokers. Anecdotally, patients treated by partial cystectomy had a greater chance of local tumour relapse if they smoked, and urethral tumours were seen more often after cystectomy in smokers. The authors went so far as to recommend that if a bladder cancer patient smoked, partial cystectomy was not indicated, and if radical cystectomy was performed, urethrectomy should also be considered.

Thus there is considerable evidence for an association between cigarette smoking and bladder cancer which is not just a general association but one that is specific and statistically significant in several details. However, as cautioned by Doll and Hill[49] it is still necessary to determine whether **association** implies **causation**. This latter crucial step in our understanding of the aetiology of bladder cancer can only be made with any certainty if there is sufficient direct evidence that constituents of tobacco smoke are bladder carcinogens.

A major component in the initiation of cancer is the conversion of chemical carcinogens to electrophilic species which react (covalently) with cellular macromolecules such as DNA to yield adducts. The production of these reactive species often involves an enzyme-mediated oxidation reaction. However, the amount of reactive species present is determined not only by the activity of the activation pathways but also by enzyme-mediated detoxication mechanisms which operate to inactivate these electrophiles. In man some of the enzymes which mediate these activation and detoxication pathways are expressed in different forms (polymorphically) such that enzyme activity shows considerable interindividual variation. Through case-control studies, genetic polymorphisms in several enzymes that metabolize foreign compounds have been shown to be important factors in determining susceptibility to a variety of cancers, including that of the bladder. The following review considers the evidence for consitutents of tobbaco smoke to act as bladder carcinogens and for genetic polymorphisms in enzymes that metabolize these compounds to determine an individual's susceptibility to bladder cancer.

CONSTITUENTS OF TOBACCO SMOKE

Tobacco smoke is a complex mixture of at least 3800 chemicals, more than 50 of which have been shown to be carcinogenic to experimental animals. The three major chemical families of carcinogens in tobacco smoke are **aromatic amines (arylamines)**, **polycyclic aromatic hydrocarbons (PAHs)** and **tobacco-specific nitrosamines (TSNA)**. Several lines of research have produced evidence to suggest that the aromatic amines are important causative agents for bladder cancer in cigarette smokers. The role of PAHs and TSNAs in this disease is less well defined. In addition to these well-recognized carcinogenic components of tobacco smoke, the carcinogenic potential of the tobacco alkaloids, particularly nicotine and its metabolites, has also been investigated.

Aromatic amines

There is considerable evidence from studies in a variety of animal species that exposure to aromatic amines (AAs; arylamines) is associated with an increased incidence of bladder tumours[100,108]. Furthermore, epidemiological studies implicate exposure to AAs as a major risk factor in the development of bladder cancer in man. As early as 1895, Rehn[101] observed an association between occupational exposure of German dye workers to AAs and bladder cancer. Since that time, exposure to AAs in a variety of occupational settings (including the dye manufacturing industry, rubber and cable making and the production of coal gas) has been shown to increase an individual's risk of bladder cancer[26,27,28,42]. From such studies the AAs, 4-aminobiphenyl (4-ABP), 2-naphthylamine (2-NA) and benzidine, have been implicated as causative agents in transitional urothelial cell carcinomas.

Since cigarette smoke contains a number of carcinogenic AAs including 2-NA, 3-aminobiphenyl, 4-ABP and *O*-toluidine[98], it has long been hypothesized that the AAs in cigarette smoke are important aetiological agents in bladder cancer.

Aromatic amine metabolism

It appears that AAs have the ability to exert their carcinogenic effects only after metabolic activation in the liver and transportation to the bladder. The initial step in this activation process is an *N*-oxidation reaction which is mediated by the hepatic cytochrome P450, CYP1A2 and results in the production of *N*-hydroxyarylamine metabolites[21]. However, the amount of *N*-hydroxylated AA produced is not solely dependent upon the activity of CYP1A2. An important hepatic detoxication pathway for AAs at this stage is mediated by the *N*-acetyltransferase (NAT) enzyme, NAT2[63].

Once produced, the *N*-hydroxy metabolite will either enter the blood where it may form covalent bonds with haemoglobin (see below), be inactivated enzymatically and excreted, or transported to the kidney and filtered into the urine where it can be reabsorbed into the bladder epithelium. At this site, NAT1 activity may mediate the final activation process leading to the formation of DNA adducts, which is a critical event for the initiation of bladder carcinogenesis[5].

Aromatic amine–macromolecule adduct formation

Haemoglobin adducts

As a consequence of their moderate reactivity and their transportation in blood, the *N*-hydroxyarylamines can form adducts with haemoglobin in erythrocytes. Since such adducts are present in both cigarette smokers and non-smokers at concentrations which are amenable to analysis, they have been used widely as biomarkers of exposure to tobacco smoke-derived AAs.

Bryant and co-workers[18,19] demonstrated that haemoglobin adducts of 4-ABP were higher in a group of cigarette smokers than in non-smokers and that there was a significant correlation between 4-ABP adduct levels and cigarette consumption. Since that time it has been shown repeatedly that smokers have higher levels of arylamine– or 4-ABP–haemoglobin adducts than non-smokers but that the amount of adduct produced in an individual is modulated by *N*-acetyltransferase activity[6,130].

Several case-control studies have demonstrated dose–response relationships between the quantity of cigarettes smoked and the relative risk for bladder cancer[62,70,89]. Similar dose–response curves have been demonstrated when levels of 4-ABP–haemoglobin adducts were related to markers of recent smoking (urinary levels of nicotine plus cotinine and number of cigarettes)[6]. Such findings lend greater weight to the hypothesis that AAs are important causative agents for bladder cancer.

4-ABP–haemoglobin adducts have been shown to be higher in smokers of black tobacco than blond tobacco[19]. This is in keeping with the greater risk of bladder cancer in smokers of black tobacco[33,45,71,131]. Furthermore, the chemical analysis of black and blond tobacco-derived smoke, indicates that aromatic amine concentrations are higher in the former[98]. Although such observations further reinforce the proposed role for the AAs as causative agents in bladder cancer, only one case-control study of 4-ABP–haemoglobin adducts and bladder cancer has been reported. In this, bladder carcinoma patients had slightly, but statistically significantly, higher levels of 4-ABP–haemoglobin adducts than controls when paired on the basis of exposure to cigarette smoke as determined by urinary cotinine levels[44]. Levels of haemoglobin adducts of arylamines other than 4-ABP have been shown to be significantly higher in tobacco smokers when compared to non-smokers[19]; however, these adducts have not found widespread use as biomarkers of tobacco smoke exposure as those of 4-ABP.

As previously stated, dose–response relationships are apparent between levels of 4-ABP–haemoglobin adducts and markers of recent smoking such that significant correlations are found between the adducts and urinary cotinine in light smokers (up to 20 cigarettes/day), individuals exposed to environmental tobacco smoke and non-smokers.

DNA adducts

Although the presence of putative carcinogen–DNA adducts in human urinary bladder tissues had been reported previously[39,74,75], it was not until the early 1990s that specific arylamine–DNA adducts were identified. Talaska and co-workers[121] investigated smoking-related carcinogen–DNA adducts in human urinary bladder biopsy samples taken from 42 men of known smoking and occupational histories undergoing cytoscopic examination. The total mean carcinogen–DNA adduct level and several specific adducts (including one that was identified as *N*-(deoxyguanosin-8-yl)-4-ABP) were significantly elevated in samples from current smokers when compared with those from never-smokers and ex-smokers. The authors concluded that the data suggested a molecular basis for the initiation of human urinary bladder cancer by cigarette smoke.

The development of methodology to monitor carcinogen–DNA adducts in exfoliated urothelial cells represented the introduction of an important non-invasive technique for human biomonitoring. Using exfoliated urothelial cells from individuals of known smoking status, Talaska and co-workers[122] demonstrated the presence of five distinct DNA adducts which were 2–9 times higher in smokers than in non-smokers. One of these adducts was chromatographically similar to *N*-(deoxyguanosin-8-yl)-4-ABP and showed the strongest correlation with the number of cigarettes smoked per day and 4-ABP–haemoglobin adduct levels. In a subsequent study of the levels of DNA adducts in biopsies of bladder cancer patients and in exfoliated urothelial cells of healthy volunteers, a dose–response relationship between smoking levels and adduct levels was present in both groups[124].

Given the wealth of evidence presented, there can be little doubt that the AAs contribute to the bladder carcinogenicity associated with cigarette smoke. However, it seems likely that other chemicals from the thousands present in cigarette smoke act in conjunction with the AAs to result ultimately in bladder cancer[36].

Polycyclic aromatic hydrocarbons

Exposure to environmental PAHs is virtually unavoidable since these compounds are produced during the incomplete combustion of a variety of natural materials such as coal and petroleum and synthetic organic materials such as plastics. Cigarette smoke represents an additional source of environmental PAHs in the form of benzo[a]pyrene (BP) and benz[a]anthracene[95,138], two well-established genotoxic (DNA damaging) agents. Studies of occupational exposure to PAHs have revealed an association with cancers of the skin, lung and bladder[34,61,67], the presence of PAHs in cigarette smoke and the fact that cigarette smoking is epidemiologically linked to bladder cancer have led to investigations of the role of cigarette smoke-derived PAHs in the aetiology of bladder cancer.

Polycyclic aromatic hydrocarbon metabolism and macromolecule adduct formation

Like other environmentally stable carcinogens, PAHs undergo metabolic activation by cytochromes P450 (CYP) to electrophilic species which have the capacity to form adducts with nucleophilic cellular macromolecules, including proteins and DNA. CYP1A1, CYP3A4 and CYP2C isozymes have all been implicated in the oxidation reactions involved[38,43,57,103,113,114]. Some PAHs are known substrates for glutathione *S*-transferase and as such this enzyme may serve an important role in the detoxication of the activated forms of cigarette smoke-derived PAHs[29,79].

In the early 1980s, studies in which human explant cultures of bladder were incubated with radiolabelled BP demonstrated BP metabolite–DNA binding which varied 70-fold between individuals[112,119]. Although valuable, this technique was unable to provide information regarding environmental exposure to the PAHs.

The advent of ^{32}P-postlabelling analysis provided the much needed means to monitor exposure of human tissue biopsy-derived DNA to the genotoxic chemicals in tobacco smoke. Further modifications of the technique enabled a degree of analytical differentiation between aromatic amine–DNA adducts and PAH–DNA adducts to be achieved[56,59,122]. Using this methodology, studies of DNA adducts in human urinary bladder tissue revealed significantly higher total mean carcinogen–DNA adducts in cigarette smokers than ex-smokers and never-smokers[123]. Furthermore, in smokers nearly half the adducts detected were similar to those derived from PAHs, while the remainder exhibited properties consistent with the nature of AAs[53,99,123].

In a recent study of the relationship between the biomarker of exposure to PAHs, urinary 1-hydroxypyrene glucuronide (1-OHPG)[120], and DNA adducts in exfoliated bladder cells, 1-OHPG was found to correlate positively with urinary nicotine plus cotinine and urinary mutagenicity. However, Vineis and co-workers state that although urinary 1-OHPG is associated with tobacco smoking, DNA damage in bladder cells appears to be mainly due to AAs[133].

Tobacco-specific nitrosamines

Although the tobacco-specific nitrosamines (TSNA) have been identified as a group of chemicals that may contribute to the induction of cancer in a variety of organs including the lung, oesophagus and nasal and oral mucosa, a role for these compounds in the development of bladder cancer has not been identified. None of the *N*-nitrosamines which are consistently found in tobacco smoke have been shown to be bladder carcinogens in experimental animals[65].

However, in a study of the influence of a history of urinary tract infections (UTI) and risk of bladder cancer, the joint effect of UTI and cigarette smoking was greater than would have been predicted from the two separate exposures and was even more pronounced in those individuals in the most extreme categories of exposure[78]. Although not reported for urine, there is evidence to suggest that endogenous TSNA formation can occur under simulated gastric conditions and in saliva when nitrosamine precursors such as the nicotine metabolite, nornicotine, and nitrosating agents are present[80,93,94,126]. Given that nornicotine is a urinary metabolite of nicotine in man, albeit a minor one, and during bladder infection, urinary nitrate is reduced to the nitrosating agent, nitrite, it seems likely that the conditions in the bladder of a smoker with a UTI are conducive to the production TSNA. Nevertheless, a causative link between these compounds and bladder cancer is yet to be established.

Tobacco alkaloids and their metabolites

Nicotine is the major alkaloid present in tobacco (85–95%) and is generally regarded as the substance that is responsible for the development of dependence on cigarettes. Depending on the species of the genus *Nicotiana* and the stalk position of the leaf, the alkaloids account for between less than 0.5% to 5% of dry tobacco. When smoked, the cigarette represents a device which ensures the rapid delivery of nicotine to the brain and other organs of the body. However nicotine is short lived ($t_{\frac{1}{2}} = 2$ h) since it is rapidly degraded to a variety of metabolites, including the major primary oxidative metabolites, cotinine (subsequently metabolized to 3′-hydroxy-cotinine) and nicotine 1′-*N*-oxide.

The excretion of nicotine in urine is pH dependent such that under conditions of acidic urine (pH < 5.5) nicotine is almost completely ionized and cannot be reabsorbed through the renal tubule. As such, urinary pH determines the amount of nicotine available for metabolism[7,11,107].

Cotinine and nicotine 1′-*N*-oxide are also present in dry tobacco at low concentrations with other alkaloids including anabasine and anatabine.

IN VIVO STUDIES

Chronic toxicity of nicotine and cotinine

As a consequence of its rapid elimination and high acute toxicity (LD_{50} in mice after intravenous administration = 0.6 mg/kg body weight), testing nicotine for carcinogenicity has proved difficult[118]. In 1935 Staemmler[117] observed that daily subcutaneous injections of nicotine to rats for up to 20 months induced hyperplasia and tumours of the adrenal medulla. Schmähl[109] showed that nicotine-containing ethanolic extracts of five different tobaccos were not carcinogenic on injection into rats but smoke condensates or tars made from the tobaccos induced sarcomas at the site of injection; however, there was no evidence that this treatment induced tumours of the bladder.

When cotinine was administered to mice by several different routes, 11 adenomas of the bladder were seen in 69 mice in which pellets of cholesterol containing 10% cotinine were implanted into the urinary bladder[14]. Although the implantation of cholesterol pellets into the urinary bladder of mice had previously been shown to be associated with increased bladder tumour incidence[2], Boyland[14] concluded that cotinine is a direct carcinogen for the epithelium of the urinary bladder. Oral administration of cotinine induced no bladder tumours[110,127].

Co-carcinogenicity of nicotine and its metabolites

The ability of nicotine, cotinine and nicotine 1′-*N*-oxide to act as a tumour promoters in the urinary bladder has been investigated in rats but studies to date have not supported this possibility[72,83].

Nicotine metabolism in bladder cancer

In a study by Gorrod and co-workers in 1974[58], urinary cotinine/nicotine 1′-*N*-oxide ratios were determined in a 24-hour period in bladder cancer patients and compared to those in healthy volunteers. All participants smoked cigarettes *ad libitum* throughout the study and no attempt was made to control for confounding effects of fluctuating urinary pH. The ratio was significantly higher in bladder cancer patients than in the control group, suggesting that the cases produced either more cotinine or less nicotine 1′-*N*-oxide than the controls.

IN VITRO STUDIES

Although bladder cancer tissue has not been investigated, the presence of opioid receptors and endogenous opioids has been demonstrated in wide variety of other human and animal tumours[140]. Both opioid and nicotine receptors were shown to be expressed in lung cancer cell lines of diverse histologies and nicotine was shown to reverse opioid-mediated growth inhibition of these cell lines[87]. Such observations have led to the hypothesis that nicotine may promote tumour formation by inhibiting the 'suppressive' effect of the endogenous opioids on cell proliferation.

It has been proposed that cell death by apoptosis may be an important mechanism to prevent tumour development. The ability of nicotine, cotinine, the tobacco alkaloid anabasine and the tobacco smoke-derived carcinogen, *N*-nitrosodiethylamine to inhibit apoptosis in normal and transformed cells derived from a variety of species and tissues was investigated. Although no bladder-derived cells were studied, nicotine, cotinine and anabasine were all shown to inhibit apoptosis, whereas *N*-nitrosodiethylamine was without effect[137]. The authors concluded that their data implicate nicotine as a potential tumour-promoting agent which may contribute to the development of tobacco-related malignancies. Using lung cancer cells, Maneckjee and Minna[87] provided supporting evidence for this hypothesis. They reported that nicotine acts through nicotinic acetylcholine receptors to suppress apoptosis and postulated that nicotinic receptor antagonists, such as hexamethonium, could have potential value as chemopreventive agents for lung cancer in cigarette smokers.

Jeremy and co-workers[73] investigated the effects of nicotine, cotinine and cigarette smoke extracts on the synthesis of prostacyclin (PGI_2), a proposed cytoprotectant of epithelia, in the urinary bladder of the rat. In contrast to cigarette smoke extracts, neither nicotine or cotinine inhibited the synthesis of PGI_2 in this *in vitro* model.

Using the Ames test (*Salmonella typhimurium*/microsome assay), nicotine, and its metabolites cotinine and nicotine 1′-*N*-oxide and the minor alkaloids, anabasine and anatabine, were shown to have no mutagenic activity, however nicotine and anatabine induced repairable DNA damage in a test with *Escherichia coli pol* A^+/*pol* A^-[102]. Similarly, nicotine and four of its major metabolites (cotinine, nicotine-*N*′-oxide, cotinine-*N*-oxide and 3′-hydroxycotinine) were shown not to be genotoxic in both the Ames test and the Chinese hamster ovary sister chromatid exchange at concentrations orders of magnitude higher than those observed in the plasma of cigarette smokers[50].

GENETIC POLYMORPHISMS IN XENOBIOTIC METABOLISM AND SUSCEPTIBILITY TO BLADDER CANCER

N-acetyltransferase

Many epidemiological studies have suggested that the expression of the cytosolic enzyme *N*-acetyltransferase (NAT) is an important determinant of susceptibility to arylamine-induced bladder cancer. Two isozymes of NAT, NAT1 and NAT2, have been identified. The latter appears to mediate an important detoxication pathway for AAs in the liver, whereas it has been suggested that NAT1, which is expressed at a higher level in bladder epithelium than NAT2, may mediate the final activation step for the *N*-hydroxyarylamine in this tissue. NAT2 can also catalyse the *O*-acetylation of the *N*-hydroxyarylamine metabolites which results in their conversion to an ultimate carcinogen that forms arylamine–DNA adducts.

Allelic variants of NAT2 are known to be responsible for the classical *N*-acetylation polymorphism discovered as a consequence of variability in metabolism of isoniazid[51]. Although 14 mutant alleles have been identified[128], seven of these plus the wild-type allele (NAT2*4) accounted for all NAT2 alleles in a group of 533 unrelated German subjects[25]. These genotypic differences give rise to two distinct phenotypes. Individuals either homozygous or heterozygous for the wild-type allele are **fast acetylators** and those homozygous for mutant alleles are **slow acetylators**. In Caucasian populations, 30–50% of individuals are fast acetylators and 50–70% are slow acetylators. However, there is considerable inter-ethnic variability with respect to this enzyme activity, with slow acetylators representing only 5% of Canadian Eskimos but 90% of North Africans.

In the early 1980s, two studies indicated that individuals with the slow acetylator phenotype (NAT2) who were occupationally exposed to arylamines were more susceptible to bladder cancer than fast acetylators[23,136]. In the UK study[23], 96% of chemical workers with bladder cancer were slow acetylators, whereas the Danish study found 65% slow acetylators in the bladder cancer group compared to 51%

among controls[135]. Such results suggested an association between the slow-acetylator phenotype and increased susceptibility to bladder cancer in individuals occupationally exposed to arylamines. Since that time, many authors have confirmed this observation by performing studies that have taken into account cigarette smoking as well as arylamine exposure as a consequence of occupation. By combining the results of 10 case-control studies, Vineis and Ronco[132] determined a 30% increased risk of bladder cancer in slow acetylators compared to fast acetylators.

Since *N*-acetylation represents a detoxication pathway for the electrophilic species generated by the *N*-hydroxylation of arylamines, the influence of acetylator status (NAT2) on the formation of adducts with cellular macromolecules has been investigated. Bartsch and co-workers[6] studied the relationship between type of tobacco (black or blond) and number of cigarettes smoked, levels of 4-ABP–haemoglobin adducts and the NAT phenotype in 97 healthy male volunteers. Slow acetylators were shown to have higher levels of 4-ABP–haemoglobin adducts for the same type and quantity of cigarettes smoked than fast acetylators. These results were confirmed by Yu *et al.*[139], who demonstrated consistently higher levels of both 4-ABP and 3-ABP–haemoglobin adducts in slow acetylators relative to rapid acetylators independent of level of smoking. An investigation of the presence/absence of DNA adducts in exfoliated bladder cells from 39 subjects of known acetylator phenotype and smoking status, revealed the presence of 12 adducts, two of which were moderately associated with smoking status[129]. One of these was qualitatively similar to *N*-(deoxyguanosin-8-yl)-4-ABP, the adduct formed in the bladders of dogs treated with 4-ABP and in biopsies of bladder cancer patients[121,122,123]. Although the presence of both adducts was associated with the slow-acetylator phenotype, this was not statistically significant.

A survey of Caucasian, Afro-American and Asian males in Los Angeles by Bernstein and Ross[12] demonstrated that the incidence of bladder cancer in Afro-American and Asian men was one-half or less the incidence in Caucasian men, although smoking habits were comparable among the three groups. In an attempt to establish whether these differences are attributable to inter-ethnic variation in acetylator status (NAT2), Yu and co-workers[139] determined the extent of variability in the NAT2-mediated metabolism of caffeine to determine acetylator phenotype in Caucasian, Afro-American and Asian males of known smoking status and related this to levels of ABP–DNA adducts. The prevalence of slow acetylators among the three ethnic groups closely paralleled their bladder cancer incidence and explained a substantial proportion of the racial differences in levels of ABP–haemoglobin adducts. Genotypic analysis of the wild-type allele and five mutant alleles of NAT2 revealed slow acetylation to be a significant risk factor for bladder cancer in smokers (OR = 1.2) such that a continuously increasing impact of the slow NAT2 trait with increasing lifetime cigarette consumption was seen, with the highest risk associated with individuals with more than 50 lifetime pack-years irrespective of occupational risk (OR = 2.2)[17]. Furthermore, this study demonstrated the greatest risk to be associated with those smokers homozygous for the NAT2*5B allele (OR = 1.8). Similarly, Okkels and co-workers[97] have demonstrated a small but significant association between the slow-acetylator genotype (NAT2) and bladder cancer risk in cigarette smokers and an overrepresentation of the NAT2*5B allele in the incident case group.

The polymorphic nature of NAT1 has only recently been delineated and information relating genetic polymorphisms in NAT1 to fast/slow NAT1 phenotype is limited at present. Bell and co-workers[9] screened a series of bladder and colon tissues samples for NAT1 phenotype and genotype to reveal twofold higher NAT1 activity in those samples from individuals with the NAT1*10 allele. This may have some consequence for bladder cancer risk since tissue from individuals heterozygous for the NAT1*10 allele had twofold higher levels of putative AA–DNA adducts than that from subjects homozygous for the wild-type allele[5]. Combined with other findings, such data lend weight to the hypothesis that NAT1 activity in bladder epithelium may mediate the final activation of urinary *N*-hydroxyarylamines to reactive *N*-acetoxy esters that form DNA adducts.

Glutathione *S*-transferase

The glutathione *S*-transferases (GST) are a family of multifunctional, cytosolic enzymes which render harmless potentially harmful electrophilic xenobiotic metabolites. The supergene family that encodes these enzymes is subdivided into four classes: α, μ, π and θ[88]. Within the GST μ family, GSTM1 is expressed polymorphically such that approximately 50% of a Caucasian population has the GSTM1 0/0 genotype and lack the enzyme activity[13,69,111].

In 1993, Bell and co-workers[8] determined the frequency of the homozygous deleted genotype, GSTM1 0/0, in a group of patients with transitional cell carcinoma of the bladder and in a group of race-, sex- and age-matched controls. The GSTM1 0/0 genotype was shown to confer a 70% increased risk of bladder cancer, and when smoking status was considered this genotype significantly increased the risk of bladder cancer in smokers but had little effect on this risk in non-smokers. Individuals with the GSTM1 0/0 genotype and smoking exposure of greater than 50 pack-years had a sixfold greater relative risk than non-smokers with either the GSTM1 +/0 or 0/0 genotype. Daly and co-workers[41] similarly observed that only 15% of a bladder tumour patient group were seen to have the GSTM1 gene compared to 40% of patient controls and 46% of healthy controls. The protective role seen for GSTM1 in smokers by Bell *et al.*[8] was not apparent. In a study of leucocytic GSTM1 activity, Lafuente and co-workers[81] demonstrated a significantly higher proportion of control smokers with measurable activity compared with bladder cancer patients and an increased susceptibility to early appearance of bladder neoplasia in heavy smokers with a genetic absence of the enzyme. The following year a comprehensive study by Brockmöller and co-workers[16], using both genotyping and phenotyping methods, revealed a statistically significant overrepresentation of GSTM1-defective patients in a group with bladder cancer. However, the impact of the GSTM1 0/0 genotype was not significantly different in individuals with occupational risk when compared to those without. Furthermore, and in agreement with the data of Daly *et al.*[41], the proportions of GSTM1-deficient individuals were not statistically different between never-, moderate and heavy smokers. Further investigation by the same group has revealed GSTM1 deficiency to be a risk factor for bladder cancer which is independent of smoking and occupation[17]. A recent study in Egyptian bladder cancer patients revealed a significantly higher frequency of the GSTM1 null genotype in the cancer patients compared to controls; however, the aetiology of bladder cancer in these subjects may differ from that in other populations given that in Egypt the majority of this disease is associated with *Schistosoma haematobium* infection[4].

Thus several studies suggest that the GSTM1 null genotype/phenotype increases the risk of bladder cancer. Two other studies do not reach this conclusion[84,141], but the use of young, healthy individuals rather than source-, age- and smoking status-matched controls in these two studies may have influenced the outcome. The results of a study by Okkels and co-workers[97] have provided further evidence that the GSTM1 deletion is not a risk factor for the development of bladder cancer, but suggest a role for this genotype in the survival of bladder cancer patients.

Also, Hirvonen and co-workers[64] compared the urinary mutagenicity of smokers with and without the GSTM1 gene. The level of urinary mutagenicity in two different bacterial strains was approximately three times greater for smokers without the GSTM1 gene than for those with it.

Taken as a whole, the data presented above suggest that the possession of the GSTM1 gene confers a protective effect on the individual with respect to the development of bladder cancer. However, whether the presence of GSTM1 activity has a protective effect on cigarette smoking- and/or occupational exposure-associated bladder cancer remains equivocal.

Cytochromes P450

The enzyme cytochrome P4502D6 (CYP2D6; debrisoquine 4-hydroxylase) is absent or inactive in 3–10% of Caucasian subjects[52]. As a consequence of their resultant inability to metabolize a variety of drugs (including the archetypal substrate debrisoquine), these subjects are referred to as ***poor metabolizers*** (PMs), a phenotype which results from the presence of two mutant alleles for the gene that codes for CYP2D6, *CYP2D6*.

In 1984 when Cartwright and co-workers[24] compared CYP2D6 activity in bladder cancer patients with that in a control group, no differences between cases and controls were observed. A later study confirmed this observation with respect to patients with non-aggressive bladder tumours but found a reduced frequency of PMs among patients with aggressive tumours[76]. A study of debrisoquine metabolism in Spanish bladder cancer patients did not confirm this but demonstrated higher CYP2D6 activity in bladder cancer patients than controls[10]. Furthermore, patients with high occupational risk for bladder cancer had significantly lower metabolic ratios than those without previous occupational exposure. In contrast, an investigation of *CYP2D6* genotype in bladder cancer patients revealed a significant increase in the frequency of mutant alleles in bladder cancer patients when compared to controls[135]. Nevertheless PMs appear less likely to suffer a recurrence of the tumour after therapy than those who were extensive metabolizers (EMs)[54]. A recent study failed to show any apparent influence of CYP2D6 genotype and phenotype on bladder cancer susceptibility[17].

Given the contradictory nature of these results, a definative role for CYP2D6 in bladder cancer susceptibility and long-term prognosis cannot be stated. Nevertheless, it has been proposed that CYP2D6 may be responsible for the conversion, in either liver or bladder mucosa, of an as yet unidentified procarcinogen to the proximate carcinogen. A correlation between the polymorphic expression of CYP2D6 mRNA in bladder mucosa and tumour tissue to *in vivo* debrisoquine hydroxylase activity has been demonstrated[105]. A recent study has demonstrated an association between high CYP2D6 activity, as determined by *in vivo* debrisoquine 4-hydroxylation, and loss of expression of or mutated retinoblastoma protein in bladder tumour tissue[106].

Cytochrome P450, CYP3A4, accounts for up to 25% of the total cytochrome P450 present in human liver and for the majority of cytochrome P450 present in human small bowel. Although considerable variability in the *in vivo* activity of CYP3A4 has been demonstrated, a genetic polymorphism has not been established.

In a case-control study of the CYP3A4-mediated activity in bladder cancer, low activity of this metabolic pathway was shown to be a susceptibility risk factor in aggressive, but not non-aggressive, bladder cancer[55]. The authors conclude that such results suggest a role for this enzyme in the detoxication of environmental procarcinogens. Subsequent investigation of these subjects revealed a history of occupational exposure and alcohol intake to be significant risk factors for bladder cancer whereas smoking habit was unimportant[30]. Recently, low CYP3A activity was significantly associated with overexpression of or mutated p53 protein in bladder tumour tissue[107].

SUMMARY

1. There is considerable epidemiological evidence suggesting an association between cigarette smoking and bladder cancer. Risk of bladder cancer appears to be elevated with increased cigarette consumption, inhalation of cigarette smoke and consumption of black tobacco, and reduced by smoking cessation.

2. DNA damage (an important event in the initiation of cancer) in bladder cells appears to be due mainly to the aromatic amine constituents of tobacco smoke rather than the polycyclic aromatic hydrocarbons and tobacco-specific nitrosamines. Although there is no evidence that tobacco alkaloids have either genotoxic or carcinogenic properties, some of these compounds can inhibit apoptosis.
3. The expression of the polymorphic *N*-acetyltransferase, NAT2, appears to be an important factor in determining an individual's susceptibility to smoking-related bladder cancer, such that slow acetylators have an elevated risk which increases with increasing lifetime cigarete consumption. NAT2 appears to mediate an important detoxication pathway for aromatic amines.
4. Possession of the gluthathione *S*-transferase M1 null genotype/phenotype appears to increase the risk of bladder cancer. However, whether this effect is further increased in smokers and individuals with occupational risk remains equivocal.
5. CYP2D6 activity may confer a risk for non-occupational exposure associated aggressive bladder cancer, whereas CYP3A4 activity may have a protective role in such cases. These associations do not appear to apply to non-aggressive bladder cancer.
6. In conclusion, although much of the evidence presented in this review supports a causal relationship between cigarette smoking and bladder cancer, whether exposure to the aromatic amine component of tobacco smoke alone is sufficient to cause bladder cancer remains unknown. Furthermore, it must be stressed that bladder cancer is a mutifactorial disease which, even in the most well-delineated of cases, such as those associated with occupational exposure to a known bladder carcinogen, is not only dependent upon the extent and nature of environmental exposure but also on the individual's inherent ability to activate/detoxicate the compound(s) responsible. The contents of this review demonstrate that such caveats also apply to bladder cancer that is associated with cigarette smoke.

REFERENCES

1. Akdas A, Kirkali Z, Bilir N (1990) Epidemiological case control study on the aetiology of bladder cancer in Turkey. *Eur Urol* **17**: 23–26.
2. Allen MJ, Boyland E, Dukes CE *et al.* (1957) Cancer of the urinary bladder induced in mice with metabolites of aromatic amines and tryptophan. *Br J Cancer*, **11**: 212–228.
3. Anthony HM, Cole P, Thomas GM and Hoover R (1971) Tumors of the urinary bladder: an analysis of the occupations of 1030 patients in Leeds, England. *J Natl Cancer Inst* **46**(5): 1111–1113.

4. Anwar WA, Abdel-Rahman SZ, El-Zein RA *et al.* (1996) Genetic polymorphism of GSTM1, CYP2E1 and CYP2D6 in Egyptian bladder cancer patients. *Carcinogenesis* **17**: 1923–1929.
5. Badawi AF, Hirvonen A, Bell DA *et al.* (1995) Role of aromatic amine acetyltransferases, NAT1 and NAT2, in carcinogen–DNA adduct formation in human urinary bladder. *Cancer Res* **55**: 5230–5237.
6. Bartsch H, Caporaso N, Coda M *et al.* (1990) Carcinogen hemoglobin adducts, urinary mutagenicity and metabolic phenotype in active and passive smokers. *J Natl Cancer Inst* **82**: 1826–1831.
7. Beckett AH, Gorrod JW and Jenner P (1972) A possible relation between pKa_1 and lipid solubility and the amounts excreted in urine of some tobacco alkaloids given to man. *J Pharm Pharmacol* **24**: 115–120.
8. Bell DA, Taylor JA, Paulson DF *et al.* (1993) Genetic risk and carcinogen exposure: a common inherited defect of the carcinogen metabolism gene glutathione *S*-transferase M1 (GSTM1) that increases susceptibility to bladder cancer. *J Natl Cancer Inst* **85**: 1159–1164.
9. Bell DA, Badawi AF, Lang NP *et al.* (1995) Polymorphism in the *N*-acetyltransferase 1 (*NAT1*) polyadenylation signal: association of *NAT1*10* allele with higher *N*-acetylation activity in bladder and colon tissue. *Cancer Res* **55**: 5226–5229.
10. Benitez J, Ladero JM, Fernández-Gundin MJ *et al.* (1990) Polymorphic oxidation of debrisoquine in bladder cancer. *Ann Med* **22**: 157–160.
11. Benowitz NL, Kuyt F, Jacob P, III (1983) Cotinine disposition and effects. *Clin Pharmacol Ther* **309**: 139–142.
12. Bernstein L and Ross RK (1991) *Cancer in Los Angeles County: A Portrait of Incidence and Mortality 1972–1987*, University of Southern California Press, Los Angeles, p. 61.
13. Board P, Coggan M, Johnston P *et al.* (1990) Genetic heterogeneity of the human glutathione transferases: a complex of gene families. *Pharmacol Ther* **48**: 357–369.
14. Boyland E (1968) The possible carcinogenic action of alkaloids of tobacco and betel nut. *Planta Med Suppl:* 13–23.
15. *British Medical Journal* (1971) Bladder cancer and occupation. *BMJ* **6 March**: 517–518.
16. Brockmöller J, Kerb R, Drakoulis N *et al.* (1994) Glutathione *S*-transferase M1 and its variants A and B as host factors of bladder cancer susceptibiliy: a case-control study. *Cancer Res* **54**: 4103–4111.
17. Brockmöller J, Cascorbi I, Kerb R and Roots I (1996) Combined analysis of inherited polymorphisms in arylamine *N*-acetyltransferase 2, glutathione *S*-transferase M1 and T1, microsomal epoxide hydrolase and cytochrome P450 enzynes as modulators of bladder cancer risk. *Cancer Res* **56**: 3915–3925.

18. Bryant MS, Skipper PL, Tannenbaum SR and Maclure M (1987) Hemoglobin adducts of 4-aminobiphenyl in smokers and non-smokers. *Cancer Res* **47**: 602–608.
19. Bryant MS, Vineis P, Skipper PL and Tannenbaum SR (1988) Hemoglobin adducts of aromatic amines: associations with smoking status and type of tobacco. *Proc Natl Acad Sci* **85**: 9788–9791.
20. Burch JD, Rohan TE, Howe GR *et al.* (1989) Risk of bladder cancer by source and type of tobacco exposure: a case-control study. *Int J Cancer* **44**: 622–628.
21. Butler MA, Iwasaki M, Guengerich FP and Kadlubar FF (1989) Human cytochrome P-450_{PA} (P-4501A2), the phenacetin *O*-deethylase, is primarily responsible for the hepatic 3-demethylation of caffeine and *N*-oxidation of carcinogenic arylamines. *Proc Natl Acad Sci* **86**: 7696–7700.
22. Carpenter AA (1989) Clinical experience with transitional cell carcinoma of the bladder with special reference to smoking. *J Urol* **141**: 527–528.
23. Cartright RA, Glashan RW, Rogers HJ *et al.* (1982) Role of *N*-acetyltransferase phenotypes in bladder carcinogenesis: a pharmacogenetic epidemiological approach to bladder cancer. *Lancet* **2**: 842–846.
24. Cartright RA, Philip PA, Rogers HJ and Glashan RW (1984) Genetically determined debrisoquine oxidation capacity in bladder cancer. *Carcinogenesis* **5**: 1191–1192.
25. Cascorbi I, Drakoulis N, Brockmöller J *et al.* (1995) Arylamine *N*-acetyltransferase (*NAT2*) mutations and their allelic linkage in unrelated Caucasian individuals: correlation with phenotypic activity. *Am J Hum Genet* **57**: 581–592.
26. Case RAM and Hosker ME (1954) Tumours of the urinary bladder as an occupational disease in the rubber industry in England and Wales. *Br J Prev Soc Med* **8**: 39–50.
27. Case RAM and Pearson JT (1954) Tumours of the urinary bladder in workmen engaged in the manufacture and use of certain dye-stuff intermediates in the British chemical industry. Part II. Further consideration of the role of aniline and of the manufacture of auramine and magenta (fuchsine) as possible causative agents. *Br J Ind Med* **11**: 213–216.
28. Case RAM, Hosker ME, MacDonald DB and Pearson JT (1954) Tumours of the urinary bladder in workmen engaged in the manufacture and use of certain dye-stuff intermediates in the British chemical industry. Part I. *Br J Ind Med* **11**: 75–104.
29. Chasseaud LF (1979) The role of glutathione and glutathione *S*-transferases in the metabolism of chemical carcinogens and other electrophilic agents. *Adv Cancer Res* **29**: 175–274.
30. Chern HD, Romkes-Sparkes M, Hu JJ *et al.* (1994) Homozygous deleted genotype of glutathione *S*-transferase M1 increases susceptibility to aggressive bladder cancer. *Proceedings of the 85th Annual Meeting of the American Association for Cancer Research* **35**: 285.

31. Chinegwundoh FI and Kaisary AV (1996) Polymorphism and smoking in bladder carcinogenesis. *Br J Urol* **77**: 672–675.
32. Claude J, Kunze E, Frentzel-Beyme R *et al.* (1986) Life-style and occupational risk factors in cancers of the lower urinary tract. *Am J Epidemiol* **124:** 578–589.
33. Clavel J, Cordier S, Boccon-Gibod L and Hemon D (1989) Tobacco and bladder cancer in males: increased risk for inhalers and black tobacco smokers. *Int J Cancer* **44:** 605–610.
34. Clavel J, Mandereau L, Limasset J-C *et al.* (1994) Occupational exposure to polycyclic aromatic hydrocarbons and the risk of bladder cancer: a French case-control study. *Int J Epidemiol* **23**: 1145–1153.
35. Clayson DB (1975) Epidemiology of bladder cancer, in *The Biology and Clinical Management of Bladder Cancer,* (eds EH Cooper and RE Williams), Blackwell Scientific, Oxford, pp. 65–86.
36. Cohen SM and Ellwein LB (1995) Relationship of DNA adducts derived from 2-acetylaminofluorene to cell proliferation and the induction of rodent liver and bladder tumours. *Toxicol Pathol* **23**: 136–142.
37. Cohen SM and Johansson SL (1992) Epidemiology and aetiology of bladder cancer. *Urol Clin N Am* **19:** 421–428.
38. Conney AH (1982) Induction of microsomal enzymes by foreign chemicals and carcinogenesis by polycyclic aromatic hydrocarbons. GHA Clowes Memorial Lecture. *Cancer Res* **42**: 4875–4917.
39. Cuzick J, Routledge MN, Jenkins D and Garner C (1990) DNA adducts in different tissues of smokers and non-smokers. *Int J Cancer* **45**: 673–678.
40. D'Avanzo B, LaVecchia C, Negri E *et al.* (1995) Attributable risks for bladder cancer in Northern Italy. *Ann Epidemiol* **5**: 427–431.
41. Daly AK, Thomas DJ, Cooper J *et al.* (1993) Homozygous deletion of gene for glutathione *S*-transferase M1 in bladder cancer. *BMJ* **307**: 481–482.
42. Davies JM (1965) Bladder tumours in the electric cable industry. *Lancet* **2**: 143–146.
43. Degawa M, Stern SJ, Martin MV *et al.* (1994) Metabolic activation and carcinogen–DNA adduct detection in human larynx. *Cancer Res* **54**: 4915–4919.
44. Del Santo P, Moneti G, Salvadori M *et al.* (1991) Levels of the adducts of 4-aminobiphenyl to hemoglobin in control subjects and bladder carcinoma patients. *Cancer Lett* **50**: 245–251.
45. De Stefani E, Correa P, Fierro L *et al.* (1991) Black tobacco, mate and bladder cancer. *Cancer* **67**: 536–540.
46. Doll R, Fisher REW, Gammon EJ *et al.* (1965) Mortality of gas workers with special references to cancers of the lung and bladder, chronic bronchitis and pneumoconiosis. *Br J Ind Med* **22**: 1–12.
47. Doll R and Hill AB (1952) A study of the aetiology of carcinoma of the lung. *BMJ* **2**: 1271–1286.

48. Doll R and Hill AB (1954) Mortality of doctors in relation to their smoking habits; a preliminary report. *BMJ* **1**: 1451–1456.
49. Doll R and Hill AB (1964) Mortality in relation to smoking: ten years' observations of British doctors. *BMJ* **1**: 1399–1410 and 1460–1467.
50. Doolittle DJ, Winegar R, Lee CK *et al.* (1995) The genotoxic potential of nicotine and its major metabolites. *Mut Res* **344**: 95–102.
51. Evans DAP, Manley KA and McKusick VA (1960) Genetic control of isoniazid metabolism in man. *BMJ* **2**: 485–491.
52. Evans DAP (1993) The debrisoquine/sparteine polymorphism (cytochrome P450 2D6), in *Genetic Factors in Drug Therapy: Clinical and Molecular Pharmacogenetics*, Cambridge University Press, Cambridge, pp. 54–88.
53. Everson RB, Randerath E, Santella RM *et al.* (1986) Detection of smoking-related covalent DNA adducts in human placenta. *Science* **231**: 54–57.
54. Fleming CM, Kaisary A, Wilkinson GR *et al.* (1992) The ability to 4-hydroxylate debrisoquine is related to recurrence of bladder cancer. *Pharmacogenetics* **2**: 128–134.
55. Fleming CM, Persad R, Kaisary A *et al.* (1994) Low activity of dapsone *N*-hydroxylation as a susceptibility risk factor in aggressive bladder cancer. *Pharmacogenetics* **4**: 199–207.
56. Gallagher JE, Jackson MA, George MH *et al.* (1989). Differences in detection of DNA adducts in the ^{32}P-postlabelling assay after either 1-butanol extraction or nuclease P1 treatment. *Cancer Lett* **49**: 7–12.
57. Gautier J-C, Urban P, Beaune P and Pompon D (1996) Simulation of human benzo[a]pyrene metabolism deduced from the analysis of individual kinetic steps in recombinant yeast. *Chem Res Toxicol* **9**: 418–425.
58. Gorrod JW, Jenner P, Keysell GR and Mikhael BR (1974) Oxidative metabolism of nicotine by cigarette smokers with cancer of the urinary bladder. *J Natl Cancer Inst* **52:**1421–1424.
59. Gupta RC and Earley K (1988). ^{32}P-adduct assay: comparative recoveries of structurally diverse DNA adducts in the various enhancement procedures. *Carcinogenesis* **9**: 29–36.
60. Hammond EC and Horn D (1958) Smoking and death rates – report on 44 months of follow up of 187,783 men. *J Am Med Ass* **166**: 1294–1308.
61. Hammond EC, Selikoff IJ, Lawther PL and Seidman H (1976) Inhalation of benzo[a]pyrene and cancer in man. *Ann NY Acad Sci* **271**: 116–124.
62. Hartge P, Silverman D, Hoover R *et al.* (1987) Changing cigarette habits and bladder cancer risk: a case-control study. *J Natl Cancer Inst* **78**: 1119–1125.
63. Hein DW, Doll MA, Rustan TD *et al.* (1993) Metabolic activation and deactivation of arylamine carcinogens by recombinant human NAT1 and polymorphic NAT2 acetyltransferases. *Carcinogenesis* **14**: 1633–1638.
64. Hirvonen A, Nylund L, Kociba P *et al.* (1994) Modulation of urinary mutagenicity by genetically determined carcinogen metabolism in smokers. *Carcinogenesis* **15**: 813–815.

65. Hoffmann D, Brunnemann KD, Prokopczyk B and Djordjevik MV (1994) Tobacco-specific *N*-nitrosamines and *areca*-derived *N*-nitrosamines: chemistry, biochemistry, carcinogenicity and relevance to humans. *J Toxicol Env Hlth* **41**: 1–52.
66. Hoffmann D, Masuda Y and Wynder L (1969) Alpha-naphthylamine and beta-naphthylamine in cigarette smoke. *Nature* **221**: 255–256
67. Hogstedt C, Andersson K, Frenning B and Gustavsson A (1981) A cohort on mortality among long-time employed Swedish chimney sweeps. *Scand J Work Env Hlth* **1**: 72–78.
68. Howe GR, Burch JD, Miller AB *et al.* (1980) Tobacco use, occupation, coffee, various nutrients and bladder cancer. *J Natl Cancer Inst* **64**: 701–713.
69. Hussey AJ, Hayes JD and Beckett GJ (1987) The polymorphic expression of neutral glutathione *S*-transferase in human mononuclear leucocytes as measured by specific radioimmunoassay. *Biochem Pharmacol* **36**: 4013–4015.
70. IARC (1986) Tobacco smoking, in *Monographs on the Evaluation of the Carcinogenic Risk of Chemicals to Humans*, IARC, Lyon, Vol. 38.
71. Iscovich J, Castelletto R, Esteve J *et al.* (1987). Tobacco smoking, occupational exposure and bladder cancer in Argentina. *Int J Cancer* **40**: 734–740.
72. Ito N, Fukushima S, Shirai T and Nakanishi K (1983) Effects of promotors on *N*-butyl-*N*-(4-hydroxybutyl)nitrosamine-induced urinary bladder carcinogenesis in the rat. *Env Hlth Perspect* **50**: 61–69.
73. Jeremy JY, Mikhailidis DP and Dandona P (1985) Cigarette smoke extracts inhibit prostacyclin synthesis by the rat urinary bladder. *Br J Cancer* **51**: 837–842.
74. Kadlubar FF and Badawi AF (1995) Genetic susceptibility and carcinogen–DNA adduct formation in human urinary bladder carcinogenesis. *Toxicol Lett* **82/83**: 627–632.
75. Kadlubar FF, Talaska G, Lang NP *et al.* (1988) in *Methods for Detecting DNA Damaging Agents in Humans: Applications in Cancer Epidemiology and Prevention*, (eds H Bartsch, K Hemminski and IK O'Neill), Int Agency Res Cancer, Lyon, No. 89, pp.166–174.
76. Kaisary A, Smith P, Jaczq E *et al.* (1987) Genetic predisposition to bladder cancer: Ability to hydroxylate debrisoquine and mephenytoin as risk factors. *Cancer Res* **47**: 5488–5495.
77. Kantoff PW (1990) Bladder cancer. *Curr Probl Cancer* **14**: 233–292.
78. Kantor AF, Hartge P, Hoover RN *et al.* (1984) Urinary tract infection and risk of bladder cancer. *Am J Epidemiol* **119**: 510–515.
79. Ketterer B, Harris JM, Talaska G *et al.* (1992) The human glutathione *S*-transferase supergene family, its polymorphism and its effects on susceptibility to lung cancer. *Env Hlth Perspect* **98**: 87–94.
80. Klimisch H-J and Stadler L (1976) Untersuchungen zur Bildung von *N*-nitrosonornikotin aus Nikotin-*N'*-oxid. *Talanta* **23**: 614–616.

81. Lafuente A, Pujol F, Carretero P *et al.* (1993) Human glutathione *S*-transferase *mu* (*GSTMU*) deficiency as a marker for the susceptibility to bladder and larynx cancer among smokers. *Cancer Lett* **68**: 49–54.
82. LaVecchia C, Boyle P, Franceschi S *et al.* (1991). Smoking and cancer with emphasis on Europe. *Eur J Can* **27**: 94–104.
83. LaVoie EJ, Shigematsu A, Rivenson A *et al.* (1985) Evaluation of the effects of cotinine and nicotine-*N*'-oxides and the development of tumours in rats initiated with *N*-[4-(5-nitro-2-furyl)-2-thiazolyl]formamide. *J Natl Cancer Inst* **75**: 1075–1081.
84. Lin HJ, Han C-Y, Bernstein DA *et al.* (1994) Ethnic distribution of the glutathione transferase Mu 1–1 (*GSTM1*) null genotype in 1473 individuals and application to bladder cancer susceptibility. *Carcinogenesis* **15**: 1077–1081.
85. Lopez-Abente G, Gonzales CA, Errezola M *et al.* (1991) Tobacco smoke inhalation pattern, tobacco type and bladder cancer in Spain. *Am J Epidemiol* **134**: 830–839.
86. McCarthy PV, Bhatia AJ, Saw SM *et al.* (1995) Cigarette smoking and bladder cancer in Washington County, Maryland: ammunition for health educators. *Maryland Med J* **44**: 1039–1042.
87. Maneckjee R and Minna JD (1994) Opioids induce while nicotine suppresses apoptosis in human lung cancer cells. *Cell Growth Differentiation* **5**: 1033–1040.
88. Mannervik B, Awasthi YC, Board PG *et al.* (1992) Nomenclature for human glutathione transferases. *Biochem J* **282**: 305–306.
89. Møller-Jensen O, Wahrendorf J, Blettner M *et al.* (1987) The Copenhagen case-control study of bladder cancer: role of smoking in invasive and non-invasive bladder tumours. *J Epidemiol Commun Hlth* **41**: 30–36.
90. Mommsen S and Aagaard J (1983) Tobacco as a risk factor in bladder cancer. *Carcinogenesis* **4**: 335–338 .
91. Momas I, Daures JP, Festy B *et al.* (1994) Relative importance of risk factors in bladder carcinogenesis: some new results about Mediterranean habits. *Cancer Causes Control* **5**: 326–332.
92. Morrison AS, Boring JE, Verhoek WG *et al.* (1984) An international study of smoking and bladder cancer. *J Urol* **131**: 650–654.
93. Nair J, Ohshima H, Friesen M *et al.* (1985) Tobacco-specific and betel nut specific *N*-nitroso compounds in saliva and urine of betel quid chewers and formation *in vitro* by nitrosation of betel quid. *Carcinogenesis* **6**: 295–303.
94. Nair J, Nair UJ, Ohshima H *et al.* (1987) Endogenous nitrosation in the oral cavity of chewers while chewing betel quid with and without tobacco, in *The Relevance of N-nitroso Compounds in Human Cancer: Exposure and Mechanisms*, (eds H Bartsch, IK O'Neil and R Schulte-Hermann), IARC, Lyon, Vol. 84, pp. 465–469.

95. National Cancer Institute (1976) *Smoking and Health Program. Toward Less Hazardous Cigarettes: The Second Set of Experimental Cigarettes.* DHEW Publ. No. (NIH) 76–7111, p. 153.
96. Nomura A, Kolonel LN and Yoshizawa CN (1989) Smoking, alcohol, occupation and hair dye use in cancer of the lower urinary tract. *Am J Epidemiol* **130**: 1159–1163.
97. Okkels H, Sigsgaard T, Wolf H and Autrup H (1995) Glutathione *S*-transferase mu as a risk factor in bladder tumours. *Pharmacogenetics* **6**: 251–256.
98. Patrianakos C and Hoffmann D (1979) Chemical studies of tobacco smoke. LXIV. On the analysis of aromatic amines in cigarette smoke. *J Analyt Chem* **3**: 150–154.
99. Phillips DH, Hewer A, Martin CN *et al.* (1988) Corrrelation of DNA adduct levels in human lung with cigarette smoking. *Nature* **336**: 790–792.
100. Radomski JL and Brill E (1970) Bladder cancer induction by aromatic amines: role of *N*-hydroxy metabolites. *Science* **16**: 992–993.
101. Rehn L (1895) Blasengeschwülste bei anilinarbeitern. *Arch Klin Chir* **50**: 588–600.
102. Riebe M, Westphal K and Fortnagel P (1982) Mutagenicity testing, in bacterial test systems, of some constituents of tobacco. *Mut Res* **101**: 39–43.
103. Roberts-Thomson SJ, McManus ME, Tukey RH *et al.* (1993) The catalytic activity of four expressed human cytochrome P450s towards benzo[a]pyrene and the isomers of its proximate carcinogen. *Biochem Biophys Res Commun* **192**: 1373–1379.
104. Robinson P, personal communication.
105. Romkes-Sparks M, Mnuskin A, Chern H-D *et al.* (1994) Correlation of polymorphic expression of CYP2D6 mRNA in bladder mucosa and tumour tissue to *in vivo* debrisoquine hydroxylase activity. *Carcinogenesis* **15**: 1955–1961.
106. Romkes M, Chern H-D, Lesnick TG *et al.* (1996) Association of low CYP3A activity with *p53* mutation and CYP2D6 activity with *Rb* mutation in human bladder cancer. *Carcinogenesis* **17**: 1057–1062.
107. Rosenberg J, Benowitz NL, Jacob P, III and Wilson KM (1980) Disposition kinetics and effects of intravenous nicotinc. *Clin Pharmacol Ther* **28**: 517–522.
108. Schieferstein GL, Littlefield NA, Gaylor DW *et al.* (1985) Carcinogenesis of 4-aminobiphenyl in BALB/cStCrlfC3Hf/Nctr mice. *Eur J Cancer Clin Oncol* **21**: 865–873.
109. Schmähl Von D (1968) Vergleichende Untersuchungen an ratten über die carcinogene Wirksamkeit verschiedener Tabakextrakte und tabakraunchkondensate. *Arzneimittelforsch*, **8**: 814–816.
110. Schmähl D and Osswald H (1968) Fehlen einer carcinogenen Wirkung von Cotinin bei Ratter. *Zeitschrift für Krebsforschung*, **71**: 198–199.

111. Seidegard J and Pero RW (1985) The hereditary transmission of high glutathione transferase activity towards *trans*-stilbene oxide in human mononuclear leucocytes. *Human Genet* **69**: 66–68.
112. Selkirk JK, Nikbakht A and Stoner GD (1983) Comparative metabolism and macromolecular binding of benzo(a)pyrene in explant cultures of human bladder, skin, bronchus and esophagus from eight individuals. *Cancer Lett* **18**: 11–19.
113. Shimada T, Martin MV, Preuss-Schwartz D *et al.* (1989) Role of individual human cytochrome P-450 enzymes in the bioactivation of benzo[a]pyrene, 7,8-hydroxy-7,8-dihydrobenzo[a]pyrene, and other dihydrodiol derivatives of polycyclic aromatic hydrocarbons. *Cancer Res* **49**: 6304–6312.
114. Shimada T, Yun C-H, Yamazaki H *et al.* (1992) Characterization of human microsomal cytochrome P-450 1A1 and its role in the oxidation of chemical carcinogens. *Mol Pharmacol* **41**: 856–864.
115. Siemiatycki J, Dewar R, Crewski D *et al.* (1994) Are the apparent effects of cigarette smoking on lung and bladder cancers due to uncontrolled confounding by occupational exposures? *Epidemiology* **5**: 57–65.
116. Sorahan T, Lancashire RJ and Sole G (1994) Urothelial cancer and cigarette smoking: findings from a regional case-controlled study. *Br J Urol* **74**: 753–756.
117. Staemmler M (1935) Die chronische Vergiftung mit Nicotin. Ergebnisse experimenteller Untersuchungen an Ratten. *Virchows Archiv* **295**: 366–393.
118. Stålhandske T (1970) Effects of increased liver metabolism of nicotine on its uptake, elimination and toxicity in mice. *Acta Physiol Scand* **80**: 222–234.
119. Stoner GD, Daniel FB, Schenck KM *et al.* (1982) Metabolism and DNA binding of benzo(a)pyrene in cultured human bladder and bronchus. *Carcinogenesis* **3**: 195–201.
120. Strickland PT, Kang D, Bowman ED *et al.* (1994) Identification of 1-hydroxypyrene glucuronide as a major pyrene metabolite in human urine by synchronous fluorescence spectroscopy and gas chromatography-mass spectrometry. *Carcinogenesis* **15**: 483–487.
121. Talaska G, Dooley KB and Kadlubar FF (1990) Detection and characterization of carcinogen–DNA adducts in exfoliated urothelial cells from 4-amimobiphenyl-treated dogs by ^{32}P-postlabeling and subsequent thin layer and high-pressure liquid chromatography. *Carcinogenesis* **11**: 639–646.
122. Talaska G, Al-Jubiri AZSS and Kadlubar FF (1991) Smoking-related carcinogen–DNA adducts in biopsy samples of human urinary bladder: identification of *N*-(deoxyguanosin-8-yl)-4-aminobiphenyl as a major adduct. *Proc Natl Acad Sci USA*, **88**: 5350–5354.
123. Talaska G, Schamer M, Skiper P *et al.* (1991) Detection of carcinogen–DNA adducts in exfoliated urothelial cells of cigarette smokers: association with smoking, hemoglobin adducts and urinary mutagenicity. *Cancer Epidemiol Biomarkers Prev* **1**: 61–66.

124. Talaska G, Schamer M, Casetta G *et al.* (1994) Carcinogen–DNA adducts in bladder biopsies and urothelial cells: a risk assessment exercise. *Cancer Lett* **84**: 93–97.
125. Thompson IM, Peek M and Rodriguez FR (1987) The impact of cigarette smoking on stage, grade and number of recurrences of transitional cell carcinoma of the bladder. *J Urol* **137**: 401–403.
126. Tricker AR, Haubner R, Spiegelhalder B and Preussmann R (1988) The occurrence of tobacco-specific nitrosamines in oral tobacco products and their potential for formation under simulated gastric conditions. *Food Chem Toxicol* **26**: 861–865.
127. Truhaut R, de Clercq M and Loisillier F (1964) Sur les toxicités aiguë et chronique de la cotinine et sur son effet cancérig‘ene chez le rat. *Path Biol* **12**: 39–42.
128. Vatsis KP, Weber WW, Bell DA *et al.* (1995) Nomenclature for *N*-acetyltransferases. *Pharmacogenetics* **5**: 1–17.
129. Vineis P, Bartsch H, Caporaso N *et al.* (1994) Genetically based *N*-acetyltransferase metabolic polymorphism and low level environmental exposure to carcinogens. *Nature* **369**: 154–156.
130. Vineis P, Caporaso N, Tannenbaum SR *et al.* (1990) Acetylation phenotype, carcinogen hemoglobin adducts and cigarette smoking. *Cancer Res* **50**: 3002–3004.
131. Vineis P, Est‘eve J, Hartge P *et al.* (1988) Effects of timing and type of tobacco in cigarette-induced bladder cancer. *Cancer Res* **48**: 3849–3852.
132. Vineis P and Ronco G (1992) Interindividual variation in carcinogen metabolism and bladder cancer risk. *Env Hlth Perspect* **98**: 95–99.
133. Vineis P, Talaska G, Malaveille C *et al.* (1996) DNA adducts in urothelial cells: Relationship with biomarkers of exposure to arylamines and polycyclic aromatic hydrocarbons from tobacco smoke. *Int J Cancer* **65**: 314–316.
134. Williams RR and Horm JW (1997) Association of cancer sites with tobacco and alcohol consumption and socio-economic status of patients: interview study from the third national cancer survey. *J Natl Cancer Inst* **58**: 525–547.
135. Wolf CR, Smith CAD, Gough AC *et al.* (1992) Relationship between the debrisoquine hydroxylase polymorphism and cancer susceptibility. *Carcinogenesis* **13**: 1035–1038.
136. Wolf H, Lower GM and Bryan GT (1980) Role of *N*-acetyltransferase phenotype in human susceptibility to bladder carcinogenic arylamines. *Scand J Urol Nephrol* **14**: 161–165.
137. Wright SC, Zhong J, Zheng H and Larrick JW (1993) Nicotine inhibition of apoptosis suggests a role in tumour promotion. *FASEB J* **7**: 1045–1051.
138. Wynder EL and Hoffmann D (1963) Ein experimenteller Beitrag zur Tabakrauchkarzinogenese. *Deutsch Med Wochensch* **88**: 623–628.

139. Yu MC, Skipper PL, Taghizadeh K *et al.* (1994) Acetylator phenotype aminobiphenyl-hemoglobin adduct levels and bladder cancer risk in white, black and Asian men in Los Angeles. *J Natl Cancer Inst* **86**: 712–716.
140. Zagon IS, McLaughlin PJ, Goodman SR and Rhodes RE (1987) Opioid receptors and endogenous opioids in diverse human and animal cancers. *J Natl Cancer Inst* **79**: 1059–1065.
141. Zhong S, Wyllie AH, Barnes D *et al.* (1993) Relationship between the GSTM1 genetic polymorphism and susceptibility to bladder, breast and colon cancer. *Carcinogenesis* **14**: 1821–1824.

Commentary

Karlheinz Kurth

Smoking and dietary elements are among the most important proven causes of cancer. The knowledge about this relationship is not limited to physicians. Tobacco smoking remains the largest single avoidable cause of premature death internationally and the most important known carcinogen to human beings[7]. The US National Youth Risk Behavior Survey reported that smoking prevalence among US high-school students increased between 1991 and 1995. This trend represented a turnabout from an earlier decline. It seemed that teenagers were less convinced of the danger of smoking than they were some years ago[1]. Whether a similar belief (less fear of smoking risks) influences the behaviour of the European youth is unknown to me, but considering that smoking is still less banned by the public in Europe than in the US (with remarkable differences from country to country), one may assume that European youth does not differ. In Poland the life expectancy at age 45 years in men has been declining for over a decade owing to the increasing premature death rates from cancer and smoking-related vascular disease[4]. Between 25 and 30% of all cancers in developed countries are tobacco related. In the member states of the European Union in 1990 there were over a quarter of a million deaths in middle age directly caused by tobacco smoking: 219 700 in men and 31 900 in women[6]. There were many more deaths caused by tobacco at older ages.

Tobacco smoke is a human carcinogen. Lung cancer is the most frequent smoking-related cancer but bladder cancer, adenocarcinoma of the oesophagus and gastric cardia, cancer of the mouth and pharynx are all related to smoking behaviour[5]. For over 30 years it has been clear that prevention of smoking would lead to substantial reductions in death associated with lung and other cancers, heart disease, bronchitis, emphysema and a number of other conditions. Tobacco smoking kills at least one in four of those who smoke 20 cigarettes or more daily and leads to an estimated 15 years' reduction in life span[4].

Bladder cancer is mainly a disease of the elderly man, meaning that the disease is diagnosed after 20 years or more exposure to tobacco smoking and/or a variety of occupational exposures. Thus the cessation of smoking probably will have little impact on further development of the natural disease, although this possibility has been discussed in the foregoing chapter. The greatest potential for improvement is to convince smokers to give up their self-damaging behaviour before rather than after developing the disease. Once they have the disease the urologist will inform them of the relationship between smoking and the development of bladder cancer. However, in my own experience, only a minority follow the advice to cease smoking.

After decades of steady increases, the age-adjusted mortality due to all malignant neoplasms decreased by 1.0% from 1991 to 1994 in the USA. The decline in mortality due to cancer was greatest among persons under 55 years of age. This trend reflects a combination of changes in death rate from specific types of cancer, with important declines due to reduced cigarette smoking. Thus, and as concluded by Bailar and Gornik, the most promising approach to the control of cancer is a commitment to prevention[2]. Barendregt *et al.* recently published a provocative report on the health care costs of smoking[3]. They stated, 'if people stopped smoking, there would be a saving in health care costs, but only in the short term. Eventually, smoking cessation would lead to increased health care costs . . . yet given a short enough period of follow up and a high enough discount rate, it would be economically attractive to eliminate smoking'.

Urologists deal with about 25% of all solid cancers in men and they therefore should have knowledge about the epidemiology, aetiology and prevention of the type of cancer they are treating. When urologists counsel their patients with bladder cancer about preventive measures they should not be afraid that it is unknown 'whether exposure to the aromatic amine component of tobacco smoke alone is sufficient to cause bladder cancer' (this chapter). Rather, they should consider that in any programme of cancer control priority should be given to control of tobacco. Even when it may be too late for the patient's bladder (after cystectomy) to experience the advantages of having given up smoking, other organs still may benefit from such advice.

COMMENTARY REFERENCES

1. American Cancer Society (1997) Trends in teenage smoking. Cancer risk report: prevention and control 1996. *J Natl Cancer Inst* **89**: 118.
2. Bailar JC and Gornik HL (1997) Cancer undefeated. *N Engl J Med* **336**: 1569–1574.
3. Barendregt JJ, Bonneux L and Van der Maas PJ (1997) The health care costs of smoking. *N Engl J Med* **337**: 1052–1057.
4. Boyle P, La Vecchia C, Masionneuve P *et al.* (1995) Cancer epidemiology and prevention, in *Oxford Textbook of Oncology*, (eds M Peckham, HM Pinedo and U Veronesi), Oxford University Press, Oxford, Vol. 1, pp. 199–273.
5. Gammon MD, Schoenberg JB, Ahsan H *et al.* (1997) Tobacco, alcohol and socioeconomic status and adenocarcinomas of the esophagus and gastric cardia. *J Natl Cancer Inst* **89**: 1277–1284.
6. Osborne M, Boyle P and Lipkin M (1997) Cancer prevention. *Lancet* **349**: 27–30.
7. Peto R, Lopez AD, Boreham J *et al.* (1992) Mortality from tobacco in developed countries: indirect estimation from national statistics. *Lancet* **39**: 1268–1278.

14

Talking with patients about bladder cancer

Reginald R. Hall, Margaret M. Charlton and Patricia Ongena

As clinicians who are dealing with bladder cancer almost every working day, it is very difficult to put ourselves in the position of a patient who has just been diagnosed with the disease. Imagine never having been in a hospital before, not feeling particularly unwell, with little idea of the whereabouts or anatomy of the urinary tract, recovered from an anaesthetic, embarrassed by so much medical and nursing attention to the genitalia, relief after coping with the entirely new and strange experience of a catheter, hoping to return to normal life and work in a day or two – and then being told that you have 'a tumour' or possibly worse, cancer. 'What? Me? It's not possible. But I feel so fit. Why? Will I need surgery? I will have to return to hospital? More surgery? Not another catheter! Can you cure it? What do you mean, I'm fortunate it's only superficial! It really is a cancer? You've cut it all away, how do you know it won't come back? Oh, it may do but you can fix it. It's a good cancer? Such things don't exist! Oh my God! So you think it will be alright? Really?'

A patient is an individual person with his or her own personality, intellect, social background, religion, age and concomitant medical problems. Their initial reaction to the diagnosis of cancer will be influenced by their previous 'cancer experience' either within their own family or circle of friends, or gleaned from reading or the media. There are, therefore, two reasons for talking with the patient about his or her bladder cancer. The first reason is that the patient is a human being, concerned about his or her well-being, livelihood and future, and in need of information about the disease that may threaten any or all of these. The extent of this need will differ from patient to patient. It is the responsibility of the doctor to assess that need and to respond to it in the most appropriate manner. An obligation to provide treatment, or at least an opinion or advice about treatment, has always been implicit in the doctor–patient relationship, but an obligation to provide information has often been overlooked. The second reason is medico-legal, to obtain the patient's written consent to undergo treatment or participate in a clinical trial. This discussion is more to do with the treatment than the disease. Although the patient has sought advice about their illness, he or she is under no obligation to undergo treatment and may choose to accept or reject the advice given. It is therefore essential that they understand what is being proposed and agree to it. Although the signing of a consent form is often seen to be protection for the doctor against any subsequent claim of improper treatment or complications arising from treatment, the prime purpose is the protection of the patient.

THE PHYSICIAN

Patients' expectations have been determined by national social patterns and tradition. For many years North American patients have been much more pro-active than European patients about disease, medicine and health-related subjects and the amount of information they have demanded has been correspondingly greater. As a result of international communication, patients' attitudes are changing and in many countries patients and their families are asking for more information. The past decade, in particular, has seen the growth of patient organizations, the proliferation of press coverage and films about cancer and the innovation of web sites, all of which have had a challenging impact on the physician–patient relationship.

In 1995 the Chief Medical Officers of England and Wales published a *Policy Framework for Commissioning Cancer Services* which set out seven principles for the improvement of cancer care[2]. Of the seven, four addressed the need to improve the doctor–patient relationship. The Expert Advisory Committee recommended:

- more public education;
- clear information to patients and families;
- treatment to be patient centred with good communication; and
- the psychosocial aspects of cancer to be considered at all stages.

It is salutary to ask why such guidance was considered necessary. Perhaps it was the presence of family physicians and a patient on the committee that succeeded in gaining official recognition of the fact that although the medical profession provides high-quality treatment for cancer, many of its members are not so good at communicating with their patients. In general, it has been the pressure, even the threat, of legal procedure that has made the medical profession aware of the need to inform patients. In North America initially it was the professional standard of what a 'reasonable physician' would expect to explain to a patient. This was followed by the reverse criterion, the 'reasonable patient' standard, based on what a reasonable patient would wish to know about the risks and benefits of a particular procedure. Most recently this has developed to become the 'subjective patient viewpoint' standard, to describe what any reasonable patient would want to know about his or her medical condition[4].

There are still some urologists who do not use the words 'cancer' or 'malignancy' with patients, preferring 'polyp' or 'growth' unless a more specific explanation is requested. Such euphemisms may be appropriate for some patients, but studies have shown that the large majority wish to know the true nature of their disease and to be helped to come to terms with the likely outcome. Gross examples of poor communication tend to be regarded as anecdotal and obsolete. For example, the senior urologist who informed the patient in a single sentence from the foot of the bed that he had inoperable bladder cancer and then continued the ward round without further discussion, or the cancer surgeon who spoke to his 'successful' surgical patients but ignored those with metastases; these would be considered to be unacceptable clinical practice. However, a not infrequent present-day equivalent is to recommend unproven or experimental treatment for advanced disease instead of explaining the true gravity of the situation to the patient who might, if properly informed, choose no further treatment. Telling a patient that his or her cancer cannot be cured is a stressful experience for most professionals and some may attempt to avoid the stress by offering treatment instead of facing the difficult facts with the patient. Ostensibly this is done for the patient's benefit but as often as not is a protective mechanism for the doctor. Inevitably, patients with advanced bladder cancer will become aware of their true situation. For most, the sooner this point is reached the better. Helping patients to discuss the fact that they are dying is usually the first step in their coping with the other problems that may arise during the intervening months or years. The giving of information about how we can deal with specific problems is an important part of treatment, but this process cannot begin if the patient is not aware of the inevitable outcome.

It is assumed that because an intelligent young person takes up medicine or nursing as a career he or she has a natural talent for communicating with patients, but this is clearly not true of many. Even among oncologists, be they nursing, surgical, radiation or medical, the reasons for specializing in the treatment of cancer are many and varied, least of which may be the desire or ability to talk with patients about their life-threatening disease. These discussions require not only sympathy and concern, that arise from personal, emotional and possibly religious feelings, but knowledge of the disease, insight and skill that are based on professionalism and experience with previous patients in this situation. These skills can be improved by training and education, but unfortunately, few pre- or postgraduate oncology or urology curricula include such training.

Most urologists choose to become surgeons because they enjoy surgery. Job satisfaction comes from completing a radical cystectomy well, even though it may offer a less than 50% chance of cure. Many may be more comfortable spending several hours in the operating room than a few minutes talking with a patient when metastases develop. No surgeon likes to be reminded of his or her failures. They can be avoided by careful patient selection. We tend to forget that less than half of all patients with muscle invasive bladder cancer are considered suitable for cystectomy[6] and the majority are left to others to palliate. When the cystectomy histology report reveals unsuspected lymph node metastases the instinctive reaction is to recommend adjuvant chemotherapy in the hope, rather than any statistical likelihood, that survival will be improved. How often in these situations are the unwelcome facts made clear to the patient? How often are they acknowledged by the urologist? Discussions of this nature are difficult, tiring, possibly painful and certainly time-consuming. The common perception of surgeons is that they are very busy, hence the tendency to rely on other members of the clinical team, nursing staff, junior medical staff, the family physician or non-surgical oncology colleagues to talk with the patient and attempt to deal with the psychological aspects of the patient's cancer. Other professional colleagues have vital roles to play in the informing, counselling and supportive process. However, in the management of bladder cancer it is the urologist who undertakes the investigation of symptoms and establishes the diagnosis, and who is regarded by the patient as the specialist on whom he or she relies for advice and the principal treatment decisions. It is thus the urologist who bears the prime responsibility for informing the patient about the disease and ensuring that the patient understands the nature of the diagnosis, the prognosis and the treatment options, in so far as it is appropriate to the needs of the individual patient.

This chapter was to be entitled 'What to tell the patient about bladder cancer', until the senior author recognized the error. Our role is not to tell but to listen to, and talk with our patients. It is they who should be helped to set the agenda for the several, possibly many discussions that will be necessary during the course of their disease.

THE PATIENT

It is hardly surprising that most patients retain little detailed information from the first occasion when they learn their diagnosis. Although the diagnosis is often expected and may have been suggested initially by cystoscopy, confirmation of the diagnosis by the biopsy result is none the less a shock for many patients. 'I remember leaving, it was like somebody had hit me with a hammer. It was incredible, the feeling. Whether just being able to talk to someone would have helped, I don't know, but I didn't want to be alone, I knew that.' Most patients will be ignorant of the significance of the diagnosis and the consequences for themselves, their family, their work and their future. Numerous questions will present themselves but this is almost certainly not the occasion to try to answer them. A second calmer, more detailed discussion will be needed preferably with the nurse and a partner or family member present. 'It would be a great idea to have a nurse sitting in when the doctor was talking. You don't take everything in so if there was someone there who could come back to you later, it would be a great help.'

The main problems faced by a patient on learning that they have cancer are (1) the knowledge that they have a life-threatening disease, and (2) fear, or ignorance, of the implications of the disease and the effects of the treatment that they will require. These problems are coped with best by giving the patient sufficient, accurate information to enable them to answer their various questions, fears and doubts. As observed by Fallowfield[3],

> Top of the list of potential problems experienced by patients is the inadequacy of information. Although the numbers of doctors who deliberately withhold the diagnosis of cancer have declined over the past decade, there is plenty of evidence to suggest that many patients with cancer in the United Kingdom feel inadequately informed about details of their illness and the treatments they must face. Worrying about the implications of symptoms, tests and treatments does not help an individual to cope with or adjust to their cancer, it just increases the anxiety and stress. No news is not good news, it is an invitation to fear.

A survey of patients in Newcastle with both superficial and muscle invasive bladder cancer (Charlton, unpublished data) confirmed the desire of the majority to be given the fullest possible information in unambiguous, non-technical language. Only one of 15 patients did not wish to be given detailed information, although the need to pace the giving of information was considered important. Most patients wanted full information as soon as possible but understanding and coming to terms with bladder cancer, be it superficial or invasive, is a slow process that can be achieved seldom in the course of one interview or with one person. In this situation knowledge is acquired most beneficially 'little and often'. Explanations are reinforced by repetition and information given by all the professionals involved in the patient's care is helpful, provided it is consistent. Nurses were regarded by most patients as important sources of information, mainly because they spent more time with patients but also because they are perceived as being more approachable than doctors. 'You see the doctor for a few minutes and then off they go, until they come back the next day. The nursing staff are dealing with you all the time; they have more idea how you are feeling.' 'The nurses were first class. They told me things the doctor hadn't told me. It was useful to get information from both.' The extent to which nurses and junior medical staff are able to provide additional information should always be clarified. Only a close working understanding between all the staff involved in a patient's care will ensure the consistency and accuracy of information that are essential. It is better that an inexperienced member of staff should say 'I do not know' rather than give inaccurate or confusing information.

Waiting for information is a source of added anxiety for two reasons. Patients realize that the results of biopsies may not be available instantly but it is 'another piece in the jigsaw'. Although maximum information may cause distress, the reassurance of knowing 'the complete picture' is most important. How and where patients receive the result of the histological examination of their tumour is important. With appropriate discussion immediately following cystoscopy and TUR, a telephone call or letter may suffice. For example,

> Your biopsy result confirms what we discussed after your operation. It shows that you had a small bladder cancer that did not invade the bladder wall and has been removed completely. It was not the aggressive type of tumour that it might have been and there is no need for any more treatment until you return for a follow up cystoscopy in three month's time. If you would like any further information in the meantime please telephone or arrange an appointment to see me.

If the news is not so good, a personal interview will be necessary. Whatever the result, it is vital to agree with the patient before they leave hospital how they wish to be informed, and who to contact if they have heard nothing after a specified date. Secondly, there is a common perception that cancers grow quickly and that any delay in making the diagnosis and commencing treatment may be harmful. For patients with Ta or T1 bladder tumours treated by TUR it is therefore important to emphasize that the surgery has been complete and that any delay in learning the result of the histology will have no adverse physical effect.

In our study, without prompting, some patients expressed a worry about the completeness of the information they had been given. Any reluctance on the part of the professionals to answer questions unambiguously, or any hint that they were avoiding discussion was interpreted as possibly withholding bad but important news. 'He didn't use the word cancer. He was avoiding the word.' 'They were hedging; was there something going on I should know about?' Patients listen very carefully to what is said and how it is said, in an effort to 'read between the lines' and obtain maximum information from what is often a relatively short discussion. Most will attempt to minimize the negative aspects of their diagnosis and increase their level of hopefulness. This is undermined by inconsistencies or the perception that information has been withheld. 'I wanted to know everything. Not knowing is the biggest problem. You hope that they are telling you everything. I believed that they were, and I found that helped me enormously.'

Despite a very full and clear discussion, patients may miss pieces of information or fail to ask questions of specific concern. 'I missed things because of my own thoughts, not because I wasn't being told. It was easy enough to assimilate the information. The hard part is accepting it.' Patients tend to blame themselves, and the suspicion that it is their fault may prevent them asking for an explanation to be repeated or to raise the matter at a subsequent meeting. Ample opportunity needs to be given to patients to understand what has been explained. A patient has not been 'fully informed' if the information has not been received or has been misunderstood.

All medical personnel use technical terms that mean nothing to patients, or use common words that will not necessarily convey the same meaning to patients. 'When he said he would examine my bladder I had no idea he meant through the penis.' A TUR is a surgical procedure but may not be considered to be an 'operation' by many patients because there is no skin incision. A patient attending for a 'three-month check-up' may be alarmed to discover that the examination includes a cystoscopy, which is another 'operation' to them. Recalling the suggestion that cystectomy would be necessary a patient explained; 'They were talking about major surgery but I didn't know what major surgery was. I'd never had an operation, except my tonsils out and that was 62 years ago.'

The words 'tumour' and 'growth' are often used synonomously with cancer and malignancy by professionals but this may not be understood by many patients. A 'growth' is often considered to be benign rather than malignant. 'Tumour' is frequently ill understood: 'It's not malignant is it?' Equally 'malignant' is more a medical than a lay word and may be thought to refer only to very advanced or metastatic cancers. Although cancer is a very emotive word and has many unhelpful associations when used outside a medical context, it is the one word that virtually every patient will understand and the large majority will use when talking about their disease. Its disadvantage is that cancer is a generic term that includes a range of diseases with widely differing life-threatening potential, but is used by the lay public to mean a disease of sinister, painful and fatal outcome. Thus, for patients with bladder cancer, many of whom will have a good prognosis and will never die of their cancer, the word must be introduced with caution, qualification and skill.

TA T1 BLADDER CANCER

The large majority of patients with superficial tumours that are not poorly differentiated (Ta, T1, G1, G2) have a very good long-term prognosis. Although they have a cancer it is curable by TUR, the risk of recurrence can be reduced by intravesical therapy, the chance of needing radical treatment is small and the likelihood of dying from their bladder disease is very small. To maximize the benefit of proven therapy and minimize the risk of serious relapse, the patient will need regular cystoscopic follow-up, perhaps repeated intravesical therapy and an occasional IVU. All of this will be inconvenient, possibly uncomfortable, rarely very painful, will require lifelong association with the medical profession or a hospital, cost some time off work, maybe some expenditure or loss of income and will probably cause some recurring anxiety. However, looking to the long-term future it will be a nuisance rather than a danger.

This would be a very reasonable clinical summary that should reassure the patient, but will almost certainly not be sufficiently reassuring for many months, possibly years. 'If it is a cancer how come it is so "easy" to treat?' 'All my family who had cancer died.' Despite repeated negative follow-up cystoscopies, and the reassurance that regular follow-up provides, patients often continue to be concerned about the long-term implications and spend a restless night before the obligatory visit to the urologist. 'You say it doesn't worry you, but it always worries you. It's always there in the back of your mind.' 'Check-ups are great, but every one of them worries you.'

Experience with patients who attend for repeated instillations of intravesical treatment have revealed the extent of anxiety suffered by some patients, even with such good-prognosis disease. TURs can easily be explained to patients without telling the diagnosis of cancer, or without their understanding that cancer has been found. The duration of hospital stay for TUR is short, the opportunities for more than one discussion are few, the histology result is very good, there should be no long-term problem and cystoscopy in 3 months is advised. Only when the patient returns for instillations of intravesical therapy does he or she realize the nature of the disease or develop the confidence to ask questions that have been causing concern for many weeks or possibly months. When attending for follow-up cystoscopy the patient is more concerned with the immediate result of the examination, and the urologist so busy explaining how healthy the bladder looks that a discussion of unanswered questions is easily overlooked. Another 3 or 6 months passes before another opportunity may arise. Each attendance for intravesical therapy provides an opportunity to explain possible side-effects, answer questions and reinforce information given previously. A telephone number and a person to contact in case of problems are all part of caring for patients with Ta, T1 bladder cancer.

Our experience with patients with good-prognosis, superficial bladder cancer suggests that they require just as much information as those with more advanced cancers, but because of the lack of clinical anxiety about the future these patients are often not provided with the information that they need. For these patients who do not spend much time in hospital a booklet to read or video to view at home may be very helpful, but such material should not be given to the patient without clear explanations from the medical or nursing staff. It is also important that the material itself should be prepared very carefully and in consultation with patients. Several years ago we published a small but comprehensive booklet to give to all our patients with bladder cancer. It included clearly marked sections dealing with different stages of the disease to meet the needs of different patients. However, it became apparent subsequently that a number of patients had been caused considerable distress and anxiety by reading about the advanced stages of the disease, cystectomy, radiotherapy and chemotherapy which they had been informed would be most unlikely to be needed in their particular case. Increasing numbers of patients are seeking information for themselves from the Internet where they will encounter a similar wide range of information. It is thus particularly important that patients with good-prognosis superficial disease receive sufficient information at the time of their initial treatment to enable them to view their cancer in its true context.

MUSCLE INVASIVE BLADDER CANCER

Invasive bladder cancer is a very different situation, although not entirely dismal. As discussed in Chapter 7, 5-year survival in some highly selected tertiary referral centres is reported to be 70%, and as much as 40% even for patients with microscopic lymph node metastases. Overall, however, muscle invasive bladder cancer is a nasty disease that will kill the majority of patients in 5 years and many within 2 years. It is therefore important to consider very carefully what is said to an individual patient. The information should be neither overly optimistic nor unduly pessimistic and in this situation prognostic factors as they apply to the individual are crucial.

Cystectomy

Some patients believe unreservedly in radical surgery and are anxious to proceed with cystectomy without delay. 'I wasn't worried or anxious, just on my mind was to get the bladder out.' 'As long as I could get rid of the pain, I wasn't worried about what happened. I said to the doctor that I am not happy, but I'll just get on with it.' In this situation the patient may not wish much information and may not be interested to discuss other treatment options. Because their decision has been made, they may not even wish to discuss the consequences of surgery, the possibility of complications or the risks. 'I didn't hesitate about accepting it. I didn't think deeply about it. I thought he knows more about it than I do so I was happy to go along with it.' However, as this patient explained subsequently, the anxiety to press on with curative surgery and the lack of due consideration can cause problems later: 'At the time the sex problem wasn't relevant. There's only one thing you are interested in; am I going to live, can I be cured? To discuss anything else at that time was irrelevant because you might not be around to have the problem. However, now the sex thing is a big problem.'

These comments illustrate the effect that the diagnosis of cancer can have on patients and emphasize the wisdom of spending time with the patient and his or her partner or a member of the family on more than one occasion before admission for surgery. Patients who have been interviewed after their treatment had been completed, repeatedly stated that because they knew so little about the disease and its treatment they did not know what questions to ask. Written information may prompt questions, as may the opportunity to talk with nursing staff or other patients. Finally, however, the responsibility lies with the urologist to understand and address the individual patient's need for information, remembering that the majority will appreciate more rather than less. Simple diagrams which the patient can keep are often very useful provided their content is explained carefully, the organs identified legibly and the parts related to the patient's own anatomy. The options of bladder reconstruction and continent or incontinent urinary diversion will be determined by individual or local cultural factors. Patients will usually have a clear preference but most will welcome the opportunity for an 'independent' opinion from a nurse counsellor or continence adviser before making their final choice.

The complexities of the surgery or bladder reconstruction are seldom of great interest, but it is essential to provide a realistic assessment of the likelihood of postoperative continence, the need to self-catheterize, the possibility of sexual activity, erectile impotence and the impact, if any, on work and leisure activities. The amount of written or verbal information about possible complications will depend upon local medico-legal practice. The request from the patient to 'spare me the gory details' should usually be ignored and at the very least the patient should be informed in writing of the organs and structures that will be removed. For women cystectomy almost always includes hysterectomy, but few patients will be aware of this and, unless it is explained explicitly, misunderstandings may arise.

Treatment options

In some countries cystectomy is the only realistic option for muscle invasive bladder cancer because radiotherapy, conservative surgery and chemotherapy are either not available, are of inadequate quality or are discounted by urological surgeons. The pros and cons of these alternative treatments have been discussed elsewhere in this book. Where they are available and are offered to patients, the possibility of bladder preservation has become a subject of considerable interest. Individual or collective bias will colour urologists' recommendations, but following the example of women with breast cancer, an increasing number of bladder cancer patients are stating their preference to avoid mutilating, radical surgery if possible. In view of the fact that apparently successful bladder-preserving treatments may fail in the long term, these patients need to understand that cystectomy may be delayed rather than avoided. There is also the possibility that if cancer recurs cystectomy may no longer be an option. In addition, they will need to understand the necessity for lifelong cystoscopic follow-up. If, despite these three qualifications, bladder preservation is of interest to the patient an opportunity to meet with the non-surgical oncologist is essential to discuss the details of radiotherapy and/or chemotherapy and for the oncologist to assess the appropriateness of non-surgical treatment for the particular patient.

Some patients are disturbed by being offered a choice of treatment. They have consulted an expert for his or her advice which they expect to be a clear opinion of what is considered to be the best treatment. Being offered a choice may be taken to imply a lack of understanding of the patient's particular case. Furthermore, some patients believe that surgery provides the greatest likelihood of cure and a proposal to consider radiotherapy may be interpreted as indicating that cure is no longer possible. The need for clarity in discussion is self-evident.

In our experience most patients welcome the opportunity to consider alternative treatments and to be involved in the decision-making process. 'I was fortunate, I had a choice. It was important that I took part in the decision.' 'I was happy about having a choice because at the end of the day everything is down to the patient. He has got to sign for whatever is going to happen to him.' 'It was important that I was involved. I had to know the options.' 'As a patient I need to be in control of my life and, as far as possible, my cancer. I need to understand the alternatives so that I can consider their impact on the other parts of my life.'

Locally advanced and metastatic bladder cancer

The finding of inoperable extravesical cancer extending to the pelvic side wall (T4b or T5) is very bad news. The following summarizes a possible discussion with such a patient.

> Your cancer has already spread outside the bladder and cannot be cured by an operation. Almost certainly you will not survive two years. However, your kidney function is good and you are otherwise very fit so we can try chemotherapy if you wish. There may be some unpleasant side effects and your life style will be limited for several months, but there is a small chance that your cancer will respond to the treatment sufficiently [for us] to be able to remove your bladder. That will mean a major operation with two or three months of convalescence. The chance of being cured is very small. Alternatively, now that we have removed part of your tumour, the bleeding has stopped and you do not have any other symptoms at present. You may therefore prefer to have no treatment at the present time and to make the most of the time that you have left without spending time in hospital. If bleeding starts again, or other symptoms develop, you could have radiotherapy or try chemotherapy at that stage. Think about it, discuss it with your wife or husband and we will talk about it again tomorrow.

This is a very abbreviated version of one side of a conversation that could take a few minutes or more than an hour, and will certainly require further discussion and supportive follow-up. Intensive systemic chemotherapy and cystectomy in patients demonstrating a complete clinical response has been reported to be successful in a few patients. However, for the large majority of these patients a fatal outcome cannot be prevented, and patients with these very advanced cancers deserve a sympathetic but realistic assessment of the likelihood of a successful outcome, and the negative impact of several courses of systemic chemotherapy on what will almost certainly be a restricted life expectancy.

In general, patients with metastases will have an even shorter prognosis. Median survival without chemotherapy is only 3–6 months, and although durable complete remissions following chemotherapy have been observed in some patients, very few with metastases survive 5 years (Chapter 11). It is impossible to anticipate a patient's reaction to this information. Some would seize upon it as a reason for hope and therefore request intensive chemotherapy. Others, particularly if asymptomatic, would not regard such an outcome to be worth the morbidity of treatment. Immediate reactions, emotions, pressure within the family or national culture may dictate intensive chemotherapy or other experimental treatment, but a more measured discussion of personal circumstances, individual expectations and the meaning of life may conclude otherwise. Some patients may wish to participate in trials of new agents as a result of a genuine desire to improve the outlook of others in the future, even though there may be little hope for themselves. These wishes are to be acknowledged and supported but should never be exploited. Before meeting a patient to discuss the finding of advanced or metastatic disease, preliminary discussion with nursing staff, the family doctor or perhaps a family member may help to anticipate the patient's reaction to such news or desire to know of the poor prognosis.

Most patients will wish to know if cure can no longer be expected, but many will not seek more specific information about the outcome, at least initially. Few patients ask directly how long they are expected to live, but this question tends to be asked more frequently by relatives. Only the unwary clinician will suggest anything other than a rather vague prediction. Specific estimates are often wrong.

> Handing out an exact prophesy is akin to issuing the patient with a death sentence. The relatives may be triumphant if the patient exceeds the time 'given' by the doctor or angry if he or she is proved to be wrong. A lie to a patient with advanced disease may seem to bring short term comfort, but does nothing to help adjustment and coping when the fact of approaching death becomes inevitable[3].

RANDOMIZED CLINICAL TRIALS

The current management of many aspects of bladder cancer has been determined by the results of clinical trials. In a very practical way today's practice guidelines were yesterday's trial protocols, and today's patients benefit from the willingness of others to participate in previous randomized trials. This does not mean that patients should feel particularly grateful nor under any pressure to join in an ongoing clinical trial, but an explanation of this background helps patients to appreciate the management of their own disease in context.

The statistical principles of clinical trials, in particular the need to avoid bias, have been discussed in Chapter 12. From the patient's perspective the principle that causes greatest concern is randomization. Even more so than offering a patient a choice of treatments, randomization implies uncertainty which, in turn, suggests quite clearly that the doctor does not know what is best. Patients, especially those with cancer, have a need to feel that they are receiving the best treatment. Experts are supposed to know and patients should be allowed to choose. Not only does randomization make it clear that the expert does not know, but it also removes the patient's freedom to choose, just at the time when they feel the greatest need for security and control. The 'lottery' of randomization may be very unnerving for some patients. It is therefore important to help patients understand that although experts may have different opinions about so-called 'best treatment' for their particular situation, the treatments offered in a randomized trial are considered to be at least as effective as current standard therapy. The need for randomized trials arises from progress and new knowledge rather than medical ignorance. They are thus a positive rather than a negative influence on patient care. Despite this, for the patient faced with the diagnosis of cancer, being asked to participate in a trial becomes a matter of immediate and serious concern. Historically, clinicians have protected patients, and to a certain extent themselves, from uncertainty by recommending one form of therapy with confidence when unequivocal evidence to support the opinion is lacking. As Buckman[1] has observed, in our professional training (and in private medical practice) we are seldom rewarded for saying that we do not know. Throughout a physician's graduate and postgraduate training his or her standing is seen to be diminished by a confession of ignorance. However, when communicating with patients honesty shown by a professional strengthens the relationship, increases trust and encourages honesty in return. Compared with some cancers we are fortunate that the treatment of many bladder cancer patients can be based upon good evidence. When this is not the case, however, how often does a urologist respond to the question 'What would you do if you were me?' by saying 'I would randomize myself in the trial I have just described'?

The need to provide detailed information to patients being recruited to a trial is often considered an 'additional' burden, when in reality the information is no more than any patient should receive. Patients who are not informed of different treatment options or of the questions posed by relevant clinical trials may well be underinformed. None the less, asking a patient to participate in a trial does introduce an extra and unexpected dimension to the discussion which, as a result, will need to be more detailed, include additional information (about trial design and randomization) and allow a period of reflection before making a final decision. Written information takes time to read and may be discussed with a relative or friend. This process may be undertaken best by a trained data manager or a clinical trial nurse, who may be regarded by patients as being independent and therefore more sympathetic to their need for simple explanation in non-technical language. Patients may also feel more comfortable asking questions or saying no to these staff than to the specialist to whom they feel indebted for their care. If patients are helped to recognize that they are participating in an open discussion, with no information withheld and no pressure to concur, many will be pleased to join a trial and welcome the opportunity to contribute to medical knowledge for the benefit of others. In addition the extra information, the relationship with the data manager or research nurse and the closer or more frequent clinical supervision required for trial purposes all provide resources that are appreciated by trial patients in coping with their cancer. For example, patients who are asked to complete quality of life questionnaires often report that this interest in other aspects of their life and well-being is a comfort, and appreciated as an expression by professionals of their concern for the patient as a person rather than a disease.

SUMMARY

Talking with patients about bladder cancer is as much a skill as a thorough endoscopic resection, a meticulous radical cystectomy, planning the dosimetry for external-beam radiation or supervising optimal chemotherapy with the least morbidity. The interview with the patient has two components[1]:

- the divulging of information by the professional to the patient; and
- therapeutic dialogue during which the professional listens to, hears and responds to the patient's reaction to the information.

> My previous experience left me unclear about the benefits of surgery and certainly fearful for the future. My consultation with the specialist largely dispelled these concerns. He was open, reassuring and ready to answer all my questions and those of my wife. On admission to hospital this openness continued. All members of staff fully explained the procedures to be carried out and the reasons for them. Although unsettled by the operation, I was much more trusting and confident. When the reasons for the trial were outlined I was confident that nothing would be done to compromise my best interests. Additionally, I was happy that the results of the trial could help others suffering from the same complaint'.

> The doctor told me everything. Anything I wanted to know I could ask him. The most important thing as far as I was concerned was I felt I could trust him.

Simpson[5] suggested that truth is like a drug. Insufficient doses are ineffective and may harm the patient's trust in the therapist, while overenthusiastic scheduling may cause symptoms of overdosage; there are also known cases of idiosyncratic reactions, tachyphylaxis and tolerance, as well as occasional individuals who appear to be resistant to it.

REFERENCES

1. Buckman R (1992) Breaking bad news – a six-step protocol, in *How to Break Bad News: A Guide for Health-Care Professionals*, Papermac, London, pp. 54–81.
2. Calman K and Hine D (1995) *A Policy Framework for Commissioning Cancer Services*. Department of Health, London.
3. Fallowfield L (1991) Counselling patients with cancer, in *Counselling and Communication in Health Care*, (eds H Davis and L Fallowfield), John Wiley & Sons, Chichester, pp. 253–269.
4. Heland KV (1993) Ethics versus law: a lawyer's road map to the Ethics Committee opinon on informed consent. *Women's Health Issues* **3**: 22–24.
5. Simpson MA (1982) Therapeutic uses of truth, in *The Dying Patient*, (ed. E Wilkes), MTP Press, Lancaster, pp. 255–262.
6. Waehre H, Ous S, Klevmark B *et al*. (1993) A bladder cancer multi-institutional experience with toal cystectomy for muscle-invasive bladder cancer. *Cancer* **72**: 3044–3051.

Index

Note: page references to tables are in *italics*; those to illustrations are in **bold**.